A chance conversation and a surprising motto penned into a sixteenth-century astronomy classic prompted Owen Gingerich to begin his Great Copernicus Chase – a search for all existing copies of Copernicus' monumental *De Revolutionibus*. This adventure is the subject of the title essay in the collection of 36 episodes in the history of astronomy.

Gingerich visits an amazing variety of geographical and chronological settings: the Alexandria of Aristarchus and Archimedes, Mogul India, Renaissance Rome, seventeenth-century London, Penobscot Bay, modern observatories of California, and Albert Einstein's 'laboratory of the mind' among others.

Originally he described these explorations in articles for *Sky & Telescope*, *Scientific American*, and other periodicals. They are collected here in popularly written, and well-illustrated chapters which explore the origin of the zodiac, the secrets of detecting fake astrolabes, how optical astronomers beat radio astronomers in the race to discover the spiral arms of our Milky Way, and much more – Gingerich even provides a 'Stonehenge decoder' to allow you to illustrate seasonal sunrise alignments.

The Great Copernicus Chase
and other adventures in astronomical history

The Great Copernicus Chase
and other adventures in
astronomical history

OWEN GINGERICH

Harvard-Smithsonian Center for Astrophysics

SKY PUBLISHING CORPORATION

Cambridge, Massachusetts

and

CAMBRIDGE
UNIVERSITY PRESS

Published by Sky Publishing Corporation
49 Bay State Road, Cambridge, Massachusetts 02138
and by the Press Syndicate of the University of Cambridge
The Pitt Building, Trumpington Street, Cambridge CB2 1RP
40 West 20th St, New York, NY 10011-4211, USA
10 Stamford Road, Oakleigh, Victoria 3166, Australia

First published 1992

Printed in the United States of America

A catalogue record for this book is available from the British Library

Library of Congress cataloging in publication data

Gingerich, Owen.
   The great Copernicus chase and other adventures in astronomical history / Owen
Gingerich.
     p.   cm.
Essays originally published in various journals.
Includes bibliographical references and index.
ISBN 0 521 32688 5
1. Astronomy—History. 2. Astrophysics—History. I. Title.
QB15.G56 1992
520.9—dc20   90–21855   CIP

ISBN 0 521 32688 5 hardback

Contents

Foreword

Over several decades Owen Gingerich has been simultaneously pursuing several different careers as a member of the faculty of Harvard University and a scientist at the Harvard-Smithsonian Center for Astrophysics. In addition to his professional work as astronomer, Professor Gingerich has been an active teacher, giving at Harvard both a general course in the physical sciences for non-science students and a series of conference courses and seminars in the History of Science Department. He has also been producing scholarly monographs on the history of astronomy, and he is chairman of the Editorial Board of the multi-volume *General History of Astronomy* being published under the auspices of the International Astronomical Union and the International Union for the History and Philosophy of Science by the Cambridge University Press. To many readers, however, Owen Gingerich is best known for his more popularly oriented articles on historical aspects of astronomy, intended for the historically literate scientist and the general historical reader.

The present collection contains thirty-six articles in this last category. They range all the way from discussions of astronomy in Egypt, Babylon, and prehistoric England, through aspects of astronomy in Greece, in Islam, and in the Middle Ages, to the astronomy of modern times. Some of these center on astronomical instruments, while another cluster deals with astronomy in the founding century of the Scientific Revolution. At the end of the collection there are half a dozen essays on astronomy and astrophysics of our own time.

These presentations display the author's skill in presenting difficult scientific topics to non-specialists. This ability has been displayed in Professor Gingerich's Harvard course on astronomy and the physical sciences, in which he has presented both the principles and data of science to non-specialists – stressing the way scientists work and think. Professor Gingerich's interest in the historical aspects of his subject is of long standing. I recall that at the start of his teaching career at Harvard, when he was still a graduate student in astronomy, he was a teaching assistant and later a lecturer in my own Harvard course, 'Natural Sciences 3,' in which I sought to use the historical development of the sciences as a means of introducing students to the essentials of scientific thinking, and to some of the problems of the scientist as a member of society.

It should be clear from these remarks that the reader of this book is in for an intellectual treat. Here are displayed, in an engaging and attractive manner, some of the major historical adventures that have produced our knowledge of the heavenly bodies and their physical properties and constitutions. In these presentations, Professor Gingerich frequently gives us an account of some of his own recent and current research, as well as that of other historians of science, which might otherwise remain the exclusive properties of those who read and write for learned journals. Additionally, he often relates his own personal adventures and experiences in exploring aspects of the history of astronomy.

Since these articles for the most part were published in journals not ordinarily read by historians of science, this anthology will make more accessible some of the original research not published elsewhere. Here historians can learn of the significance of Halley's daytime observations of Mercury, how the 'circle of invisibility' gives a misleading clue for dating Ptolemy's star catalog, or why a cartographic error may have placed Harvard's first eclipse expedition outside the path of totality. They can discover what might have prompted Bunsen and Kirchhoff to turn their spectroscope to the stars, or how swiftly Robert Trumpler came to his conclusions about the absorption of light in space.

It is to be hoped that the gathering together of these essays, published in a variety of sources, may make astronomers, historians of science, and general readers better aware of the exciting scholarly discoveries in which Professor Gingerich participated.

I. Bernard Cohen
Harvard University

Preface

The time: August, 1982
The place: Federal District Court, Washington, DC

The defendant, a conservatively dressed man who held a position that required a security clearance, had been indicted for 'knowingly carrying stolen property worth more than $5000 across a state line.' The stolen property in question was a 1566 edition of Copernicus' *De Revolutionibus*, a book that had been missing for eight years from the Franklin Institute in Philadelphia.

As the first witness, I explained to the jury who Copernicus was, why his book was important for the foundation of modern science and worth more than $5000 even in a second edition, and how the volume could be identified. Contrary to the mass-produced volumes of a modern edition, which are like so many peas in a pod, sixteenth-century books were almost always sold unbound, and each buyer added his own distinctive binding. For some years I had been making a census of copies of Copernicus' book, recording details of each one, including the edition at the Franklin Institute. Hence, when I saw this copy described in a bookseller's catalog, I immediately suspected that it was the *De Revolutionibus* that had meanwhile vanished from its Philadelphia home.

I testified, over the objections of the defense, that copies of this edition typically sold for $6000–$9000, and that one in London was being currently offered for $12 000. In the cross examination, I was asked if I had ever taken a course in book appraising, and I answered in the negative, adding that although I had never taken a course in the history of science, I was a professor of history of science at Harvard. 'Just answer my question,' snapped the attorney for the defense, to which the judge remarked, *sub voce*, 'He's trying to!'

In the end the defendant, a former Frankin Institute employee, was convicted, lost his job and his security clearance, as well as his wife, who had apparently walked out in disgust.

This episode is just one of many curious adventures that have occurred in the course of my chasing after all possible sixteenth-century copies of *De Revolutionibus*. Were I a more imaginative writer, I could frame a novel based on international book thievery and my adventures with Interpol. As it is, this present collection of popular articles on researches in the history of astronomy takes its title from just a single Copernican essay, one that originally appeared in *The American Scholar*. Perhaps eventually I'll write up more of the Copernicus adventures.

The selections in this collection, written over a period of 20 years, originally appeared in a variety of publications. Most of them, however, were written for *Sky and Telescope* between 1981 and 1986. After the unexpected death of editor Joseph Ashbrook, I was invited to take over his popular 'Astronomical Scrapbook' column. At that time I was swamped with other assignments, but I knew that my files contained a wide assortment of half-completed researches, and that such a series would provide a perfect opportunity to finish them.

Although the essays were published as general-interest popularizations, most of them contain novel results or insights from original researches in the history of astronomy. My professional colleagues tend to overlook such unfootnoted reports, which fall outside the 'serious' history of science journals. It is my hope that having the essays in book form will make it somewhat easier for historians to find them. I have taken this opportunity to make a few changes in the articles to bring them up to date, to correct errors, or to replace material that had been cut out for requirements of layout. Instead of footnotes, most of the articles include references at the end, wherein the sources are revealed and documented.

I must record here my gratitude to three consecutive directors of the Smithsonian Astrophysical Observatory who have given me the encouragement and resources to pursue these historical studies: Fred Whipple, George Field, and Irwin Shapiro. And very many of the articles comprising this anthology would never have been written without the enthusiasm and cooperation of Leif Robinson and his fellow editors at *Sky and Telescope*. I also wish to express my appreciation to the various other publishers, indicated in the table of contents, who have graciously and freely given me permission to reprint these essays. Finally, I must thank Barbara L. Welther, whose careful eye has prevented many infelicities and whose research skills have helped settle countless fine points; she is, in fact, the joint author of Chapters number 4 and 30 in this collection.

Owen Gingerich
Cambridge, Massachusetts

1 *Ancient Egyptian sky magic*

'Only in the east, where six hundred million human beings live, is it possible to found great empires and realize great revolutions.' Thus wrote Napoleon Bonaparte, and in the spring of 1798 he set out for Egypt with a force of 38 000 men and 175 civilian 'savants.' From this otherwise rather disastrous expedition came the first great work on Egyptology, the huge, multivolume *Description de l'Egypte*.

The obelisks, the Sphinx, the great pyramids – these excited the curiosity of the Europeans who were for the first time really encountering the mysteries of ancient Egypt. And then there was Dendera, an enormous, partly buried temple lying just west of the Nile about 300 miles south of Cairo. (It should not be confused with the much smaller temple of Dendur, which is now preserved in New York City's Metropolitan Museum of Art.) Napoleon's troops were overwhelmed by the spectacle when they reached Dendera on 25 May 1799. Its walls were chiseled with bas-reliefs and even more mysterious hieroglyphics, but on its ceilings was something wondrously recognizable and familiar: the signs of the zodiac.

A great circular zodiac, about $1\frac{1}{2}$ m in diameter and forming the principal part of the ceiling of one of the associated chambers, aroused some of the greatest interest, for it had all the makings of a celestial planisphere. The savants carefully sketched it, and their large engraving is one of the finest plates in the *Description* as well as one of the most handsome (though not perfectly accurate) renditions ever made of this zodiac.

When the existence of the Egyptian zodiacs became known, and particularly after the circular zodiac itself was brought in 1820 to the Louvre in Paris, a flood of articles debating the work's great antiquity swept through the learned journals of Europe. Some argued from the zodiacal signs shown in the solstitial positions that the stone dated from 4000 BC. Others identified it with the star chart supposedly seen by Eudoxus in Egypt, and dated it 1300 BC. Abbé Halma, the redoubtable translator of Ptolemy into French, thought he saw within the zodiac evidence for a pair of solar and lunar eclipses dating from 364 BC.

It must have been quite a shock when in 1822 Jean-Francois Champollion announced that he could read the hieroglyphs for AOTKRTR in association with the circular zodiac temple, and that the title 'Autocrator' had been used only by the emperors Claudius and Nero in the first century AD. In 1828 Champollion had a chance to visit the temple itself. So eager was his party to go ashore that they pushed their way through the grass and brush under the light of the full moon. His excitement registers even in the calmness of his words: 'I will not try to describe the impression that the temple, and particularly its portico, made on us. . . . We stayed there two hours, filled with ecstasy.' Champollion quickly confirmed his earlier conjectures: The temple was not 'extremely ancient' as the Egyptian commission

Figure 1.1 The great circular zodiac of Dendera. This engraving by the scholars of the Napoleonic expedition to Egypt was published around 1815. The zodiacal signs can be easily traced, and within that band five planets are depicted as gods holding staffs: Mercury, just above Cancer; Venus, between Pisces and Aquarius; Mars, standing on top of Capricorn; Jupiter, in the space between Gemini, Cancer, and Leo; and Saturn, following Virgo and just before Libra.

had concluded, but less than 2000 years old. Although begun in the Middle Kingdom and extended by the pharaohs Thutmose and Ramses, the temple had received its final form only after Alexander the Great.

Despite its status as one of the first recognizable ancient Eygptian depictions, and despite the magnificent deciphering of the hieroglyphics by Champollion and his successors, the Dendera zodiac has proved to be one of the most intractable pieces of lore. It is still puzzling Egyptologists. Although the zodiacal constellations are quite evident, most of the other symbols remain undecipherable.

Within the zodiac the hippopotamus and the front leg of a bull (or ox) are most conspicuous; beyond it to the south are a variety of gods, animals, and an outer boundary of 36 timekeeping groups. The last are called the decans (from the Greek *deka* for 10), since each one is roughly 10° wide. In 1856 the German scholar Heinrich Brugsch successfully read the names of the five naked-eye planets, which are scattered throughout the Dendera zodiac according to their astrologically most propitious places. Since then further progress has been tediously slow.

The Dendera zodiac proves to be one of the most recent in Egypt. On earlier Egyptian depictions of the sky the zodiac is entirely missing. The explanation is natural enough: The 12 zodiacal signs are a Mesopotamian invention, coming to Egypt from Babylon a few centuries before Christ. The superstitious artists at Dendera added the

Figure 1.2 *The northern constellations in the tomb of Seti I (about 1300 BC) in the Valley of the Kings. Here Meskhetiu (the Big Dipper) is represented by an entire bull. The sky goddess Hippo has not been identified with a specific constellation but may well represent a large zone of the summer evening sky including the Summer Triangle of Vega, Deneb, and Altair. The band of stars across the bottom may represent the Milky Way.*

new constellations but didn't throw anything else away, leading to the result that some parts of the sky were doubly represented. It is no wonder that attempts to match the Dendera circular zodiac directly against a modern star chart are doomed to failure.

The liberties taken by the ancient Egyptians in arranging their sky figures are easily seen by examining the unfinished tomb of the nobleman Senmut (about 1473 BC). Its ceiling displays an astronomical representation that is the oldest work of its kind. Here stands a fantastic array of figures: Hippo, Man, Lion, Croc, and others. Strikingly, the crocodile can be seen in two versions: a finished pose next to the smiting man's fist (see Figure 1.3), and an earlier sketch, barely peeking through, showing Croc in a horizontal pose. In other words, the artist chose a symbolic representation, not a detailed mapping of the sky.

It has always seemed to me astonishing that a major way of looking at the sky for at least two millennia, by one of the great civilizations of the past, has now been almost entirely lost to us. On the other hand, when we realize how inexact the depictions of the constellations are, perhaps we can feel lucky that even part of this tradition has been securely interpreted.

What have we learned about this ancient cosmology? The Egyptian sky can be divided into two main sections. The northern constellations are those north of the ecliptic, or possibly north of the Milky Way. The southern constellations include the series of timekeeping decans that have given rise to our 24-hour division of the day. According to the monumental *Egyptian Astronomical Texts* of O. Neugebauer and Richard A. Parker, only three ancient Egyptian configurations can be identified with any certainty: the Big Dipper, shown as the foreleg of a bull and named Meskhetiu; Sirius, represented as the goddess Isis; and Orion, represented as the god Osiris.

Figure 1.3 The so-called northern constellations in the unfinished tomb of Senmut; a careful examination shows that the artist originally drew the crocodile at left in a horizontal position. The outline of the deleted version is vaguely visible at the bottom of this picture.

The remarkable symbolism of the Egyptian sky is well represented by Meskhetiu. On Middle Kingdom coffin lids and on the very late Dendera zodiac, this figure is shown as the foreleg of a bull. Indeed, the stars of the Big Dipper can be envisioned as having such a pattern. On Senmut's ceiling, however, the constellation is depicted as the entire front part of a bull with four very atrophied legs.

The shape of the Big Dipper is also like that of the adze, a sculptor's tool with a sharp blade at right angles to its handle. In Egyptian society the sculptor occupied a special place. With his tool he could make statues, and with the proper magical incantations such images could be brought to life in the afterlife. Thus, a standard part of the Egyptian death ritual was accomplished by a god wielding an adze before the mummy, the 'opening-of-the-mouth' ceremony. The scene is beautifully shown in the famous tomb of Tutankhamon, where the dead pharaoh's successor, Ay, holds the adze before the mouth of Tutankhamon's mummy. The symbolism continues on the stand in front of Ay, where an adze is neatly stacked with the foreleg of a bull.

The equation between the sculptor's magical adze and the foreleg of the bull is even more explicit in a number of other tomb scenes. One example occurs in the tomb of Rekhmere, where one panel shows the adze used in the opening-of-the-mouth ceremony, and the next panel is identical except that a foreleg replaces the adze. Clearly the artists were covering all bets, making sure that the sky symbol would be brought correctly into the tomb, if not in one way, then in the other. Incidentally, I have been told by archaeologists in Upper Egypt that model adzes made of meteoritic iron have been found in some tombs.

Of all of the modern Egyptologists who have probed the mysteries

Figure 1.4 In this frieze on the wall of Tutankhamon's tomb the shrouded pharaoh stands in the guise of Osiris while his successor, King Ay, performs the 'opening-of-the-mouth' ceremony with the magical adze (stylized image of the Big Dipper). Other celestial symbols rest on the stand below.

of these ancient depictions, Virginia Lee Davis has perhaps been the most successful and certainly the most ingenious. For example, she has noticed on the stand in the Tutankhamon mural not only the two representations of the Meskhetiu (the foreleg and the adze) but also the feather, the sign of Maat. This goddess, who represents the natural order of the universe, was the sun-god's starry daughter. Her hieroglyph is a trapezoid, symbolic of the banks of the Nile and also the shape of the constellation Canis Major, the asterism that stands on the banks of the Milky Way. Can this be another magical sky symbol in the Egyptian scheme? She has also noticed that Isis (Sirius) and Osiris (Orion) are often depicted as standing in boats, perhaps on the Milky Way Nile, and she suggests an identification between the Ship decan and the Great Square of Pegasus.

"Of the northern constellations, the texts say that they form 'a ring of fighting-faced characters,'" she writes. "Artistic convention tends to square up their configuration, but their ferocious forms remain." The lion, the crocodiles, the smiting man, the bull, and the scorpion are, indeed, a ferocious lot! "The texts often mention two claws or adzes or fingernails that 'hack up the celestial mansion.'" One of these, Davis points out, is transformed into a foreleg and then into a complete bull; could not the other be similarly transformed from a jackal's small digging hoe, to a falcon's claw, and then to a falcon-headed man? Thus, she argues, both the Big and Little Dippers could be the celestial adzes chasing each other around the celestial pole (which was in Draco 4000 years ago), and the falcon-headed man could be Draco with Ursa Minor, the Little Dipper. Using similar lines of argument, she identifies the Smiter with Gemini, the Man-with-the-Ropes with Bootes, and the Lion with Leo.

Although it is presently impossible to verify any of these fascinating but speculative identifications, they seem plausible. Further interpretation of texts not now taken as astronomical may eventually bring a much greater understanding of the Egyptian sky mysteries.

Notes and references

A useful history of Egyptology is C. W. Ceram's classic popularization *Gods, Graves, and Scholars* (New York, 1953). The definitive treatment and a set of illustrations of Egyptian constellations are found in the two-part Volume III of O. Neugebauer and Richard A. Parker, *Egyptian Astronomical Texts* (London, 1969). I have profited from discussions with Virginia Lee Davis and from some of her manuscript material. Concerning the Egyptians' use of symbolism, see Dr Davis' 'Pathways to the Gods,' pp. 154–67 in the National Geographic Society's *Ancient Egypt*.

2 *The origin of the zodiac*

About midway through the fifteenth century, agents working for the Grand Duke Cosimo de' Medici obtained a curious and remarkable manuscript. Attributed to Hermes Trismegistus (scribe of the gods), the document excited great intellectual interest, for this so-called thrice-greatest Hermes was presumed to antedate not only the ancient Greek philosophers but even Moses. Hence, there was every expectation that the manuscript would reveal secrets of a long-forgotten golden age. Cosimo turned the manuscript over to the celebrated Greek scholar Marsilio Ficini, with instructions to stop translating Plato and to start straightaway on the older, presumably more fundamental text.

The Hermetic text was not a model of clarity. Rather, it was a confused pot-pourri of magic, omens, and portents. Not until a century and a half later did a linguistic analysis by Isaac Casaubon show that a colossal mistake had been made, and that the Hermetic corpus was an eclectic batch of texts compiled at the beginning of our own era. Nevertheless, the damage had been done, for the manuscript, misconstrued as it was, had given a renewed respectability to the study of astrology.

Among modern devotees of astrology, we encounter from time to time the firmly held notion that astrology represents an extremely ancient source of wisdom. This idea is somewhat akin to the Renaissance belief in the hoary antiquity of the Hermetic manuscripts. Clearly, an astrology based on planetary positions could not have existed before it became possible to predict those positions, that is, sometime in the middle of the first millennium BC. Furthermore, zodiacal astrology could not have come before the creation of the zodiac itself.

When and whence did the zodiac and the other constellations originate? Many of the 88 constellations recognized by astronomers today, particularly those in the southern sky such as Tucana (the Toucan) and Fornax (the Chemical Furnace), were invented in modern times, so we can be quite precise about their origin. The basic patterns of the older constellations were firmly established by the star catalog in Ptolemy's *Almagest* (or *Syntaxis*), a great handbook of mathematical astronomy written about AD 150.

Ptolemy described 48 constellations, including the 12 signs of the zodiac. He named a few individual stars – Arcturus, Capella, Antares, Procyon, and Canopus among them – but mostly he carefully designated each star by its position within the mythological figure. Thus, the most conspicuous star in Cygnus, known today as Alpha Cygni, was called by Ptolemy 'the bright star in the tail' of the bird. This description was subsequently translated into Arabic along with the rest of the *Almagest*, and the Arabic word *Deneb*, for 'tail,' has stuck as the proper name of this star. Most of the traditional constellation descriptions thus derive from Ptolemy's descriptions, which passed through the Islamic world into the late medieval Latin tradition, from there to the magnificent woodblocks of Albrecht Dürer, and thence to Johannes Bayer's splendid engraved *Uranometria* of 1603.

Ptolemy's description of the zodiacal constellations offers a curiosity with respect to Libra, which he described under the name of Chelae, the Claws. Libra, the Balance, is today anomalous in the zodiac, the only sign that is neither human nor beast. Ptolemy, however, describes each of Libra's principal stars with respect to the anatomy of the adjacent Scorpion: in the south claw, the north claw, and so on. We thereby get a hint of a more ancient and larger scorpion spread across the southern sky, which was split when the twelvefold division of the zodiac was made.

On the other hand, the constellation of the Balance occurs in one of the oldest known lists of zodiacal constellations, the first Babylonian tablet in the so-called *mul-Apin* series, dating from around 700 BC. This tabulation of 15 star groups along the moon's path doesn't match Ptolemy's (and our own) dozen very well. It contains separate entries for the Pleiades and the Hyades or Bull, and includes such asterisms as the Great Swallow (southwestern Pisces plus Epsilon Pegasi) and the Lady of the Heavens (northeastern Pisces plus the central part of Andromeda).

The earliest cuneiform text using the standard 12 signs dates only to around 400 BC. Scholars agree that here all of the signs, including the Balance, were strictly equal in length; in fact, the recorded positions of the planet's would not agree with modern computations if the situation were otherwise. The Egyptian Zodiac of Dendera, which comes from the beginning of our era and which is full of Babylonian influence, clearly depicts a balance (see Chapter 1). Indeed, Ptolemy himself used the Babylonian Balance interchangeably with the Greek Claws elsewhere in his work. Thus, at the time of Ptolemy, two somewhat different traditions were being merged.

From Mesopotamia, the evidence points to the concept of the zodiac gradually evolving during the first millennium BC, particularly as increasing attention was given to the motions of the sun, moon, and planets. In Greece, meanwhile, the mere handful of available sources seems to suggest that the zodiac was introduced almost full-blown. Although the Greeks had no great tradition of planetary observations, the idea of the ecliptic and the zodiac seems to have been almost universally known to them by 400 BC.

In the *Phaenomena* of Eudoxus of Cnidus, written soon after 400 BC, we have the earliest known description of our present constellation patterns. Although the work is now lost, its contents are at least partly preserved through a long astronomical verse of the same name (from around 250 BC) by the poet Aratus. From the *Phaenomena* we know that the Greek use of 'Chelae,' or the Claws of the Scorpion, goes back at least to the fourth century BC.

Of the constellations listed in the three different *mul-Apin* tablets, many match their classical Greek counterparts (such as the Bull, the Twins, the Lion, and the Scorpion), but others do not (such as the

Figure 2.1 Designated VAT 4924, this cuneiform text at the Berlin State Museum contains the earliest known use of the 12 zodiacal signs. Line 5 reads, 'Jupiter and Venus at the beginning of Gemini, Mars in Leo, Saturn in Pisces. 29th day: Mercury's evening setting in Taurus.' These positions correspond to April, 419 BC.

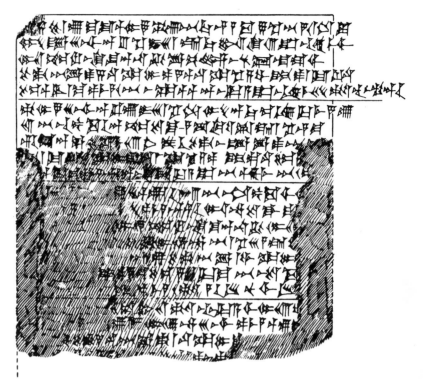

Balance, the Griffin – Cygnus and the lower part of Cepheus – and the Great Swallow). However, because of the very considerable similarities and because of the way in which the evolution of the system can be seen in Babylon, it is possible to conclude that the concept of the zodiac and horoscopic astrology developed in Mesopotamia. "Not without reason were astrologers called 'Chaldeans' throughout the Roman Empire!" concludes the contemporary mathematician B. L. van der Waerden in one of his brilliant essays on Babylonian astronomy.

Even in the absence of specific texts, it is possible to press the deductions still further. Because of the similarities between the Greek and Babylonian systems, we can imagine that some common earlier tradition played a significant role in shaping these two ways of looking at the sky. Furthermore, because many of the Babylonian constellation names have roots in the earlier Sumerian language, it seems clear that some of our present constellations have ancient origins that go back beyond the first millennium BC, the time when the zodiacal path with its 12 equal signs was first established.

Such a conclusion was drawn by the late Willy Hartner, who argued that three conspicuous constellations, Taurus, Leo, and Scorpius, whose names are particularly old, had a highly visible role in the rhythmic pattern of the seasons. Uniquely recognizable are the Pleiades, whose Babylonian name *mul-Mul* means 'The Star,' and associated with this cluster are the other stars of Taurus, including its sparkling red-orange eye, Aldebaran.

Also along the zodiac are two other important configurations. One

Figure 2.2 On this Mesopotamian boundary stone from about 1100 BC, now in the British Museum, are depicted a scorpion and a lion. Both are still represented among the zodiacal constellations. At the top appear Venus, the moon, and the sun.

is Scorpius, two conspicuous curves of stars lying across the Milky Way and encompassing the bright red star Antares. The other, not so brilliant in itself but outstanding against its relatively sparse surroundings, is Leo the Lion. Like Scorpius, Leo contains a first-magnitude star, Regulus. It is interesting to note that these key groups, Taurus, Scorpius, and Leo, include three of the four first-magnitude stars lying along the ecliptic (the fourth being Spica, in the rather amorphous constellation of Virgo, the Virgin).

At present, Taurus lies near the northernmost extreme of the sun's annual track, and its stars emerge in the eastern dawn sky around the beginning of summer. Because of the slow precessional movement of the stars, in Sumerian times (5000–6000 years ago) the predawn visibility of the Pleiades and Taurus would have marked the coming of spring. Eleven months later, in the dead of winter, Leo would have stood high in the sky as Taurus vanished into western twilight.

Hartner believed that the recurrent artistic motif of the lion–bull combat, found throughout ancient Near Eastern art, reflects an awareness of these key constellations and their changing relationships as winter overtook autumn and as spring, in turn won out over the winter. The antiquity of these artistic artifacts helped convince him that these key constellations must have been well recognized as early as 3500 BC.

It is possible to push back the origins of any of the constellations still further? Probably the two most persistently recognized asterisms are the Big Dipper and the Pleiades. The fuzzy, clustery nature of the Pleiades makes them quite unmistakable. I remember trying to get my bearings once on a dark road in a Central American jungle; the sky was mostly overcast, but even through the patches of thinner clouds it was plain that something unique was overhead: the Pleiades cluster. Besides being peculiarly recognizable, this cluster lies within a few degrees of the ecliptic. Not surprisingly, in many preliterate cultures throughout the world the Pleiades have played and continue to play essential calendrical roles, but these could have been independently discovered over and over again.

Somewhat different is the case of the Big Dipper, which is seen as a Great Bear by many different Indian tribes of North America. It is difficult to be sure how old their oral traditions are. Could the Indians have got the European notion of Ursa Major from the early explorers? I think this is unlikely. It seems more plausible that a very early tradition of a celestial bear crossed the Bering Straits with ancient migrants, especially so since the same identification is found across Siberia, as has been pointed out to me by Alexander Marshak. Such an early tradition could well have diffused throughout the world from the ancient cave dwellers of Europe.

The identification of a bear with the dipper stars may have migrated not only to ancient Greece, but even farther southward. The oldest representation of these stars in Egypt is as a ferocious set of claws but since there are no bears in the land of the Nile, such a depiction gradually evolved into a sharp adze and then into the foreleg of a bull. In the widespread mythological connection of the

Figure 2.3 The so-called Farnese Atlas, now in the Archaeological Museum in Naples, displays the constellations and celestial circles much as they are described in Aratus' Phaenomena. *The statue dates from the beginning of our era but is presumably copied from an earlier Greek original.*

dipper stars with a Great Bear (Ursa Major) we have a hint that a few of the constellations may date back as far as the Ice Ages.

The hypothesis that our present constellations are a long-evolved mixture including elements from very ancient cultures stands in marked contrast to some earlier speculations. For example, in 1913 E. W. Maunder wrote, 'When the idea had once arisen of identifying stars by giving them places in imaginary figures, the whole work of constellation making could easily have been carried out in the course of a single year. It is not a case where we have to invoke a long, slow evolution; Aratus has only some 46 constellations in his list, so that if the primitive astronomer had devoted a full week to the construction of each, he would still have been able to take quite a long vacation during his year of work.' Maunder declined to speculate here, however, on the source 'both ancient and foreign' from which Aratus drew his inspiration.

In a provocative lecture given in 1965, Michael W. Ovenden took the argument further, attempting to convince his audience that the constellations were logically designed (or at least organized in their present arrangement) at a definite time and place that could be established. Using the statements in Aratus' *Phaenomena* and the great precessional clock, Ovenden argued that we see a fossilized arrangement of constellation figures dating back to 2800±300 BC. For example, he noted that one ring of nonzodiacal figures, including Auriga, Perseus, Hercules, and Bootes, was symmetrical about the north celestial pole at that time.

While the ingenuity of Ovenden's argument is admirable, I do not find it compelling in the light of the evidence for a relatively recent zodiac and for a system that evolved gradually. Nevertheless, the clearly interrelated nature of many constellation groups with respect to their mythology (such as the Andromeda–Perseus–Pegasus–Cetus legend, or the great flood story) strongly suggests that major parts of the panoramic sky depictions derive from a single lineage.

Thus, Maunder and Ovenden (and others) may well be close to the truth in speculating that at least parts of our present constellations fell into place rather swiftly. But, I would insist, any such process must have been a merger of both innovative and traditional elements, giving us a sweeping pattern of something old, something new, something borrowed, and something going back to the Ice Ages.

Notes and references

B. L. van der Waerden's 'History of the Zodiac' appears in *Archiv für Orientforschung*, **16** (1953), 216–30; see also his *Science Awakening II, The Birth of Astronomy* (Leiden and New York, 1974). W. Hartner's 'The Earliest History of the Constellations in the Near East and the Motif of the Lion-Bull Combat' is in *Journal for Near Eastern Studies*, **24** (1965), 1–16 and Plates I–XVI. E. W. Maunder's 'The Origin of the Constellations' is found in *Observatory*, **36** (1913), 329–34, and M. W. Ovenden's lecture of the same title is in *The Philosophical Journal*, **3**

(1966), 1–18. After preparing this chapter, I found Joseph Campbell's *The Way of the Animal Powers* (San Francisco, 1983), the first volume of the splendid new *Historical Atlas of World Mythology*. His sections on 'The Master Bear,' though lacking astronomical discussion, lend credence to my suggestion that the Great Bear constellation dates from the Ice Ages.

3 *The basic astronomy of Stonehenge*

About 20 years ago, when I first saw Stonehenge, I was taken by surprise. Somehow, in mind's eye, the trilithons and stone circle had assumed truly monumental proportions. In contrast, the real stones were set on a disappointingly small scale. My confession is not intended to belittle the world's most famous megalithic site, but simply to report an honest reaction to one of the most romantic of all prehistoric ruins.

My initial response, that Stonehenge seemed unexpectedly small, is in fact relevant to the viewpoint that I should here like to defend: that Stonehenge is not so much an ancient megalithic observatory as the *monument to an earlier observatory*. By this I mean that any astronomical sighting lines at Stonehenge must have been well established centuries before they were fossilized into such a heavy, immobile configuration, and that the organization of the monumental stones is primarily dictated by the aesthetic symmetry along their principal axis and not by a secondary series of lunar sightlines, as some have proposed. At best, Stonehenge was a ritual center commemorating bygone discoveries, not a site where new knowledge of the heavens was actively sought. It was a stable monument to the eternal order and regularity in the sky, and as such its alignments still synchronize with the sun's rhythmic march through the seasons.

The statistics of Stonehenge are impressive enough. The stone ring, about 30 m in diameter, originally comprised 30 megaliths capped by 30 lintels. Within this so-called sarsen circle stood the five huge trilithons in a horseshoe pattern; each was a pair of uprights with a lintel held in place by a mortise and tenon carved in the stone. The largest of the megaliths, one of the trilithon uprights, must have weighed 50 tons; it is the largest prehistoric hand-worked stone in Britain. By comparison, the stones of the sarsen ring are a mere 25 tons each. The sarsen stones – originally huge natural boulders but dressed by pounding with stone hammers – came from the Marlborough Downs, nearly 20 back-breaking miles north of Stonehenge.

Today nearly half of the sarsen ring has been quarried away, and only three of its stones remain untouched in their original positions. However, 16 are now in place and six have regained their lintels with an assist from the archaeologists. The trilithons have escaped the ravages of time somewhat more successfully. The two southeast groups never fell. When the largest group, on the central axis, tumbled down in ages past, one of the stones broke in two; the surviving upright has apparently been replaced 2 or 3 m too close to the center of the monument. Nevertheless, the original orientations of the rings and the outlying, so-called Heel Stone, about 80 m to the northeast, can be established reasonably well.

Before we can understand the unique geometry of Stonehenge and the special orientations of other ancient monuments, we must grasp a

Figure 3.1 The early Stonehenge – or rather, the early Stonehenges, for the illustration shows several stages of construction at the site. The first of these – 'Stonehenge I' – is an earthwork ring about 100 m in diameter and 2 m high. Its completeness was broken (as of about 2400 BC) by a single gap directed in the approximate direction of an outlying marker called the Heel Stone; and in this gap, excavation has uncovered a grid of post holes: the remains, it seems, of an effort to mark the northernmost excursion of the moon. Note that the Heel Stone lies slightly away from a line drawn from the center of the earthwork ring to the horizon point marking the midsummer (solstitial) sunrise; in 2400 BC the Heel Stone was presumably more erect, and thus the alignment was more nearly perfect. Stonehenge I also included a circle of chalk-filled holes now named after John Aubrey. At some later time, Stonehenge II was added. It comprises two mounds of earth, covering some of the chalk-filled holes, and also the so-called station stones. As shown in the illustration, these additions to the site mark out the corners of a rectangle whose sides and diagonal align with various risings and settings of the sun and moon. In about 2100 BC, Stonehenge III was constructed at the center of the site (shown by the circle of dashes). Stonehenge III is the megalithic structure that draws our attention to the site today.

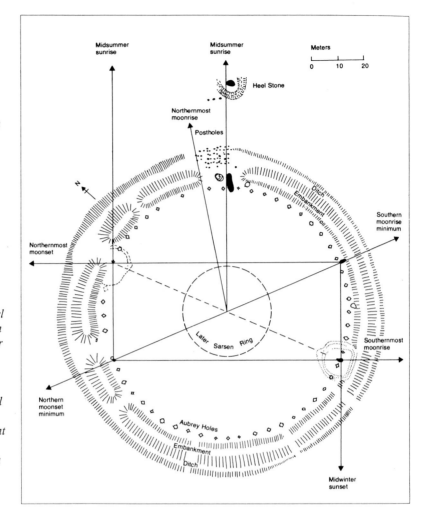

basic idea about the cyclic motions of the sun, moon, and stars. I shall use Stonehenge as a specific example, but the general rules will apply whether the monument is at the latitude of Teotihuacan in Mexico, Karnak in Egypt, or Moose Mountain in Saskatchewan. As we shall see, however, there is a dependence on latitude that makes certain phenomena much more striking in northern realms such as the British Isles or the Great Plains of Canada.

Without any doubt, the relevant celestial motions are much more easily explained visually than to readers of flat printed pages. As I contemplated the difficulties of making this presentation on paper, I could all too easily imagine my readers laying aside their books and quietly sneaking off to their refrigerators.

And then inspiration struck. As a model of the celestial sphere, a cylindrical cola can is a fair approximation. The celestial equator, which splits the sky into its northern and southern hemispheres, nicely girdles the can. Cocked at a 23° angle to the equator is the ecliptic, the other great celestial circle that concerns us when considering the basic astronomy of Stonehenge.

To convince any skeptical readers that I am really serious about this

Figure 3.2 The later Stonehenge – Stonehenge III. Dotted lines show what one time stood at the site: a horseshoe of five trilithons within a ring of 30 sarsen stones. Eighty meters from the center of the ring is an outlying marker, the Heel Stone; in combination with the central trilithon, it marks the central axis of the monument. The illustration shows several putative alignments of the monument to various risings and settings of the sun and moon. The two solstitial alignments associated with the central axis are far less problematical than the alignments associated with the off-center trilithons. The illustration fails to show the various fallen stones and fragments. Yet no rock of any size was originally present: the soil beneath the grassy surface is exclusively chalk.

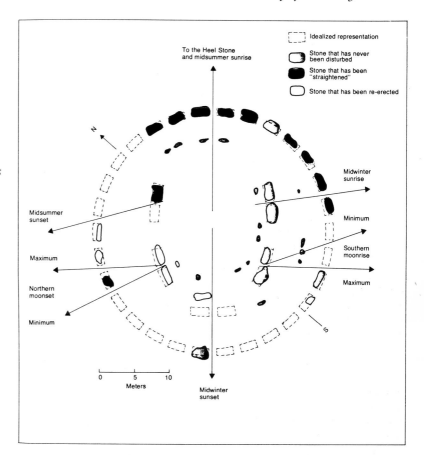

analogy, I have provided (see Figure 3.3) a wrap-around that can easily be xeroxed and then cut out and taped to a beer or soft drink can. (If you lack either of these, a Campbell's soup can has the same diameter.) To be thorough, I should have included a round paste-on for the top of the can with a dot in the center labeled 'North Pole,' but I decided that some things could be left to the imagination.

With the wrap-around taped to the cola can, the sun's daily motion around the celestial sphere can be imitated simply by rotating the can about its cylindrical axis. The sun's *yearly* motion among the stars, in the opposite direction, is shown by the succession of monthly images along the ecliptic. You can think of the sky as a giant celestial roulette wheel, with the heavens spinning eastward one revolution per day and the little ball, our sun, going around the other way, making one passage through the ecliptic each year.

The sun reaches its northernmost position near the end of June, but notice that it is almost equally far north in May or July. By our present calender the sun reaches its highest point around 21 June, the summer solstice, also known as midsummer day. It isn't midsummer by our seasonal conventions, but rather the *beginning* of summer. The arrival of the sun at this northernmost position is traditionally associated with some sort of a holiday; and Philip Morrison assures me that 4 July is, in fact, our national holiday for midsummer. When I objected, saying that everyone knows the Fourth of July marks the

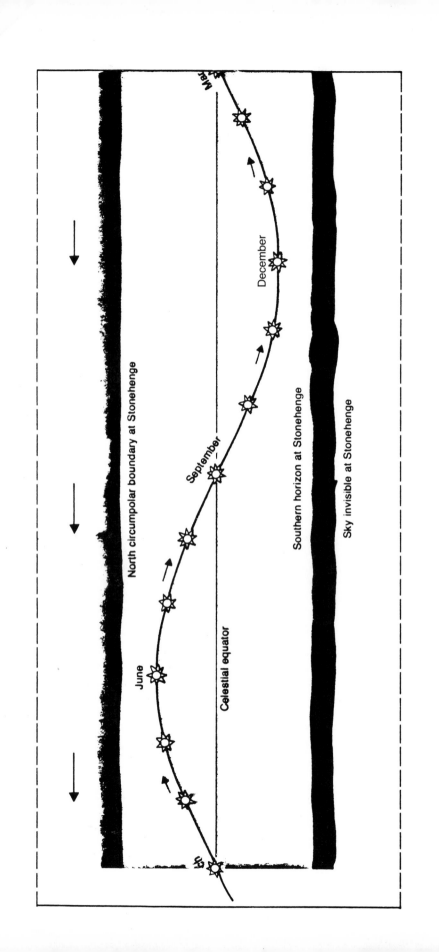

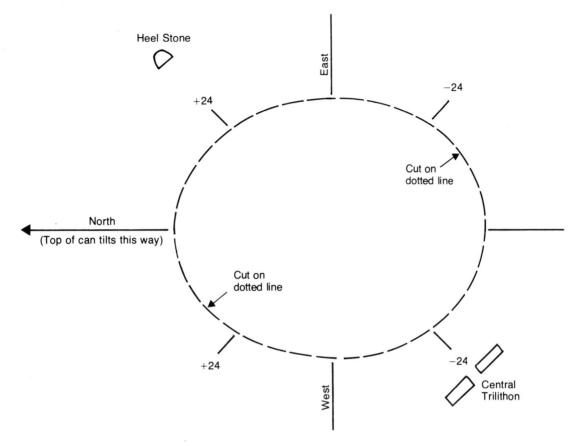

Figure 3.3 The Stonehenge decoder: a cola-can universe, with accompanying sleeve, to show the year-round motion of the sun over Stonehenge. To construct the device: Cut out the strip of paper printed on p.16 and tape it to a 12-ounce beer or cola can. The can now serves to represent the motion of the sun through the celestial sphere. In particular, that motion can be decomposed into two parts. The first of these, caused by the earth's turning on its axis once each day, is simulated by turning the can on its axis in a clockwise direction. The second, caused by the earth's revolution around the sun each year, moves the sun along the so-called ecliptic, the curved line on the paper strip, one complete cycle each
year. To visualize the superposition of these motions, the reader is urged to imagine a roulette wheel spinning in one direction as the sun moves along its track in the other. Now cut out the oval-shaped hole above. Position the cola can inside the hole in such a way that the edge of the hole closely meets the wall of the can, and tilts in the direction shown on the decoder. When positioned properly, the uppermost point on the oval's edge will touch the strip of paper along the line marked 'north circumpolar boundary,' and the lowermost point on the edge will touch it along the line marked 'southern horizon.' To find the motion of the sun over Stonehenge on any given day, simply choose a
point on the ecliptic and follow its motion as the can is turned clockwise, always keeping the can properly tilted relative to the oval. At spring or fall equinox (corresponding to sun-images labeled 'March' and 'September'), the sun rises due east and sets due west. At summer solstice (sun labeled 'June'), the monument reveals its solar alignment: on this day, when the sun is highest over the celestial equator, it rises over the Heel Stone and in alignment with the central axis of Stonehenge, as marked out by the Heel Stone and the central trilithon. At winter solstice, the sun (labeled 'December') sets in alignment with the other end of the axis.

Figure 3.4 This dramatic view of the sun shining through the narrow slit of a trilithon was taken shortly after the winter solstice. The archaeoastronomical alignments deal with rising or setting phenomena, however, and not spectacular configurations like this that can be arranged at will simply by walking into the shadow of the megaliths.

signing of the Declaration of Independence, Professor Morrison replied that 'if it weren't so, there would have been some other holiday. There *has* to be a holiday at midsummer, and just at this moment in our national life, 4 July has captured the position.'

As for the stars on the cola-can universe, I have omitted them, because the positions of the constellations today are quite different from those in the third millennium BC, when Stonehenge was built. Nowadays, for example, when the sun is at its northernmost solstitial point it crosses the northwest corner of Gemini. But a few centuries before Christ, when the constellations were being named, Cancer occupied this place – hence the name 'Tropic of Cancer.' And when the anonymous men of prehistory laid out Stonehenge, the sun went nearby the star Regulus, in Leo, at the time of summer solstice. In another 9000 years Regulus will be about as far south of the equator as it was north in 2000 BC. The slow shift in the constellations is called precession. As a consequence of this shift, the zodiacal figures last matched their present positions 26 000 years ago, and during the next 26 000 years they will cycle all the way around the ecliptic and come back to their current coordinates. It follows that the rising direction of a bright star is a comparatively ephemeral orientation. If the pyramids had been built to line up with the rising of the star Sirius, they would now be quite askew with respect to that star.

On the other hand, although the stars gradually change their orientation, the 23° (almost 24°) tilt of the ecliptic stays pretty nearly constant, so the north-south excursion of the sun is nearly the same year after year, millennium after millennium. Stonehenge, built over 4000 years ago to line up with the northernmost rising of the sun, still lines up with the northernmost rising of the sun. In order to see *where* this northernmost point lies along the horizon, I have designed a horizontal plane with an elliptical hole in which to rotate the celestial cylinder. (See Figure 3.4; after xeroxing, cut out the ellipse, and I would recommend pasting the sheet on a thin piece of cardboard to give a greater planar stability.) The can must be tilted in the hole to fill the ellipse exactly; two black strips on the wrap-around will help maintain the proper slant while you turn the can.

The particular slope of the hole – and the corresponding tilt of the cylinder – depend, of course, on the geographic latitude one wishes to depict. For example, to represent an observer at the North Pole, the can would be straight up and down, and the horizontal plane would have simply a circular hole cut into it. The sky north of the equator would always be visible, the part to the south never, so that clearly the sun would be in the visible hemisphere only during the spring and summer; the sun would circle the sky uninterruptedly throughout those six months. So much for monument building at the Pole.

For an observer at the equator the can would be on its side, and over the course of the year the sun would shuttle between 23° north of east and 23° south of east in its daily risings. At noon the sun would always be high in the sky, never more than about 23° from the zenith; and at noon-time twice a year it would be directly overhead. The annual excursion of the sun becomes larger and therefore increasingly interesting as an observer moves from the equator and tropics to more northern latitudes. On the model, the can has less and less tilt.

At an angle of 51.2° (the latitude of Stonehenge) the sun swings back and forth over an arc of nearly 90°. If you rotate the can in the cut-out ellipse (which is designed to represent the latitude of Stonehenge), you will discover how the rising and setting positions of the sun oscillate north and south along the horizon, and you will see that the swing is quite pronounced. Precisely how large is it? You can measure it on the model, or you can put it on your pocket calculator: $\sin D = \sin \delta / \cos \phi$, where δ, the sun's maximum angle above the equator (its maximum declination) is about 23.5° and the latitude ϕ is 51.2°. The maximum deviation D from the east-west line is just about 40°.

Enough for the basic mathematics. Let us next apply a thought-experiment to prehistory. Ancient man, living on the island that is now called Britain, must have been aware of the rising sun's endless excursions back and forth along the horizon. Eventually someone must have marked by trial and error the direction of the northern rising of the summer sun. How fascinating, then, to watch day after day on some later year as the sun worked its way toward its northern limit. On each successive day the sun would take a smaller step than the day before, preventing any intuitive extrapolation. Would the sun reach the same limit? Or would it go beyond? What an exciting discovery to find that each year the sun reached *exactly* the same northernmost alignment! The weather might vary in a capricious manner, but the sun had a faithful regularity. Surely a discovery worth a monument!

The grand edifice at Stonehenge indeed points to the northernmost rising of the sun, for from the center of the stone ring, the sun can be seen to rise above the Heel Stone at the time of June solstice. In some of his more dogmatic moments, the distinguished archaeologist of Stonehenge, R. J. C. Atkinson, has denied this, and he was made to look rather silly some years ago when a CBS television crew filmed the sun rising majestically past the Heel Stone at the summer solstice. Still, the sun in truth rises to the left of the Heel Stone and moves to a position above the tip of that boulder. Hence, as Atkinson correctly claimed, the shadow of the Heel Stone cannot fall into the center of the ring at the solstice. In 2000 BC the sun would have risen a degree farther north than it does now, because the tilt of the ecliptic does vary slowly. Probably, though, the Heel Stone was more upright then; and in any event, a sightline of only 80 m from the Heel Stone to the center of the sarsen circle means that the observation is not very precise: the sunrise would appear almost the same for a week before and after solstice. Certainly for ritual purposes the principal axis of the monument points to the rising sun at the summer solstice. By symmetry the opposite direction points southwest, to the southernmost setting of the sun at the winter solstice.

The real question about Stonehenge is, does it point to anything else? More specifically, would it have required a Stone Age Einstein to think of asking whether the *moon* also had a northernmost rising point? I am not sure. What would have been needed for investigation of the moon was a sufficient time free from the exigencies of food gathering. But a special sort of curiosity would have been necessary as well. Each month, in one phase or another, the moon will rise once in the vicinity of the Heel Stone, yet in general this alignment will not be very exact. It thus could be ignored as one of the vagaries of that fickle object. Furthermore, not all of these risings would be visible; the solar brilliance at midsummer would overpower a thin crescent rising soon after the sun, for instance. Finally, because the moon moves about 12° a day, it could skip past the very northern-most position on any given cycle.

The reason that the moon's northernmost rising is only in the vicinity of the Heel Stone, and not dead on, is that the moon's path is askew to the sun's. (If it were not, there would necessarily be a lunar and solar eclipse every month.) The moon's path in fact cuts the ecliptic at two opposite points in the sky, and wanders off by 5° north of the ecliptic on one side of the sky, then 5° south on the other. Moreover, the place where the maximum wandering occurs is slowly changing. Hence the moon's path can go 5° north of the 23.5° north-ernmost point of the ecliptic, but nine years later it will go 5° south of this northernmost point. In the first situation the moon would move in the sky between 28.5° north and 28.5° south, and in the second between only 18.5° north and 18.5° south. With the handy formula we used earlier in this chapter, we get an extreme swing along the horizon of 50° north of east, well past the Heel Stone. But nine years later, we find that the moon at most would rise only 30° north of east. In other words, our Stone-Age astronomer would not only have to have the insight to ask if the moon, like the sun, had a northernmost rising, but he would also have needed plenty of time to be sure – 18 years to complete the observation of just one complete cycle.

Would it be easy to think of trying this? I don't know, *but we can be rather sure that it was done*. Excavations at Stonehenge have revealed a series of post holes at angles northward of the Heel Stone in just the positions we would expect if someone was trying to establish the moon's northernmost limit by trial and error.

In its earliest stages the Stonehenge site was surrounded by an embankment approximately 2 m high and about 100 m in diameter. This earthwork ring had one entrance, in the general direction of the Heel Stone, but not symmetrically aligned with it. The opening was just wide enough to accommodate the post holes lined up with the northernmost excursion of the moon. Because similar grids are found at other neolithic sites, there is good reason to suppose that these represent lunar alignments.

Gerald Hawkins, who more than anyone else has drawn attention to the astronomical nature of Stonehenge, claims that solar and lunar alignments can also be associated with the four trilithons flanking the main axis of the monument. These are shown on Figure 3.2, where the megaliths are represented as idealized rectangles. The idealized representation is in itself instructive, because we notice at once that not all the sightlines would be possible on account of the awkward slanting views required. To be sure, the actual megalithic boulders are

Figure 3.5 Stonehenge from the east. The photograph was taken by the author in 1958; since then, some further reconstruction of the site has been done. Near the middle of the silhouetted image of the monument is a stone with a small nub on top. The stone is the remaining megalith of the central trilithon, and the nub is a tenon – a protrusion that once fitted into a corresponding hole in a lintel stone above.

more rough-hewn, so the sightlines are in fact feasible. Yet it gives pause to me, at least, to suppose that a grand edifice of this sort was specially designed so that the sightlines would depend on the imperfections or asymmetrical sizes of the rocks. I would also worry if the southern moonrise extremes and the northern moonset extremes were megalithically marked, and not the other pair of events (northern moonrise and southern moonset).

Up till now I have virtually ignored one important historical aspect of the Stonehenge site: the monument we now admire is Stonehenge III, built around 2100 BC; the sarsen ring and trilithons stand in the center of a much larger and older site whose development occurred hundreds of years earlier. And it is the early Stonehenge that has a greater claim as a research observatory, if for no other reason than that the sightlines are considerably longer and more flexible than those proposed by Hawkins for the sarsen ring and trilithons. The early Stonehenge included the previously mentioned embankment, as well as a ring of 56 chalk-filled holes, now named after the antiquarian John Aubrey. At some point in the development of the site there was added onto the ring of Aubrey Holes two mounds and two additional marker stones, known as the station stones (see Figure 3.1). These four features form the corners of a rectangle centered on the sarsen ring, although the sarsen ring had not yet been built at that time. The short sides of the rectangle are parallel to the direction to the Heel Stone; hence they point to the extreme excursion of the sun. The long sides point to the extreme excursion of the moon. (Only at the latitude of Stonehenge are these extremes joined by a right angle; if this is significant, then these relationships were not discovered here, but the observatory was erected at this specific latitude to take advantage of a previous discovery.)

The longer lunar alignments (if so they be) in the early Stonehenge do have sightlines an order of magnitude greater than those from the trilithons to the sarsen ring. Even so, they are short compared to the megalithic sightlines of 18 and 27 miles that Alexander Thom has surveyed at Ballochroy and Kintraw in Scotland. Since Stonehenge

Figure 3.6 Part of the sarsen circle at Stonehenge. Framed between the two visible sarsen stones is the Heel Stone, 80 m from the center of the ring. The smaller stones in front of the sarsen stones are two of the so-called Blue Stones, not discussed in the article. Nothing is known about their function, except that they are placed along a circle within the sarsen circle, and that the Stonehengers themselves rearranged them. The stones in the foreground and at the left are fragments; many such stones are present at the site.

seems to be an older site, perhaps the astronomy spread from there to the more accurate observatories elsewhere in Britain.

Indeed, Stonehenge cannot be considered in isolation, for the patient surveying of Alexander Thom and his associates has rather convincingly established the existence of lunar sightlines in conjunction with many of the more elaborate stone circles in the British Isles, a fact about megalithic society now almost commonly accepted by the archaeologists. Yet in Stonehenge III we find no elaborated observing site with longer baselines and more precise markers than the earlier Stonehenge constructions. Instead there is a monumental commemoration in stone of something long since discovered and perhaps already on its way to being forgotten. There is a striking parallel to this in eighteenth-century India where the ruler Jai Singh created five impressive stone observatories, all completely anachronistic (considering that the telescope had been introduced into India decades earlier), yet comfortable to the monarch's aspirations. Indeed, Jai Singh's instruments remain impressive till this day.

The combination of lunar and solar sightlines embodied in a great ritual center at Stonehenge suggests a well-organized primitive cult, possibly with the sun and moon as male and female in some grand fertility rites. Such suggestions are mere speculation, for here the stones are even more silent.

Nevertheless, the idea is no more fragile than the proposition that Stonehenge was a Stone Age eclipse calculator. There is a lunar eclipse cycle of 56, albeit not a very good one, but remember that some of the 56 Aubrey Holes were already covered by the station mounds when the sightlines were getting established. I beg the reader's pardon for not trying here to explain either the cyclic behaviour of eclipses or the imaginative suggestions by Hawkins and by Fred Hoyle for using the Aubrey Holes to calculate eclipses; the references at the end of the chapter will give a full account.

Suffice it to say that I remain skeptical. There are certain aspects that stagger the imagination. To get the idea that you could predict eclipses in some cyclical fashion, you would have to have some long record of observations and some kind of motivation for recording them in the first place. Such a record would presumably have to be oral. Today we cannot begin to conceive of the significance of oral records. We have too much cluttery detail to remember, and we are not very good at memorizing things. I am sure that memorization must have played a much more significant role for ancient people than it does for us, because we are so dependent on the written record. Even so, such a route to the prediction of eclipses seems incredible to me.

On the other hand, for a people so concerned with capturing the northernmost position of the sun and moon, the conception of the moon's nodes (the points at which its path crosses the path of the sun and where consequently eclipses can take place) may not be terribly far behind. In other words, it may well have been possible for the Stonehengers to have correlated eclipses with the celestial geometry of the solar and lunar paths rather than with cylces of eclipses derived from a communal memory of events long past. To me, this seems to be a fabulous jump for neolithic man to have made, but there is nothing to have prevented a Stone Age genius from finding the correlations simply with sticks and stones. So perhaps one of these days we will have to revise our notions concerning the sophistication of megalithic astronomy in the third millennium BC. Until then, in the words of Aubrey Burl, 'This ravaged colossus rests like a cage of sand-scoured ribs on the shores of eternity, its flesh forever lost.'

Notes and references

See Gerald S. Hawkins, *Stonehenge Decoded* (New York, 1965), Fred Hoyle, *On Stonehenge* (San Francisco, 1977), Aubrey Burl, *Rings of Stone* (London, 1979), Euan W. MacKie, *Science and Society in Prehistoric Britain* (New York, 1977), and Alexander Thom, Archibald S. Thom and Alexander S. Thom, Stonehenge, *Journal for the History of Astronomy* **5** (1974), 71–90.

4 *Some puzzles of Ptolemy's star catalog* *

Anyone who wishes to understand the origin of the constellations must sooner or later consult Ptolemy's star catalog. Dating from the second century AD, this ancient list is our oldest systematic source of star descriptions, brightnesses, and numerical positions.

Ptolemy's catalog occupies the major part of Books VII and VIII (out of 13) of his great treatise, the *Almagest*. The Alexandrian astronomer arranged his stars into 21 groups north of the ecliptic, 12 zodiacal constellations, and 15 south of the ecliptic. Within the groups he worked systematically through each constellation pattern, leaving to the end a description of the so-called 'unfigured' stars that did not help delineate the pattern.

In all, Ptolemy tabulated 1022 stars in 48 constellations. Three of the stars were duplicated, but all were marked as such: Fomalhaut is the mouth of the Southern Fish and simultaneously the splash from the Water Bearer (Aquarius); Alnath, the lower-right star of Auriga's pentagon, is both the ankle of the Charioteer and the tip of the Bull's horn; and the right foot of Hercules matches the crook of herdsman Bootes.

Clearly for Ptolemy, as for his predecessors, the arrangement of the figures had primacy over the numerical coordinates of the stars. Furthermore, he made no attempt to represent *all* the visible stars, so that many, especially the fainter ones, were omitted. However, he was very interested in knowing if the stars gradually shifted their relative positions, and thus he gave an extensive list of stars lying in straight lines, or nearly so, for future tests of stellar motions. Probably such a list was compiled observationally by the simple method of using a string to detect straight-line patterns.

Star positions in Ptolemy's catalog are given not only *descriptively* (in terms of the mythological figures), but also *numerically* by longitude and latitude, that is, with respect to the ecliptic. Unfortunately, however, there is a problem with the longitudes: they are all systematically too small by about a degree.

Ptolemy knew that stellar longitudes slowly increase over time, a phenomenon discovered by Hipparchus and known as precession of the equinoxes. Hipparchus had given a rough value for this effect, 1° per century, a conveniently round number but corresponding to a precession much too slow. Ptolemy accepted Hipparchus' degree-per-century precession and also his value for the length of the year, which was a little too long. These two errors dove-tailed, so that when Ptolemy extrapolated for the 265 years from Hipparchus to his own time, they produced deceivingly consistent results. Since Ptolemy blundered badly in measuring the time of the equinoxes in his own day, he failed to recognize these systematic errors, which then permeated his entire coordinate framework.

*Coauthored with Barbara Welther.

Now if the faulty value of the precession is applied backward to the faulty star positions in Ptolemy's catalog – lo and behold – the positions come out correctly for Hipparchus' time. Needless to say, the suspicion arose many years ago that Ptolemy may have cribbed a perfectly good catalog from Hipparchus to which he then added to each stellar longitude a standard 2° 40′ correction for precession for the $2\frac{2}{3}$ centuries elapsed since Hipparchus (instead of the true value of 1° per 72 years or 3° 40′ precession).

Is there any evidence that Ptolemy's catalog was taken over *en bloc* from Hipparchus? Ptolemy generally gives credit to his predecessors, and, in fact, most of what we know about Hipparchus comes from the information in the *Almagest*. In this case Ptolemy mentions that he has arranged the catalog differently from his predecessors, referring to the organization within the constellations rather than to the numerical positions. According to direct evidence, Hipparchus neither systematically specified the stars by longitude and latitude nor did he observe as many as 1000, though he did describe about 850 stars in a commentary on a famous work by the Greek poet Aratus.

There is, however, a curious puzzle in the way Ptolemy specified the fractions of a degree for the coordinates. Usually he used 10′ intervals specified as $\frac{1}{6}$°, $\frac{1}{3}$°, and so forth. Occasionally $\frac{1}{4}$° intervals are found. The fact that some of the latitudes were measured in quarters rather than sixths led the astronomer C. H. F. Peters to conclude, almost a century ago, that two different instruments had been used. In 1917, J. L. E. Dreyer expanded the argument to suggest that, while part of the catalog had perhaps come from Hipparchus, surely the $\frac{1}{4}$° stars had been Ptolemy's own additions. Nevertheless, Robert R. Newton has argued in his *The Crime of Claudius Ptolemy* that Dreyer's reasoning was spurious. His argument runs as follows.

If the stars are randomly distributed, there must be as many stellar coordinates with fractional parts at 5′ as at 10′, or at 15′, or at 20′, and so on. However, Ptolemy could specify 10′ as $\frac{1}{6}$°, 15′ as $\frac{1}{4}$°, 20′ as $\frac{1}{3}$°, but he did not specify any star at 5′ since he didn't use $\frac{1}{12}$°. Although it is equally likely for the fractional part to be in any 5′ interval, Ptolemy chose not to use the fractions 5′, 25′, 35′, or 55′. Therefore, stellar positions falling in such intervals were rounded up or down, resulting in a theoretical distribution like the one shown outlined in Figure 4.1. The higher distribution at 0′ and 30′ results from some of the fractional parts in the adjacent 5′ bins being rounded to those values.

Now, when the fractional parts of Ptolemy's stellar *latitudes* are plotted, they come out very close to the expected distribution, as shown by the gray bars in Figure 4.1(*a*). Consequently, R. R. Newton concluded that 'this distribution proves clearly that we are dealing with a homogeneous body of coordinates,' making it highly unlikely that a different instrument was used on the $\frac{1}{4}$° stars. However, mathematically minded readers should be able to show rather easily that a similar distribution, but fitting the fractional parts of the latitudes even better, can be made by assuming that 726 stars were observed with an instrument whose latitude scale was calibrated in sixths of a

Figure 4.1 An analysis of Ptolemy's star positions in graphical form. In (a) the number of stars in each fractional latitude *interval is plotted, with the theoretical distribution outlined in black, Ptolemy's in gray. Similarly, the* longitudes *are given in (b) and (c); the values listed by Ptolemy are the same in both drawings, while at bottom the theoretical ones have been increased by 40'. The large number of longitudes at 40' in Ptolemy's catalog suggests a 40' shift from whole degrees.*

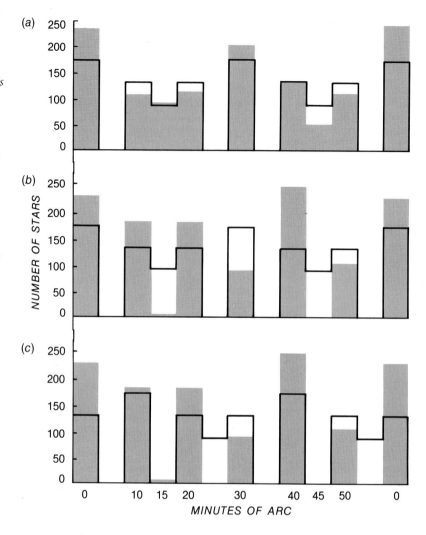

degree whereas 300 others were observed with a scale calibrated in quarters of a degree.

If we now plot the fractional parts of the *longitudes* (in gray in Figure 4.1(*b*)), the distribution is quite different from the expected one. At least part of the reason for the discrepancy is that Ptolemy rarely uses $\frac{1}{4}°$ or $\frac{3}{4}°$ for the longitudes. Is the lack of entries at $\frac{1}{4}°$ and $\frac{3}{4}°$ telling us something more? Suppose that Ptolemy started with longitudes for the time of Hipparchus, and arbitrarily precessed them all by 2° 40'? Then the precessed theoretical distribution (outlined in black) as compared with the actual distribution (gray) would give the result shown in Figure 4.1(*c*). Note that the 40' shift produces no stars at $\frac{1}{4}°$ or $\frac{3}{4}°$ but does cause many to fall at the 'not-allowed' positions of 25' and 55'.

R. R. Newton has also examined this problem and, determined to convict Ptolemy, concluded that the Alexandrian astronomer not only stole Hipparchan positions and precessed them with the faulty 2° 40', but that he rounded all those stars at 25' downward and all those at 55' upward. This produces a striking match with the shifted theoretical distribution, but we must be wary of such a circular argument. There is absolutely no reason to suppose that Ptolemy would have rounded the fractions in such a capricious manner unless there is already evidence on other grounds that he precessed the longitudes from an earlier epoch.

In any event, Ptolemy could have used a Hipparchan framework for convenience, reobserved the stars (using some key stars to get started, a stretched string as an observing aid, and a celestial globe for marking the results), and then afterward precessed these relative coordinates to his own date. It does seem plausible that, if a major earlier catalog existed, then he would have used it, at the very least, as the starting point for his own efforts. Like so many of the circumstances surrounding the catalog, this aspect still remains a mystery.

A second puzzle concerns the arrangement of the patterns in Ptolemy's catalog, or rather, the arrangement of the stars *not* included. Around the southern pole of the sky is a zone forever invisible from Alexandria's latitude of 31° north. However, the missing region of the catalog is larger than necessary and rather asymmetrically placed, as may be seen in the plot of Ptolemy's southern stars (Figure 4.2(*a*)).

Over a period of some centuries the pattern of invisible stars slowly shifts – another consequence of precession. One of the most conspicuous effects of this phenomenon is to change the north pole star. Our current pole star, Polaris, will actually pass closest to the true north polar position around AD 2100. It came close enough to the pole to become a good navigational star in the fifteenth century, just at the beginning of the 'Age of Exploration.' When the Egyptian pyramids were built, an obscure star in the constellation Draco, now named Thuban, served as an approximate pole star, and about 12 000 years from now the first-magnitude star Vega will come within 5° of the pole, becoming the 'Polaris' of that distant time.

Is it possible that Ptolemy's zone of southern invisibility is centered about an ancient pole, and therefore reflects a fossilized list of constellations made at an earlier epoch? If so, this might tell us something about the origin and age of the constellations.

In 1932 Knut Lundmark plotted Ptolemy's zone of invisibility and concluded that it corresponded to a circle centered on the south celestial pole of 4800 BC. Our own plots do not corroborate his conclusion. In the first place, it is difficult to define unambiguously a circle of invisibility, partly because Ptolemy's star positions greatly deteriorate in quality near his southern horizon. For example, the brightest stars of Centaurus and the Southern Cross contained within it are barely recognizable because of the distortions in their positions. On the other side of the zone, Eridanus ends in a spurious first-magnitude star. Had Ptolemy heard travelers' reports of Achernar, which lay forever below his horizon?

Even more dramatic is the plot of the stars *not* included in Ptolemy's catalog (Figure 4.2(*b*)). If we overlay a circle to include the largest number of omitted stars, it is quite different from the previous zone. We immediately notice how the concentration of stars in the Milky Way makes the whole situation quite unsymmetrical, and we

Figure 4.2 Two plots of bright stars in the south celestial hemisphere. (a) The stars from the Almagest *and a gray circle indicating the part of the sky where stars are missing: the precessional path of the south celestial pole is indicated. (b) Stars to magnitude 5.0 not given by Ptolemy are shown for AD 140 from data in the* Yale Bright Star Catalogue. *Here the circle of invisibility has been repositioned to include the greatest number of stars.*

(a)

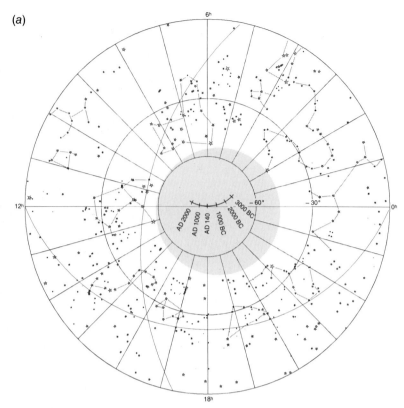

(b)

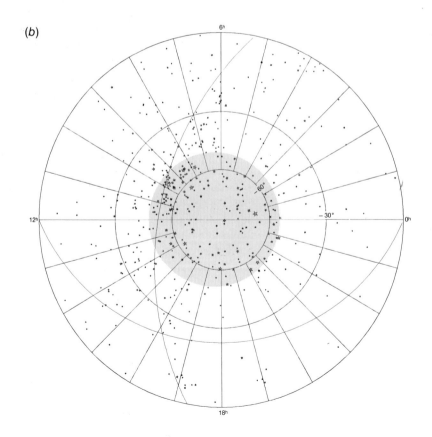

*Figure 4.3 Claudius
Ptolemy, second-century
Alexandrian astronomer,
holds a cross-staff (a later
invention) for measuring star
separations.*

also see that Ptolemy has in fact omitted relatively few stars on the 0^h side of the zone of invisibility. In looking at the second chart, we have to conclude that there is no way to derive a very early date for the particular stars listed or not listed in the Ptolemaic catalog. In fact, Ptolemy's zone of invisibility can no longer be considered a substantial puzzle.

Nevertheless, there remains at least one more mystery in the catalog. Ptolemy singles out six stars as red in color (though he does not call them by these generally more modern names): Antares, Aldebaran, Sirius, Pollux, Arcturus, and Betelgeuse. Anyone familiar with the sky will quickly notice a glaring anomaly in the list, the white Dog Star, Sirius.

There exists a considerable literature on this point alone. To date, no historian has been able to demonstrate any mistake in the transmission of Ptolemy's text with regard to the color of Sirius, nor has any astrophysicist come up with a plausible scenario that could have Sirius or its white-dwarf companion as a red star a mere two millennia ago! Nor is there any unambiguous corroboration of Ptolemy's listing. Perhaps future scholars will have better luck clearing up this puzzle of the *Almagest*'s star catalog.

Notes and references

G. J. Toomer's newly translated and annotated edition of *Ptolemy's Almagest* (London, 1984) contains a fine list of modern identifications in the catalogue portion of the text. J. L. E. Dreyer's 'On the Origin of Ptolemy's Catalogue of Stars' may be found in *Monthly Notices of the Royal Astronomical Society*, **77** (1917), 528–39, and **78** (1918), 343–49. Robert R. Newton's discussion of the fractional parts in the star positions begins on page 245 of *The Crime of Claudius Ptolemy* (Baltimore, 1977). Knut Lundmark's diagram of Ptolemy's 'zone of invisibility' appears on page 233 of 'Luminosities, Colours, Diameters, Densities, Masses of the Stars' in *Handbuch der Astrophysik*, Vol. 5, Part 1, 1932. Since this chapter was written, an important article has appeared, James Evans, 'Ptolemy's Star Catalogue,' *Journal for the History of Astronomy* **18** (1987), 155–72, **19** (1988), 233–78. Concerning the color of Sirius, see Kenneth Brecher's 'Sirius Enigmas,' pages 91–115 in *Astronomy of the Ancients* (Cambridge, Mass., 1979), edited by Kenneth Brecher and Michael Feirtag, and also Hugh M. Johnson's *Leaflet* 383 of the Astronomical Society of the Pacific, May, 1961.

5 Ptolemy and the maverick motion of Mercury

Of all the naked-eye planets, wily Mercury is the most difficult to observe. Generally obscured by the sun's glare, it can be seen by the unaided eye for only a few days around the time of its evening or morning elongations. So elusive is the swift-footed messenger of the gods that Copernicus is reputed never to have seen it. Such a statement is in all likelihood wrong, as I shall show. But to understand Mercury's tricky observational situation we should start 14 centuries earlier, with Claudius Ptolemy.

This Alexandrian astronomer is most famous for his *Almagest*, its Arabic title meaning literally 'The Greatest.' Written around AD 150, it is indeed the greatest surviving astronomical work from antiquity. What Ptolemy did was to show for the first time (as far as we know) how to convert observational data into the numerical parameters for his planetary models. With those models he was able to construct some ingenious tables from which planetary positions could be calculated for any given time.

Ptolemy accomplished his task within the geocentric framework almost universally accepted in his day. Of course, as a consequence of the revolutionary work of Copernicus, Kepler, Galileo, and Newton, the earth-centered Ptolemaic system has now been tossed into the trash can of discarded theories. Why, then, should we take time to consider the *Almagest*? Because, more than any other book, it demonstrated that natural phenomena, though complex in appearance, could be described mathematically to express relatively simple underlying regularities. It was this process that allowed for specific quantitative predictions.

Planetary motion appears complicated primarily because of retrogression – when, for a few weeks, a planet moves westward against the background stars, contrary to its normal eastward motion. Ptolemy realized that for any particular planet the lengths of these retrograde loops varied, depending on where the planet was relative to the stars. He also recognized that, apart from the retrogressions, the apparent speed of a planet varied in its course through the zodiac.

In order to account for all these features of the observed motions, Ptolemy introduced a model with three devices. First, in order to generate the retrograde motion, he used an *epicycle* riding on a larger circle called a *deferent*. The combined motion of the planet in the epicycle and the motion of the epicycle on the deferent caused the planet to swing in close to the earth. At such times the planet appears to move temporarily backward (see Figure 5.2(*a*)). Second, to account for the varying forward speed of the planet, he moved the deferent off center from the earth (to produce an eccentric circle), so the planet would appear to go faster when it was closer to us.

However, the epicycle and eccentric together were inadequate to

Figure 5.1 Ptolemy wrote: 'When I trace at my pleasure the windings to and fro of the heavenly bodies I take my fill of ambrosia, food of the gods.' His complacent self-confidence is elegantly captured in this fifteenth-century wood carving in the Ulm Cathedral in Germany.

reproduce correctly the varying size of the retrograde loops, so Ptolemy was obliged to seek yet a third device. He succeeded in solving this difficult problem with the *equant*, a locus of uniform angular motion within the deferent equal and opposite to the position of the earth (see Figure 5.2(*b*)); because the equant works so brilliantly in predicting the planetary motions, this invention by Ptolemy must be considered one of his greatest achievements.

Ptolemy had no trouble fitting planetary observations to his epicyclic model for the superior planets (Mars, Jupiter, and Saturn) and for Venus. Unfortunately, however, it was not so easy to get adequate observations of Mercury. Ptolemy would have been much better off if he had assumed that all the planets had the same mechanism, and then simply forced the available Mercury observations to fit as well as possible. What apparently happened, however, was that Ptolemy took the inadequate observations much too seriously, and thereby came up with an unnecessarily complicated model for Mercury.

One reason why Ptolemy might have been misled was that he believed Mercury traveled in the space between the moon and the other planets. Since he had been genuinely obliged to come up with an additional circle to represent the moon's intricate behavior, he probably imagined that Mercury, as an intermediate object, also required extra complexity.

Today we know that Mercury has the largest orbital eccentricity (0.21) of any naked-eye planet, tempting us to think of its path as a rather squashed ellipse. Actually, however, the orbit of Mercury is very well represented by an off-center circle – the orbit's elliptical shape is so slight that it is virtually unnoticeable in any ordinary diagram. It is essential to keep this point in mind, because the egg shape that Ptolemy came up with has absolutely nothing to do with the real configuration of Mercury's orbit.

Figure 5.3(*a*) shows the Copernican orbits of the earth and Mercury. Although it is anachronistic to discuss Ptolemy's investigation of Mercury in heliocentric terms, this arrangement makes it a little easier for us to envision the situation. What we think of as the eccentric orbit of Mercury is represented in Ptolemy's scheme as a purely circular epicycle with the sun inside; therefore, the off-centered nature of the planet's heliocentric orbit must be accounted for elsewhere. Let us now look at the observations Ptolemy reports.

Ptolemy usually gives only the minimum number of observations that he needs to determine the parameters for each planet; for each of the superior planets this is five. In the case of Mercury, however, he wants first to explore the general orientation of Mercury's orbit; thus, he gives a total of 16 Mercurian longitudes (positions measured along the ecliptic). The results of some of these observations have been included in the diagram, where the elongation from his mean sun is marked at several positions.

Two remarkable and somewhat puzzling things become apparent from this diagram. First, it is obvious that Ptolemy was unable to observe Mercury at an evening elongation in October or at a morning elongation in April. Second, from his incomplete roster of data, he has erroneously concluded that Mercury has two perigees (the positions where the epicycle appears the largest). In both February and

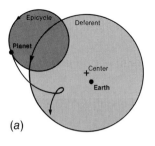 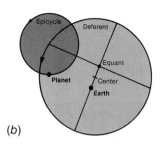

(a)   (b)

Figure 5.2 (a) The epicycle, bearing the planet, both rotates and revolves on the larger deferent circle, thereby causing apparent retrograde motion each time the planet swings close to the earth. Note that the earth is offset from the center of the deferent. (b) One of Ptolemy's greatest accomplishments was the perfection of the equant device. Like the hands of a clock, the equant turns uniformly, forcing the epicycle to move more quickly at perigee when it is close to the earth and far from the equant point. The equant is a good approximation of Kepler's second law of motion, whereby a planet sweeps out equal areas of its orbital plane in equal intervals of time.

June the combined elongations total $47\frac{3}{4}°$, whereas he deduced by assuming symmetry that in April the combined elongations would have been only $46\frac{1}{2}°$.

Why did Ptolemy fail to get the critical missing observations? Anyone who has systematically tried to observe elusive Mercury will quickly understand the reason. In early autumn evenings from the Northern Hemisphere, when the setting sun is near the equator, the ecliptic rises out of the west between the equator and the horizon. Thus, Mercury is located at an unfavorably low altitude in the sky. On early spring evenings the situation is quite the opposite, as shown in Figure 5.4(*a*). Mercury is then in its most favorable position to be sighted. For a northern naked-eye observer similar circumstances prevail with respect to the morning elongations, which are favorable in October and almost impossible in April.

The discontinuity of observations played havoc with Ptolemy's model for Mercury, for he mistakenly assumed that the unmeasured morning elongation in April would have equaled the observed evening elongation. Thus his purported double perigee was a fiction created by the lack of observations. But, because the epicycle of Mercury seemed largest at two different points, Ptolemy went ahead and incorporated into his scheme a 'crank' mechanism that pushed and pulled the epicycle (see Figure 5.3(*b*)). As the epicycle changed its distance from the earth, its apparent size varied, giving the double maximum but a single minimum. The resulting effective path of Mercury's epicycle was nearly an ellipse, but an ellipse having nothing to do with the planet's real orbit.

Such a scheme was eventually converted into a direct heliocentric equivalent by Copernicus, who also lacked the observations to contradict it. Copernicus wrote:

The foregoing method of analyzing this planet's motion was shown to us by the ancients. But they were helped by clearer skies where the Nile (it is said) does not give off such mists as does the Vistula for us. We inhabitants of a more severe region have been denied that advantage by nature. The less frequent calmness of our air, in addition to the great obliquity of the sphere, allows us to see Mercury more rarely, even when it is at its greatest elongation from the sun. For, Mercury's rising in the Ram and Fishes is not visible to us nor, on the other hand, is its setting in the Virgin and Balance. . . . This planet has accordingly inflicted many perplexities and labors on us in our investigation of its wanderings.

Note that in April the sun is in Aries, the Ram, as would be Mercury as it rises before dawn. In October the sun is in Libra, the

(a)   (b)

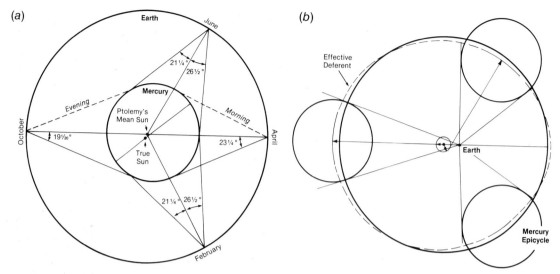

*Figure 5.3 (a) Shown
anachronistically in a
heliocentric arrangement are
some of Ptolemy's elongations
for Mercury as sighted from
the earth's orbit (outer
circle). As explained in the
text, it was impossible to
observe the morning
elongation in April, so
Ptolemy assumed (incorrectly)
that it would be the same as
the April evening elongation.
(b) In the Ptolemaic model
for Mercury, the deferent
moves on a small central
circle (here enlarged for
clarity) that produces a near-
elliptical orbit and two closest
approaches (perigees) for the
planet per orbit.*

Balance, as would be Mercury as it sets in the evening dusk. In other words, Copernicus does not say that he could never observe Mercury, but that it was impossible for him to observe it at precisely the same critical times when Ptolemy failed to measure it. I would personally find it hard to believe that Copernicus never saw the planet Mercury.

It is rather interesting to look more closely at the accuracy of Ptolemy's Mercury longitudes. I have tabulated some of them here, listing the modern dates and the mean elongation west or east, followed by the error in the elongation from the mean sun:

Date		Elongation	Error
4 June	138	26°30' E	+0°45'
5 June	130	21°15' W	+1°27'
5 April	135	23°15' E	−1°02'
2 Feb.	132	21°15' E	−1°27'
2 Feb.	141	26°30' W	+1°17'
3 Oct.	134	19°03' E	−0°32'

It is plain from the table that Ptolemy's observational base was much too coarse to justify his conclusion about the double perigee. In particular, he could hardly assume that the unmeasured morning elongation in April would be precisely $23\frac{1}{4}°$. Indeed, the observations are so poor that it would be a highly unlikely coincidence for the symmetry to have come out as perfectly as the diagram indicates.

It boggles the mind to believe that Ptolemy really had sets of matching observations in such complete agreement. Because of this and other similar problems in the *Almagest*, one contemporary critic has branded Ptolemy the greatest fraud in the history of science.

While it is difficult to believe that Ptolemy reported the Mercurian positions precisely as he measured them, there nevertheless seems to be a strong observational constraint in the way he has modeled Mercury's orbit. Had he been making up the observations, he could have saved a lot of trouble (and would probably have been more

Figure 5.4 (a) In April, when the sun is in Aries, the ecliptic stands almost perpendicular to the western horizon at sunset as seen from low northern latitudes. Thus, Mercury at eastern elongation is in a dark sky and easily visible. (b) Just the contrary is true at the eastern horizon and sunrise; Mercury is not high enough at an April western elongation for it to be sighted before morning twilight brightens the sky.

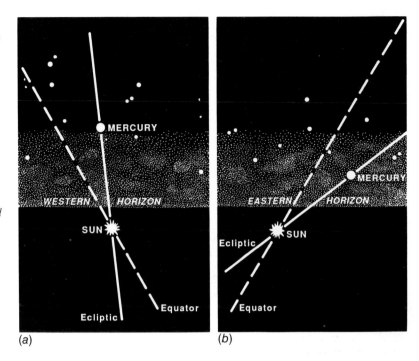

(a)   (b)

successful as well!) if he had assumed that the Mercury epicyclic model was the same as for the other planets.

It seems likely to me that, instead of inventing data out of whole cloth, Ptolemy must have placed too much faith in his own crude observations. But then, anxious to show how the model might in principle be most easily justified, and keen not to perpetuate data with observational errors, he must have also tried in some way to clean up his reported positions. Reprehensible as this may be by modern standards, we would err in casting judgment on Ptolemy's motivations in an epoch quite different from ours, when quantitative science was in its infancy.

Notes and references

The passage from Copernicus is found in Book 5, Chapter 30, of *De Revolutionibus* and was translated by Edward Rosen. A debate as to whether Ptolemy was a fraud is found in three articles by me and by R. R. Newton in the *Quarterly Journal* of the Royal Astronomical Society, **21** and **22** (1980–1).

6 How astronomers finally captured Mercury

Mercury, the innermost planet of the solar system, never moves farther than 28° from the sun in the sky. Consequently it can be seen only rarely with the naked eye, when it is near greatest elongation. From antiquity through the time of Copernicus, it was difficult to pin down Mercury's orbital motion from such bunched observations.

A theoretical model for the orbital motion of Mercury could have an appreciable error in heliocentric longitude that would not be easily detected at *elongation*, when the planet is moving pretty much along our line of sight. On the other hand, if a *transit* across the face of the sun could be observed, this would give a much stronger fix on the planet's position, as can be seen in Figure 6.1. However, without an adequate Mercury theory, an astronomer would be very lucky to predict a transit accurately enough for it to be observed. Furthermore, some kind of optical aid would be required for the observation.

Both a better theory and the telescope became available early in the seventeenth century. As Johannes Kepler carried out his investigations of the planetary system, he became convinced that some physical mechanism, common to all of the planets, had to be at work. Although he had at his disposal nearly 100 observations of Mercury made by Tycho Brahe, these also had the limitations described above. Nevertheless, Kepler demanded that Mercury have an elliptical orbit just as he had demonstrated for Mars, even though his supposition was an article of faith rather than an observational deduction. Happily, such a conviction was essentially correct, and it furnished a kind of interpolation scheme to carry Mercury past the unobserved portions of its orbit.

Kepler himself supposed that he observed a transit of Mercury on 28 May 1607. Between clouds, he caught a glimpse of a dark spot on the solar disk as it was projected through a small crack in the roof of his house in Prague. He was so excited that he ran all the way up the castle hill so that Emperor Rudolf II could be informed! Eventually, Kepler understood that a Mercurian transit was impossible on that date, for one can occur only when the earth has about the same heliocentric longitude as the nodes of Mercury's orbit. (The nodes are the places where the planet, moving north or south, crosses the plane of the earth's orbit. In the sky this plane is called the ecliptic.) This happens around 7 November and 5 May in the Gregorian calendar (or around 28 October and 25 April in the Julian calendar in the seventeenth century).

By 1618, older and wiser, Kepler realized that what he had seen was undoubtedly a sunspot – thereby anticipating Galileo's epochal discovery by several years, though he was unaware of what he had really glimpsed. Kepler finally admitted his blunder in the introduction to his *Ephemerides* for 1617: 'Did I pass off a spot I saw as

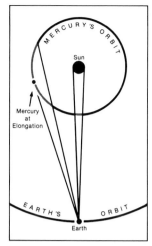

Figure 6.1 This schematic diagram shows why a transit of Mercury gives a relatively precise determination of the planet's heliocentric longitude. Since the sun's disk subtends half a degree, if a transit occurs the planet's position in the sky is known at least to that precision. (The uncertainty in heliocentric longitude is only slightly greater, because the planet is closer to the sun than to the earth.) At an elongation, however, the same measurement error would have an uncertainty many times larger, as indicated by the gray portions of Mercury's orbit.

Mercury? Then lucky me, the first in this century to observe sunspots.' (Kepler knew that a Mercury transit had been reported in the time of Charlemagne, and he must have guessed that it, too, was in reality a sunspot.)

Ultimately increasingly precise planetary theory allowed him to predict a real transit for 7 November 1631, and in 1629 he issued a small tract alerting astronomers to the event. Unfortunately, Kepler died a year before the transit took place, but in Paris the astronomer Pierre Gassendi waited patiently with his telescope, fighting the clouds and hoping to catch sight of the elusive Mercury. In a letter written to Kepler's old Tübingen University friend, Wilhelm Schickard, Gassendi related his results allegorically, designating the fleet-footed messenger of the gods by a reference to his mythological birthplace:

That sly Cyllenius introduced a fog to cover the earth and then appeared sooner and smaller than expected so that he could pass by either undetected or unrecognized. But accustomed to the tricks he played even in his infancy, Apollo favored us and arranged it so that, though he could escape notice in his approach, he could not depart utterly undetected. It was permitted me to restrain a bit his winged sandals even as they fled. . . . So, to speak briefly, I am more fortunate than so many of those Hermes-watchers who looked for the transit in vain, and I saw him where no one else has seen him so far, as it were, 'in Phoebus' throne, glittering with brilliant emeralds.'

November 7th had dawned mostly cloudy in Paris, and not until nearly 9 a.m. was the sun bright enough for Gassendi to glimpse the planet. But, since Mercury was unexpectedly small, he assumed that it was a sunspot. Clouds intervened again. But when the sun reappeared, he carefully measured the 'spot,' expecting to use it as a reference mark in case Mercury should appear. Still later, during another break in the clouds, he recognized the movement of the dark spot. Gassendi wrote: 'I could hardly persuade myself that it was Mercury, as my expectation of a larger magnitude bothered me, and I wondered if I had been deceived in the previous measurement. But when the sun shone again, I discovered further movement, and only then did I conclude that Mercury had come in on his splendid wings.'

Gassendi's letter to Schickard was published in 1632 in a small pamphlet, *Mercurius in sole visus*, in which he praised Kepler's astronomy. Schickard replied in his own pamphlet, saying that just as the names of Timocharis and Hipparchus had been preserved in the *Almagest*, so would Gassendi's name be memorialized in future years. Martin Hortensus, professor of mathematics at Amsterdam, was equally enthusiastic. Although clouded out, Hortensus expressed his joy that Gassendi had succeeded.

As a result of Gassendi's confirmation of Kepler's theory, at least one ephemeris maker, the Frenchman Noël Durret, announced that he was abandoning the rival but now obviously less accurate tables of Lansberg. He had used these for the first six years of his ephemerides. His volume for 1637–51 was reprinted in London with a conspicuous italicized notice at the beginning about the 'wonderfully small discrepancy' in Kepler's prediction. This publicity certainly helped win acceptance of Kepler's somewhat complex *Rudolphine Tables*.

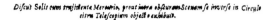

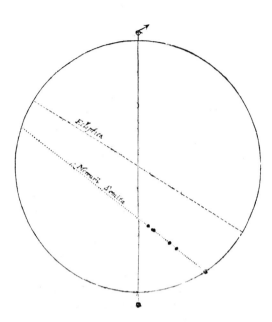

Figure 6.2 Gassendi's small tract on the 1631 Mercury transit was reprinted with his Institutio Astronomica *in 1656, where for the first time he included a diagram of the transit phenomenon.*

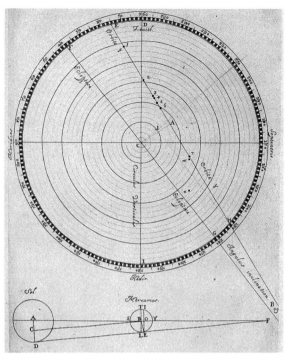

Figure 6.3 Hevelius' observations of the 1661 transit are given in a large plate in his monograph Mercurius in sole visus Gedani, *published in that city in 1662.*

As it happens, transits of Mercury take place approximately 13 times a century. Thus, astronomers would not have had to wait long for another observational opportunity had they been able to travel to other geographical longitudes. (In fact, Jeremy Shakerley, a young English astronomer, took off for India to see the transit of 3 November 1651. Later, he wrote to his friends that he succeeded, but he was never heard from again.)

The next transit visible from Europe did not occur until May, 1661, when the Gdansk astronomer Johannes Hevelius recorded it. In his lavishly produced book, he described his observations in detail. His *Mercurius in sole visus Gedani* of 1662 had a wider circulation than Gassendi's smaller tract; hence it helped convince astronomers about the small size of Mercury. In fact, Hevelius found that the planet was even smaller than Gassendi had estimated.

Although Hevelius was the only one to write extensively about the transit of 1661, it was also observed by the celebrated Dutch scientist Christiaan Huygens, who was spending a few weeks in London and who recorded the event in his diary. Huygens observed from the shop of the instrument maker Richard Reeves. With them was the 'mathematical practitioner' Thomas Streete, who was just about to issue his *Astronomia Carolina*, a new set of planetary tables named in honor of the new king, Charles II. In fact, the transit occurred on the coronation day, and Streete considered his successful prediction the consummation of his work on Mercury. Huygens had asked his brother Constantin to send to London one of their lenses, and perhaps it arrived in time for this observation. In his book Streete briefly describes using 'a good telescope, with red glasses for saving our eyes' to observe the transit for about an hour before clouds intervened.

Huygens mentioned the transit in his correspondence with the French astronomer Ismael Boulliau, who sent his congratulations, remarking sadly that 'I was on the road from Danzig [Gdansk] to Warsaw, and was thus deprived of seeing this rare and curious spectacle.' Perhaps as a consolation Hevelius dedicated his book to Boulliau. But there was also another reason: Hevelius found that of all the leading astronomical tables, only Kepler's *Rudolphines* and Boulliau's *Philolaics* (which also employed elliptical orbits) actually predicted a transit on that day.

Three years later another transit was predicted. By this time members of the newly formed Royal Society had become very interested, and the secretary sent letters to a variety of correspondents (including Gov. John Winthrop of Connecticut) giving Streete's predictions of the phenomenon. As secretary Henry Oldenburg wrote to Hevelius, 'Since the greater part of this conjunction will be hidden from us, if there be no mistake in the calculation, by the sun's being below the horizon, the Royal Society has commended its observation to some astrophiles in the English colonies in America.'

Although no successful observations were recorded, at least one attempt is documented by the English virtuoso and diarist John Evelyn. He wrote, 'I went to visit Mr Boyle (now here), whom I found with Dr Wallis and Dr Christopher Wren in the tower of the schools [that is, at Oxford], with an inverted tube or telescope, observing the discus of the sun for the passing of Mercury that day before it; but the latitude was so great that nothing happened.'

Oldenburg, writing to Boyle, confirmed this lack of success: 'I feare very much that you as well as we have taken pains in vaine about the Mercuriall conjunction. Our Virtuosi did observe both at Gresham and Mr Reeve's, but found nothing of what they looked for. I hope some of our ingenious friends in the American plantations . . . have made good observations.' But, apparently, even the American astrophiles failed.

By 1677 astronomers realized that accurate observations of a Mercury transit would provide an important test for the several planetary tables that by then had superseded Kepler's. Among those particularly eager to get precise data was Streete. Robert Hooke, then secretary of the Royal Society, noted the forthcoming event in his own diary: '*Thursday, October 25th* – Prepared experiment of weighing. Shewd the Royal Society. . . . I read from Hamborough a letter bringing severall Ephemerides about the Eclipses of ♄ by ☽ . . . . Took of Jonathan chocolat lb. $2\frac{1}{2}$ sh. . . . Mr. Street[e] here about ☿ in ☉ Sunday next.'

Unfortunately, the Sunday entry closed with the terse remarks, 'To Jonathan's with Street[e], thence with Chamberlaine to Barron. The

Figure 6.4 Huygens' diagram of the 1661 transit was not published until modern times. Note that he started to write the Julian date 'April,' used in London at that time, but overwrote 3 May of the Gregorian calendar used at his residence in France.

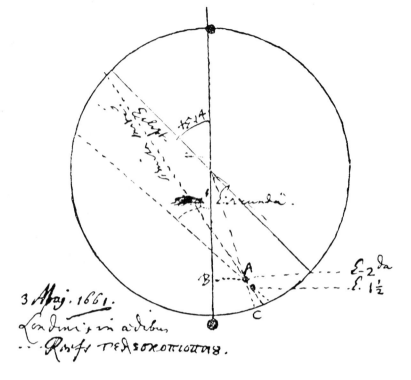

eclipse of ☉ by ☿ appeared not.' This was apparently owing to bad weather, for John Flamsteed reported an observation from the amateur instrument maker Richard Townley in Lancastershire. Edmond Halley, on his astronomical expedition to St Helena, managed to record both the ingress and egress. However, it was Halley's further observational work over the next few years (from 1682 to 1684) that made transits of Mercury far less significant for planetary theory. From his home in Islington (near London), the future Astronomer Royal was able to observe Mercury in broad daylight, even to within 15° of the sun, and these positions offered additional confirmation of the remarkable accuracy of Streete's tables.

Streete, despite naming his *Astronomia Carolina* after Charles II, 'never received a farthing' from the monarch, mostly because he refused to give a gratuity to the courtiers in order to have an introduction to the king. John Aubrey, who noted this detail in his *Brief Lives*, added that 'no man living has deserved so well of Astronomie.' Halley, ever an admirer of Streete's, echoed the sentiment when he revised and reprinted the latter's tables in 1710. Halley wrote that Streete served the science of astronomy so well that his Memory should 'still be precious with those that have any Relish for the Knowledge of the Celestial Motions.'

Yet it was Halley's own relatively short period of observing with his $5\frac{1}{2}$-foot sextant, recorded in an appendix to his reprint of Streete's volume, that finally gave astronomers the crucial missing observations of Mercury near its aphelion (its most distant point from the sun). These were observations that 'had never before been seen in

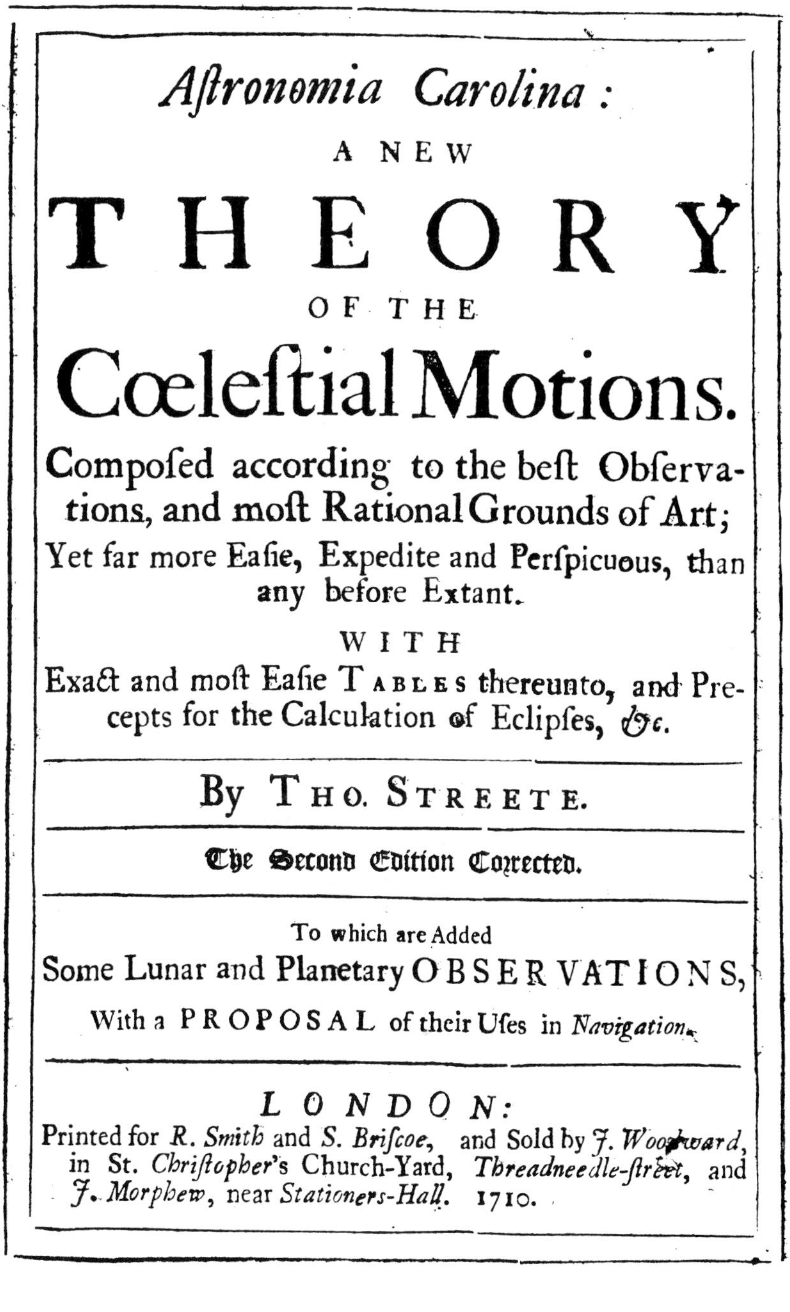

Figure 6.5 The title page of Halley's 1710 edition of Streete's tables. Halley's early lunar and planetary observations are newly appended at the end. The Astronomia Carolina was originally published in 1661.

Aſtronomia Carolina :

A NEW

THEORY

OF THE

Cœleſtial Motions.

Compoſed according to the beſt Obſerva-
tions, and moſt Rational Grounds of Art;
Yet far more Eaſie, Expedite and Perſpicuous, than
any before Extant.

WITH

Exaɛt and moſt Eaſie T ᴀʙʟᴇ s thereunto, and Pre-
cepts for the Calculation of Eclipſes, *&c.*

By T ʜ ᴏ. S ᴛ ʀ ᴇ ᴇ ᴛ ᴇ.

𝕿𝖍𝖊 𝕾𝖊𝖈𝖔𝖓𝖉 𝕰𝖉𝖎𝖙𝖎𝖔𝖓 𝕮𝖔𝖗𝖗𝖊𝖈𝖙𝖊𝖉.

To which are Added
Some Lunar and Planetary O B S E R V A T I O N S,

With a P R O P O S A L of their Uſes in *Navigation.*

L O N D O N:
Printed for *R. Smith* and *S. Briſcoe,* and Sold by *J. Woodward,*
in St. *Chriſtopher*'s Church-Yard, *Threadneedle-ſtreet,* and
J. Morphew, near *Stationers-Hall.* 1710.

these parts of Europe.' Although Halley modestly said that they 'abundantly evince the Certainty of Mr. *Streete's Mercurial Astronomy,*' they were also his own personal triumph in solving the two-millennium-old observational problem of where Mercury really was at all points in its orbit around the sun. The fleet messenger of the gods had been captured at last!

Notes and references

I would like to thank Barbara L. Welther for tracking down some of the more obscure references and Ann Wegner for a preliminary translation, here somewhat abridged, of Gassendi's report of the 1631 transit. The hypothesis that Huygens used one of his own lenses for observing the transit in 1661 was put forward by Gerrit Moll in *Astronomische Nachrichten*, **10** (1832), columns 201–3. The passage from the Evelyn diary of 24 October 1664, is cited in *The Observatory*, **17** (1894), 362. Various letters concerning the 1664 transit are found in *The Correspondence of Henry Oldenburg*, Vol. II, 1663–1665 (Madison, Wisconsin, 1966), edited and translated by A. Rupert and Marie B. Hall. Further information on the accuracy of Streete's calculations will be found in Owen Gingerich and Barbara L. Welther's 'The Accuracy of Historical Ephemerides,' *Memoirs of the American Philosophical Society*, **59S**, 1983.

7 Islamic astronomy

Historians who track the development of astronomy from antiquity to the Renaissance sometimes refer to the time from the eighth through the fourteenth centuries as the Islamic period. During that interval most astronomical activity took place in the Middle East, North Africa and Moorish Spain. While Europe languished in the Dark Ages, the torch of ancient scholarship had passed into Muslim hands. Islamic scholars kept it alight, and from them it passed to Renaissance Europe.

Two circumstances fostered the growth of astronomy in Islamic lands. One was geographic proximity to the world of ancient learning, coupled with a tolerance for scholars of other creeds. In the ninth century most of the Greek scientific texts were translated into Arabic, including Ptolemy's *Syntaxis*, the apex of ancient astronomy. It was through these translations that the Greek works later became known in medieval Europe. (Indeed, the *Syntaxis* is still known primarily by its Arabic name, *Almagest*, meaning 'the Greatest.')

The second impetus came from Islamic religious observances, which presented a host of problems in mathematical astronomy, mostly related to timekeeping. In solving these problems the Islamic scholars went far beyond the Greek mathematical methods. These developments, notably in the field of trigonometry, provided the essential tools for the creation of Western Renaissance astronomy.

The traces of medieval Islamic astronomy are conspicuous even today. When an astronomer refers to the zenith, to azimuth or to algebra, or when he mentions the stars in the Summer Triangle – Vega, Altair, Deneb – he is using words of Arabic origin. Yet although the story of how Greek astronomy passed to the Arabs is comparatively well known, the history of its transformation by Islamic scholars and subsequent retransmission to the Latin West is only now being written. Thousands of manuscripts remain unexamined. Nevertheless, it is possible to offer at least a fragmentary sketch of the process.

The foundations of Islamic science in general and of astronomy in particular were laid two centuries after the emigration of the prophet Muhammad from Mecca to Medina in AD 622. This event, called the Hegira, marks the beginning of the Islamic calendar. The first centuries of Islam were characterized by a rapid and turbulent expansion. Not until the late second century and early third century of the Hegira era was there a sufficiently stable and cosmopolitan atmosphere in which the sciences could flourish. Then the new Abbasid dynasty, which had taken over the caliphate (the leadership of Islam) in 750 and founded Baghdad as the capital in 762, began to sponsor translations of Greek texts. In just a few decades the major scientific works of antiquity – including those of Galen, Aristotle, Euclid, Ptolemy, Archimedes, and Apollonius – were translated into Arabic.

Figure 7.1 This photograph shows an astrolabe, a two-dimensional brass map of the sky that was widely used by medieval Islamic astronomers. The pointers on the open network (the rete) indicate the positions of prominent stars. By rotating the rete about the central pin one could simulate the apparent daily motions of the stars around the north celestial pole; with the help of a celestial coordinate system engraved on an underlying solid plate, one could locate the stars with respect to the horizon and the meridian. Among other things, the astrolabe made it possible to tell time by day or by night. The device was invented by the ancient Greeks, but the oldest dated specimen is the one shown here, made by a Muslim named Nastulus in AD 927–28 and now in the Kuwait National Museum. During the Middle Ages, Islamic astronomers preserved and transformed Greek astronomy, and it was from Islam that astronomy – and the astrolabe – later passed to Renaissance Europe.

The work was done by Christian and pagan scholars as well as by Muslims.

The most vigorous patron of this effort was Caliph al-Ma'mūn, who acceded to power in 813. Al-Ma'mūn founded an academy called the House of Wisdom and placed Ḥunayn ibn Isḥaq al-ʿIbādī, a Nestorian Christian with an excellent command of Greek, in charge. Ḥunayn became the most celebrated of all translators of Greek texts. He produced Arabic versions of Plato, Aristotle, and their commentators, and he translated the works of the three founders of Greek medicine, Hippocrates, Galen, and Dioscorides.

The academy's principal translator of mathematical and astronomical works was a pagan named Thābit ibn Qurra. Thābit was originally a money changer in the marketplace of Harran, a town in northern Mesopotamia that was the center of an astral cult. He stoutly maintained that the adherents of this cult had first farmed the land, built cities and ports, and discovered science, but he was tolerated in the Islamic capital. There he wrote more than 100 scientific treatises, including a commentary on the *Almagest*.

Another mathematical astronomer at the House of Wisdom was al-Khwārizmī, whose *Algebra*, dedicated to al-Ma'mūn, may well have been the first book on the topic in Arabic. Although it was not particularly impressive as a scientific achievement, it did help to introduce Hindu as well as Greek methods into the Islamic world. Sometime after 1100 it was translated into Latin by an Englishman, Robert of Chester, who had gone to Spain to study mathematics. The translation, beginning with the words *'Dicit Algoritmi'* (hence the modern word algorithm), had a powerful influence on medieval Western algebra.

Moreover, its influence is still felt in all mathematics and science: it marked the introduction into Europe of 'Arabic numerals.' Along with certain trigonometric procedures, the Arabs had borrowed from India a system of numbers that included the zero. The Indian numerals existed in two forms in the Islamic world, and it was the Western form that was transmitted through Spain into medieval Europe. These numerals, with the explicit zero, are far more efficient than Roman numerals for making calculations.

Yet another astronomer in ninth-century Baghdad was Aḥmad al-Farghānī. His most important astronomical work was his *Jawāmi'*, or *Elements*, which helped to spread the more elementary and nonmathematical parts of Ptolemy's earth-centered astronomy. The *Elements* had a considerable influence in the West. It was twice translated into Latin in Toledo, once by John of Seville (Johannes Hispalensis) in the first half of the twelfth century, and more completely by Gerard of Cremona a few decades later.

Gerard's translation of al-Farghānī provided Dante with his principal knowledge of Ptolemaic astronomy. (In the *Divine Comedy* the poet ascends through the spheres of the planets, which are centered on the earth.) It was John of Seville's earlier version, however, that became better known in the West. It served as the foundation for the *Sphere of Sacrobosco*, a still further watered-down account of spherical astronomy written in the early thirteenth century by John of Holywood (Johannes de Sacrobosco). In universities throughout Western Christendom the *Sphere of Sacrobosco* became a long-term best seller. In the age of printing it went through more than 200 editions before it was superseded by other textbooks in the early seventeenth century. With the exception of Euclid's *Elements* no scientific textbook can claim a longer period of supremacy.

Thus from the House of Wisdom in ancient Baghdad, with its congenial tolerance and its unique blending of cultures, there streamed not only an impressive sequence of translations of Greek scientific and philosophical works but also commentaries and original treatises. By AD 900 the foundation had been laid for the full flowering of an international science, with one language – Arabic – as its vehicle.

A major impetus for the flowering of astronomy in Islam came from religious observances, which presented an assortment of problems in mathematical astronomy, specifically in spherical geometry.

At the time of Muhammad both Christians and Jews observed holy days, such as Easter and Passover, whose timing was determined by

Figure 7.2 Star map of the constellation Perseus is a medieval Islamic copy of a drawing made in the tenth century by the Persian astronomer ʿAbd al-Raḥmān al-Ṣūfī. Al-Ṣūfī revised the star catalogue compiled in the second century by Ptolemy. The foremost astronomer of ancient times. The illustration is a page of a manuscript that now belongs to the Egyptian National Library in Cairo.

the phases of the moon. Both communities had confronted the fact that the approximately 29.5-day lunar months are not commensurable with the 365-day solar year: 12 lunar months add up to only 354 days. To solve the problem Christians and Jews had adopted a scheme based on a discovery made in about 430 BC by the Athenian astronomer Meton. In the 19-year Metonic cycle there were 12 years of 12 lunar months and 7 years of 13 lunar months. The periodic insertion of a thirteenth month kept calendar dates in step with the seasons.

Apparently, however, not every jurisdiction followed the standard pattern; unscrupulous rulers occasionally added the thirteenth month when it suited their own interests. To Muhammad this was the work of the devil. In the Koran (Chapter 9, Verse 36) he decreed that 'the number of months in the sight of God is 12 [in a year] – so ordained by Him the day He created the heavens and the earth; of

them four are sacred: that is the straight usage.' Caliph ʿUmar I
(634–44) interpreted this decree as requiring a strictly lunar calendar,
which to this day is followed in most Islamic countries. Because the
Hegira year is about 11 days shorter than the solar year, holidays
such as Ramadan, the month of fasting, slowly cycle through the
seasons, making their rounds in about 30 solar years.

Furthermore, Ramadan and the other Islamic months do not begin
at the astronomical new moon, defined as the time when the moon
has the same celestial longitude as the sun and is therefore invisible;
instead they begin when the thin crescent moon is first sighted in the
western evening sky. Predicting just when the crescent moon would
become visible was a special challenge to Islamic mathematical astronomers. Although Ptolemy's theory of the complex lunar motion was
tolerably accurate near the time of the new moon, it specified the
moon's path only with respect to the ecliptic (the sun's path on the
celestial sphere). To predict the first visibility of the moon it was
necessary to describe its motion with respect to the horizon, and this
problem demanded fairly sophisticated spherical geometry.

Two other religious customs presented problems requiring the
application of spherical geometry. One problem, given the requirement for Muslims to pray toward Mecca and to orient their mosques
in that direction, was to determine the direction of the holy city from
a given location. Another problem was to determine from celestial
bodies the proper times for the prayers at sunrise, at midday, in the
afternoon, at sunset and in the evening.

Solving any of these problems involves finding the unknown sides
or angles of a triangle on the celestial sphere from the known sides
and angles. One way of finding the time of day, for example, is to
construct a triangle whose vertexes are the zenith, the north celestial
pole and the sun's position. The observer must know the altitude of
the sun and that of the pole; the former can be observed, and the
latter is equal to the observer's latitude. The time is then given by the
angle at the intersection of the meridian (the arc through the zenith
and the pole) and the sun's hour circle (the arc through the sun and
the pole).

The method Ptolemy used to solve spherical triangles was a clumsy
one devised late in the first century by Menelaus of Alexandria. It
involved setting up two intersecting right triangles; by applying the
Menelaus theorem it was possible to solve for one of the sides, but
only if five other sides were known. To tell the time from the
sun's altitude, for instance, repeated applications of the Menelaus
theorem were required. For medieval Islamic astronomers there was
an obvious challenge to find a simpler trigonometric method.

By the ninth century the six modern trigonometric functions – sine
and cosine, tangent and cotangent, secant and cosecant – had been
identified, whereas Ptolemy knew only a single chord function. Of
the six, five seem to be essentially Arabic in origin; only the sine
function was introduced into Islam from India. (The etymology of the
word sine is an interesting tale. The Sanskrit word was *ardhajya*,
meaning 'half chord,' which in Arabic was shortened and trans-
literated as *jyb*. In Arabic vowels are not spelled out, and so the word

Figure 7.3 Altazimuth coordinates map the celestial hemisphere visible at a given latitude. The sky is divided by lines of equal altitude above the horizon and lines of equal azimuth around the horizon. The latter converge toward the zenith, the point directly overhead.

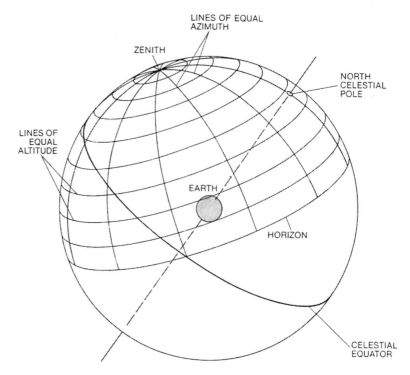

Figure 7.4 Solving spherical triangles was a fundamental problem faced by Islamic astronomers. To tell time from the altitude of the sun, for example, they had to find the hour angle formed at the north celestial pole by the meridian (the great circle through the zenith and the poles) and the sun's hour circle (the great circle through the sun and the poles). At noon, when the sun crosses the meridian, its altitude is maximum and its hour angle is zero.

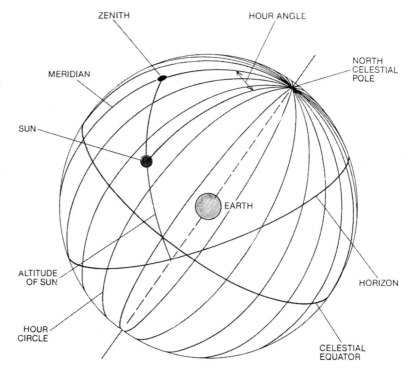

was read as *jayb*, meaning 'pocket' or 'gulf.' In medieval Europe it was then translated as *sinus*, the Latin word for gulf.) From the ninth century onward the development of spherical trigonometry was rapid. Islamic astronomers discovered simple trigonometric identities, such as the law of sines, that made solving spherical triangles a much simpler and quicker process.

One of the most conspicuous examples of modern astronomy's Islamic heritage is the names of stars. Betelgeuse, Rigel, Vega, Aldebaran, and Fomalhaut are among the names that are directly Arabic in origin or are Arabic translations of Ptolemy's Greek descriptions.

In the *Almagest* Ptolemy had provided a catalog of more than 1000 stars. The first critical revision of the catalog was compiled by ʿAbd al-Raḥmān al-Ṣūfī, a tenth-century Persian astronomer who worked in both Iran and Baghdad. Al-Ṣūfī's *Kitāb ṣuwar al-kawākib* ('Book on the Constellations of Fixed Stars') did not add or subtract stars from the *Almagest* list, nor did it remeasure their often faulty positions, but it did give improved magnitudes as well as Arabic identifications. The latter were mostly just translations of Ptolemy.

For many years it was assumed that al-Ṣūfī's Arabic had established the stellar nomenclature in the West. It now seems that his fourteenth- and fifteenth-century Latin translators went to a Latin version of the Arabic edition of Ptolemy himself for the star descriptions, which they combined with al-Ṣūfī's splendid pictorial representations of the constellations. Meanwhile the Arabic star nomenclature trickled into the West by another route: the making of astrolabes.

The astrolabe was a Greek invention. Essentially it is a two-dimensional model of the sky, an analog computer for solving the problems of spherical astronomy. A typical astrolabe consists of a series of brass plates nested in a brass matrix known in Arabic as the *umm* (meaning 'womb'). The uppermost plate, called the *'ankabūt* (meaning 'spider') or in Latin the *rete*, is an open network of two or three dozen pointers indicating the position of specific stars. Under the rete are one or more solid plates, each engraved with a celestial coordinate system appropriate for observations at a particular latitude: circles of equal altitude above the horizon (analogous to terrestrial latitude lines) and circles of equal azimuth around the horizon (analogous to longitude lines). By rotating the rete about a central pin, which represents the north celestial pole, the daily motions of the stars on the celestial sphere can be reproduced.

Although the astrolabe was known in antiquity, the earliest dated instrument that has been preserved comes from the Islamic period (see Figure 7.1). It was made by one Nastulus in 315 of the Hegira era (927–8), and it is now one of the treasures of the Kuwait National Museum. Only a handful of tenth-century Arabic astrolabes exist, whereas nearly 40 have survived from the eleventh and twelfth centuries. Several of these were made in Spain in the mid-eleventh century and have a distinctly Moorish style.

The earliest extant Arabic treatise on the astrolabe was written in Baghdad by one of Caliph al-Ma'mūn's astronomers, ʿAlī ibn ʿĪsā. Later members of the Baghdad school, notably al-Farghānī, also

Figure 7.5 The astrolabe simplified astronomical calculations, including the telling of time. Its nested brass plates are a projection of the celestial sphere onto two dimensions. The top plate, called the rete, is an open network of pointers indicating the positions of prominent stars. The off-center circle on the rete is the ecliptic. Under the rete is a solid plate on which is engraved a celestial coordinate system for a particular latitude: lines of equal altitude, lines of equal azimuth and hour circles. By turning the rete around the central pin, which represents the north celestial pole, one could reproduce the daily motions of the stars. The first step in telling the time was to determine the altitude of the sun (or of a star) with the help of a sighting bar and a degree scale on the back of the astrolabe. Next the rete was turned until the position of the sun on the ecliptic (or the star pointer) was at the correct altitude line. The hour angle was then given by the angle at the central pin between the meridian line and a line through the object's position. The astrolabe shown here was made in England in the fourteenth century and belongs to Merton College of the University of Oxford. Astrolabes were the chief means by which Arabic star names were transmitted to the West.

wrote on the astrolabe. Al-Farghānī's treatise was impressive for the mathematical way he applied the instrument to problems in astrology, astronomy, and timekeeping.

Many of these treatises found their way to Spain, where they were translated into Latin in the twelfth and thirteenth centuries. The most popular work, which exists today in about 200 Latin manuscript copies, was long mistakenly attributed to Māshā'allāh, a Jewish astronomer of the eighth century who participated in the decision to found Baghdad; it probably is a later pastiche from a variety of sources. In about 1390 this treatise was the basis for an essay on the astrolabe by the English poet Geoffrey Chaucer. Indeed, England seems to have been the gateway for the introduction of the astrolabe from Spain into Western Christendom in the late thirteenth and fourteenth centuries. It is possible that scientific activity centered at Oxford at the time contributed to the surge of interest in the device. Merton and Oriel colleges of the University of Oxford still own fine fourteenth-century astrolabes.

On them one finds typical sets of Arabic star names written in Gothic Latin letters. Included on the Merton College astrolabe, for example, are Arabic names that have evolved into standard modern nomenclature: Wega, Altahir, Algeuze, Rigil, Elfeta, Alferaz, and Mirac. Thus as a result of the astrolabe tradition of Eastern Islam, transmitted through Spain to England, most navigational stars today

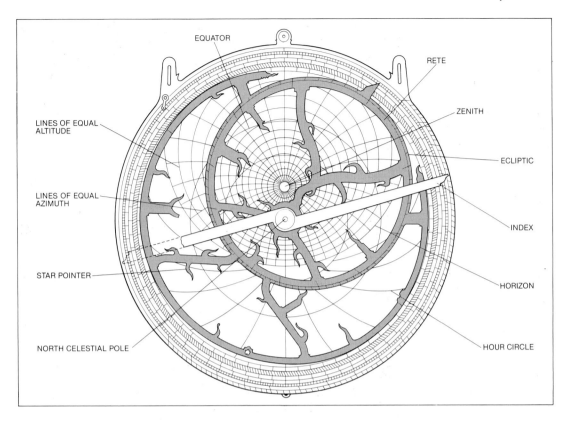

EQUATOR

RETE

ZENITH

LINES OF EQUAL
ALTITUDE

ECLIPTIC

LINES OF EQUAL
AZIMUTH

INDEX

STAR POINTER

HORIZON

NORTH CELESTIAL POLE

HOUR CIRCLE

have Arabic names, either indigenous ones or Arabic translations of
Ptolemy's Greek descriptions.

It would be wrong to conclude from the preponderance of Arabic
star names that Islamic astronomers made exhaustive studies of the
sky. On the contrary, their observations were quite limited. For
instance, the spectacular supernova (stellar explosion) of 1054,
which produced the Crab Nebula, went virtually unrecorded in
Islamic texts even though it was widely noted in China. Modern
astronomers struck by this glaring gap often do not realize that
Islamic astronomers failed to document most specific astronomical
phenomena. They had little incentive to do so. Their astrology, unlike
that of the Chinese, depended not so much on unusual heavenly
omens as on planetary positions, and these were quite well described
by the Ptolemaic procedures.

The planetary models that Ptolemy devised in the second century
AD had the sun, the moon, and the planets moving around the earth.
A simple circular orbit, however, could not account for the fact that a
planet periodically seems to reverse its direction of motion across the
sky. (According to the modern heliocentric viewpoint, this apparent
retrograde motion occurs when the earth is passing or being passed
by another planet on its way around the sun.) Hence Ptolemy had
each planet on an epicycle, a rotating circle whose center moved
about the earth on a large circle called the deferent. The epicycle,
together with other geometric devices invented by Ptolemy, gave a
fairly good first approximation to the apparent motion of the planets.

As a great theoretician, Ptolemy must have been fairly confident of the particular geometry of his models, since he never described how he settled on it.

On the other hand, the idea of applying mathematics to a specific numerical description of the physical world was something rather novel for the Hellenistic Greeks, quite different from the pure mathematics of Euclid and Apollonius. In this part of his program Ptolemy must have realized that improved values for the numerical parameters of his models were both desirable and inevitable, and so he gave careful instructions on how to establish the parameters from a limited number of selected observations. The Islamic astronomers learned this lesson all too well. They limited their observations, or at least the few they chose to record primarily to measurements that could be used for rederiving key parameters. These included the orientation and eccentricity of the solar orbit and the inclination of the ecliptic plane.

An impressive example of an Islamic astronomer working strictly within a Ptolemaic framework but establishing new values for Ptolemy's parameters was Muḥammad al-Bāttanī, a younger contemporary of Thābit ibn Qurra. Al-Bāttanī's *Zīj* ('*Astronomical Tables*') is still admired as one of the most important astronomical works between the time of Ptolemy and that of Copernicus. Among other things, al-Bāttanī was able to establish the position of the solar orbit (equivalent in modern terms to finding the position of the earth's orbit) with better success than Ptolemy had achieved.

Because al-Bāttanī does not describe his observations in detail, it is not clear whether he adopted an observational strategy different from that of Ptolemy. In any case his results were good, and centuries later his parameters for the solar orbit were widely known in Europe. His *Zīj* first made its way to Spain. There it was translated into Latin early in the twelfth century and into Castilian a little more than 100 years later. The fact that only a single Arabic manuscript copy survives (in the Escorial Library near Madrid) suggests that al-Bāttanī's astronomy was not as highly regarded in Islam as it was in Europe, where the advent of printing ensured its survival and in particular made it available to Copernicus and his contemporaries. In *De Revolutionibus orbium Coelestium* ('*On the Revolutions of the Heavenly Spheres*') the Polish astronomer mentions his ninth-century Muslim predecessor no fewer than 23 times.

In contrast, one of the greatest astronomers of medieval Islam, ᶜAlī ibn ᶜAbd al-Raḥmān ibn Yūnus, remained completely unknown to European astronomers of the Renaissance. Working in Cairo a century after al-Bāttanī, Ibn Yūnus wrote a major astronomical handbook called the *Ḥākimī Zīj*. Unlike other Arabic astronomers, he prefaced his *Zīj* with a series of more than 100 observations, mostly of eclipses and planetary conjunctions. Although Ibn Yūnus' handbook was widely used in Islam, and his timekeeping tables survived in use in Cairo into the nineteenth century, his work became known in the West less than 200 years ago.

Throughout the entire Islamic period astronomers stayed securely within the geocentric framework. For this one should not criticize them too harshly. Until Galileo's telescopic observations of the phases of Venus in 1610, no observational evidence could be brought against the Ptolemaic system. Even Galileo's observations could not distinguish between the geo-heliocentric system of Tycho Brahe (in which the other planets revolved about the sun but the sun revolved about the earth) and the purely heliocentric system of Copernicus (see Chapter 14). Furthermore, although Islamic astronomers followed Ptolemy's injunction to test his results, they did not limit themselves simply to improving his parameters. The technical details of his models were not immune from criticism. These attacks, however, were invariably launched on philosophical rather than on observational grounds.

Ptolemy's models were essentially a mathematical system for predicting the positions of the planets. Yet in the *Planetary Hypotheses* he did try to fit the models into a cosmological system, the Aristotelian scheme of tightly nested spheres centered on the earth. He placed the nearest point of Mercury's path immediately beyond the most distant point of the moon's path; immediately beyond the farthest excursion of Mercury lay the nearest approach of Venus, and so on through the spheres for the sun, Mars, Jupiter, and Saturn.

To reproduce the observed nonuniform motions of the planets, however, Ptolemy adopted two purely geometric devices in addition to the epicycle. First, he placed the deferent circles off-center with respect to the earth. Second, he made the ingenious assumption that the motion of celestial bodies was uniform not around the earth, nor around the centers of their deferents, but instead around a point called the equant that was opposite the earth from the deferent center and at an equal distance. Eccentric deferents and equants did a good job of representing the varying speeds with which planets are seen to move across the sky, but to some minds they were philosophically offensive.

The equant in particular was objectionable to philosophers who thought of planetary spheres as real physical objects, each sphere driven by the one outside it (and the outermost driven by the prime mover), and who wanted to be able to construct a mechanical model of the system. For example, as was pointed out by Maimonides, a Jewish scholar of the twelfth century who worked in Spain and Cairo, the equant point for Saturn fell right on the spheres for Mercury. This was clearly awkward from a mechanical point of view. Furthermore, the equant violated the philosophical notion that heavenly bodies should be moved by a system of perfect circles, each of which rotated with uniform angular velocity about its center. To some purists even Ptolemy's eccentric deferents, which moved the earth away from the center of things, were philosophically unsatisfactory.

The Islamic astronomers adopted the Ptolemaic–Aristotelian cosmology, but eventually criticism emerged. One of the first critics was Ibn al-Haytham (Alhazen), a leading physicist of eleventh-century Cairo. In his *Doubts on Ptolemy* he complained that the equant failed to satisfy the requirement of uniform circular motion, and he went so far as to declare the planetary models of the *Almagest* false.

Only one of Ibn al-Haytham's astronomical works, a book called *On the Configuration of the World*, penetrated into Latin Europe in the Middle Ages. In it he attempted to discover the physical reality underlying Ptolemy's mathematical models. Conceiving of the heavens in terms of concentric spheres and shells, he tried to assign a single spherical body to each of the *Almagest's* simple motions. The work was translated into Castilian in the court of Alfonso the Wise, and early in the fourteenth century from Castilian into Latin. Either this version or a Latin translation of one of Ibn al-Haytham's popularizers had a major influence in early Renaissance Europe. The concept of separate celestial spheres for each component of Ptolemy's planetary motions gained wide currency through a textbook, *Theoricae novae planetarum*, written by the Viennese Georg Peurbach in about 1454.

Meanwhile, in the twelfth century in the western Islamic region of Andalusia, the astronomer and philosopher Ibn Rushd (Averroës) gradually developed a somewhat more extreme criticism of Ptolemy. 'To assert the existence of an eccentric sphere or an epicyclic sphere is contrary to nature . . .,' he wrote. 'The astronomy of our time offers no truth, but only agrees with the calculations and not what exists.' Averroës rejected Ptolemy's eccentric deferents and argued for a strictly concentric model of the universe.

An Andalusian contemporary, Abu Isḥāq al-Biṭrūjī, actually tried to formulate such a strictly geocentric model. The results were disastrous. For example, in al-Biṭrūjī's system Saturn could on occasion deviate from the ecliptic by as much as 26° (instead of the required 3°). As for the observed motions that led Ptolemy to propose the equant, they were completely ignored. In the words of one modern commentator, al-Biṭrūjī 'heaps chaos on confusion.' Nevertheless, early in the thirteenth century his work was translated into Latin under the name *Alpetragius*, and from about 1230 on his ideas were widely discussed throughout Europe. Even Copernicus cited his order of the planets, which placed Venus beyond the sun.

At the other end of the Islamic world a fresh critique of the Ptolemaic mechanisms was undertaken in the thirteenth century by Naṣīr al-Dīn al-Ṭūsī. One of the most prolific Islamic polymaths, with 150 known treatises and letters to his credit, al-Ṭūsī also constructed a major observatory at Maragha (the present-day Marāgheh in Iran).

Al-Ṭūsī found the equant particularly dissatisfactory. In his *Tadhkira* ('*Memorandum*') he replaced it by adding two more small epicycles to the model of each planet's orbit. Through this ingenious device al-Ṭūsī was able to achieve his goal of generating the nonuniform motions of the planets by combinations of uniformly rotating circles. The centers of the deferents, however, were still displaced from the earth. Two other astronomers at the Maragha observatory, Muʾayyad al-Dīn al-ʿUrdī and Quṭb al-Dīn al-Shīrāzī, offered an alternative arrangement, but this system too retained the philosophically objectionable eccentricity.

Finally a completely concentric rearrangement of the planetary mechanisms was achieved by Ibn al-Shāṭir, who worked in

Damascus in about 1350. By using a scheme related to that of al-Ṭūsī, Ibn al-Shāṭir succeeded in eliminating not only the equant but also certain other objectionable circles from Ptolemy's constructions. He thereby cleared the way for a perfectly nested and mechanically acceptable set of celestial spheres. (He described his work thus: 'I found that the most distinguished of the later astronomers had adduced indisputable doubts concerning the well-known astronomy of the spheres according to Ptolemy. I therefore asked Almighty God to give me inspiration and help me to invent models that would achieve what was required, and God – may He be praised and exalted – did enable me to devise universal models for the planetary motions in longitude and latitude and all other observable features of their motions, models that were free from the doubts surrounding previous ones.') Yet Ibn al-Shāṭir's solution, along with the work of the Maragha astronomers, remained generally unknown in medieval Europe.

Ibn al-Shāṭir's forgotten model was rediscovered in the late 1950s by E. S. Kennedy and his students at the American University of Beirut. The discovery raised an intriguing question. It was quickly recognized that the Ibn al-Shāṭir and Maragha inventions were the same type of mechanism used by Copernicus a few centuries later to eliminate the equant and to generate the intricate changes in the position of the earth's orbit. Copernicus, of course, adopted a heliocentric arrangement, but the problem of accounting for the slow but regular changes in a planet's orbital speed remained exactly the same. Since Copernicus agreed with the philosophical objections to the equant – like some of his Islamic predecessors, he apparently believed celestial motions were driven by physical, crystalline spheres – he too sought to replace Ptolemy's device. In a preliminary work, the *Commentariolus*, he employed an arrangement equivalent to Ibn al-Shāṭir's. Later, in *De Revolutionibus*, he reverted to the use of eccentric orbits, adopting a model that was the sun-centered equivalent of the one developed at Maragha.

Could Copernicus have been influenced by the Maragha astronomers or by Ibn al-Shāṭir? No Latin translation has been found of any of their works or indeed of any work describing their models. It is conceivable that Copernicus saw an Arabic manuscript while he was studying in Italy (from 1496 to 1503) and had it translated, but this seems highly improbable. A Greek translation of some of the al-Ṭūsī material is known to have reached Rome in the fifteenth century (many Greek manuscripts were carried west after the fall of Constantinople in 1453), but there is no evidence that Copernicus ever saw it.

Scholars are currently divided over whether Copernicus got his method for replacing the equant by some unknown route from the Islamic world or whether he found it on his own. I personally believe he could have invented the method independently.

Nevertheless, the whole idea of criticizing Ptolemy and eliminating the equant is part of the climate of opinion inherited by the Latin West from Islam. The Islamic astronomers would probably have been astonished and even horrified by the revolution started by Copernicus. Yet his motives were not completely different from theirs. In

Figure 7.6 Double-epicycle system proposed in the fourteenth century by Ibn al-Shāṭir eliminated Ptolemy's equants and put the sun, moon and planets in concentric orbits around the earth. This diagram, from a manuscript now at Oxford, shows positions of the moon.

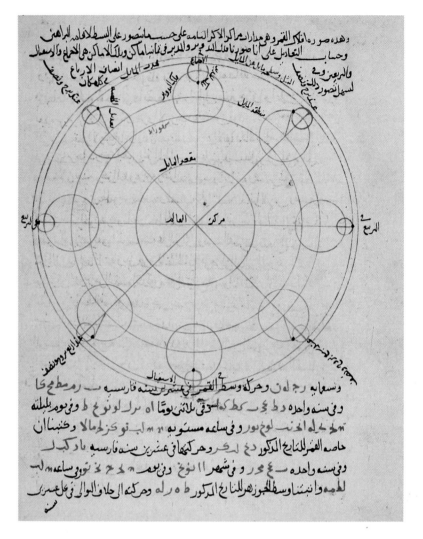

eliminating the equant, and even in placing the planets in orbit around the sun, Copernicus was in part trying to formulate a mechanically functional system, one that offered not only a mathematical representation but also a physical explanation of planetary motions. In a profound sense he was simply working out the implications of an astronomy founded by Ptolemy but transformed by the Islamic astronomers. Today that heritage belongs to the entire world of science.

Notes and references

For further reading, see Aydin Sayilli, *The Observatory in Islam and Its Place in the General History of the Observatory* (Istanbul, 1960); A. I. Sabra, ''The Scientific Enterprise'' in *The World of Islam: Faith, People, Culture*, edited by Bernard Lewis (New York, 1976); E. S. Kennedy, *Studies in the Islamic Exact Sciences* (Beirut, 1983); J. D. North, ''The Astrolabe'' in *Scientific American* (January, 1974).

8 *The astronomy of Alfonso the Wise*

Seven centuries ago there reigned in Spain one of the Middle Ages' most enlightened patrons of the arts and sciences. Alfonso X, who ruled the kingdoms of Leon and Castile from 1252 to 1284, was not much of a political success, but he is far better remembered than most medieval kings. Alfonso wrote verse, supported the last of the troubadours, collected music, and reformed the legal code. His image is emblazoned in the US House of Representatives as one of the great lawgivers of the world.

But in the late Middle Ages and Renaissance, Alfonso the Wise was known primarily as the author of the most useful astronomical tables. Never mind that he was not actually their author; he was intensely interested in their production and supported both the making of the tables and a collection of astronomical works.

In their medieval Castilian Spanish form, the astronomical tables of Alfonso no longer exist. However, a copy must have reached Paris early in the fourteenth century, for beginning in the late 1320s manuscript copies with Latin instructions spread from there throughout Europe. In 1483 the first printed edition appeared, followed in the next six decades by four more editions.

Unlike the *Alfonsine Tables*, the collection of astronomical texts (mostly translations of Arabic works) did not receive wide distribution in the Middle Ages. It was not even printed until the last century when, in a splendid burst of Castilian nationalism, Manuel Rico y Sinobas published an important section of it in five unwieldy folio volumes entitled *Libros del Saber de Astronomia* ("*The Books of the Knowledge of Astronomy*").

What, in fact, do Alfonso's books contain? First, there is a large four-part section describing the constellations star by star based upon *The Book of the Fixed Stars* by al-Sūfī, a tenth-century Persian astronomer. Each of the 15 books that follow deal with a different astronomical instrument or device. The first is a treatise on the use of the celestial globe, translated by Yehuda ben Moses, a Jewish scholar who figured in several of Alfonso's projects, and Guillen Arremon Daspa, apparently a Christian collaborator.

Then there follows a work on the spherical astrolabe, a rare instrument that derives directly from the celestial globe. The preface to this treatise is by Alfonso himself, who says that no suitable book could be found for translation, and therefore Rabiçag or Rabbi Zag had been commissioned to create a work specially for the occasion. This statement gives insight into the king's role in the undertaking. Clearly, he wanted an encyclopedic corpus, and he took an active part in formulating its requirements. Some of these astronomical translations must have begun soon after his coronation in 1252, but the idea of an integrated body of work achieved its final form sometime after 1277, near the end of his troubled reign.

In contrast to the spherical astrolabe, the plane astrolabe is a more

Figure 8.1 This medallion of
Alfonso X (1221–84) is
among a series in the
chamber of the U.S. House
of Representatives that depicts
great lawgivers.

Figure 8.1 This medallion of Alfonso X (1221–84) is among a series in the chamber of the U.S. House of Representatives that depicts great lawgivers.

convenient, more sophisticated, and also more common instrument. Relatively common, too, were instructions for its use. Thus, about a century later when Geoffrey Chaucer set out to compose a treatise on the astrolabe in Middle English, he had no difficulty modifying a Latin work from Spain that was popularly attributed to Messahalla, an eighth-century Jewish astronomer from Basra. What treatise Alfonso's translators actually used is not stated; perhaps they combined several of the many available examples.

The universal astrolabe is the logical generalization of the plane astrolabe, and thus it is not surprising to find it treated next in the collection. In fact, two separate versions of it appear in consecutive books. Next follows a book on the armillary sphere, and once more Alfonso states that it was necessary to have one specially written by Rabbi Zag.

The astrolabe in its various forms allows an astronomer to depict the sky for a particular time and place, to show exactly which stars are rising or setting, and so on. It is the direct ancestor of the familiar modern planisphere. However, the astrolabe works only for the fixed stars. To locate the wandering planets, another complementary device is required. This instrument, called the planetary equatorium, is to the armillary sphere what the astrolabe is to the celestial globe.

Next in Alfonso's collection is a treatise on the quadrant, yet another case where he commissioned a new work from Rabbi Zag to fill a gap. The following five books deal with timekeeping, and they likewise are written or edited by the rabbi. They deal with sundials, water clocks, a mechanical device using quicksilver, standard candles, and the 'palace of the hours,' which seems to be a sundial building.

Of the 16 surviving volumes from the royal Alfonsine scriptorium, one is a copy of this collection of astronomical texts, kept today in the library of the University of Madrid. This codex provided the foundation for Rico's edition of 1863–7. Another astronomical volume from the royal scriptorium is found in the Bibliothèque Arsenal in Paris. Its 142 folios contain Castilian versions of the tables of al-Bāttanī and those of al-Zarqali, plus a work on the 'single quadrant.'

These Alfonsine manuscripts were rarely recopied, probably because Castilian was never a common scholarly language. In contrast, the Latin version of the *Alfonsine Tables* exists in a few hundred manuscript copies as well as thousands of printed examples. In order to understand the role of these tables, we must realize that they embodied a practical computing scheme for finding the positions of the sun, moon, and planets according to the Ptolemaic system, in which the earth was at the center of the universe.

If we wish to calculate actual planetary positions from Ptolemy's epicyclic model, we need to have in hand the numbers that specify the actual geometry. We must know, for example, the size of the epicycle with respect to its circular deferent, the period of motion of the planet in the epicycle and of the epicycle along its deferent, and so on. Altogether there are seven such parameters, of which five are independent for each planet.

Given these parameters and the geometry of the model, we can find the celestial longitude of a planet for any time, but in the days before pocket calculators this would have been rather hard work. Even *with* a small calculator the procedure is tedious enough to ruin a morning. So this is where astronomical tables come in. By pre-calculating certain fundamental parts of the planet's motion, the whole operation becomes much more manageable.

Basically, the procedure works like this. Suppose we want to know where Mars was in 1984. Let us assume that we know where Mars was on 1 January 1252. (In fact, the positions of all the planets are so specified at the beginning of the tables, because 1252 was the year of Alfonso's coronation.) Since then, 733 full years have elapsed. If we know the average motion of Mars, we could add 733 years' worth to the starting position, and we would then know approximately where the center of Mars' epicycle was on New Year's Day in 1984.

For Mars' position later in the year, we would need to know how much it moved in the months, days, and hours since 1 January. The *Alfonsine Tables* provide for this too. But in their common form as issued from Paris, they basically give the motion in days and sixtieths of a day rather than in years and months. However, the arithmetic is made very easy by chronological conversion tables.

Furthermore, the *Alfonsine Tables* not only give the average daily motion of Mars (which is exactly the same as the average motion of the center of the epicycle) but they also tabulate the motion within the epicycle. Once you know where the center of the epicycle is, then you need some correction back or forth in longitude to account for

Figure 8.2 An armillary sphere was one of the instruments depicted in the thirteenth-century Alfonsine Libros del Saber de Astronomia, *as published in 1863 by Manuel Rico y Sinobas.*

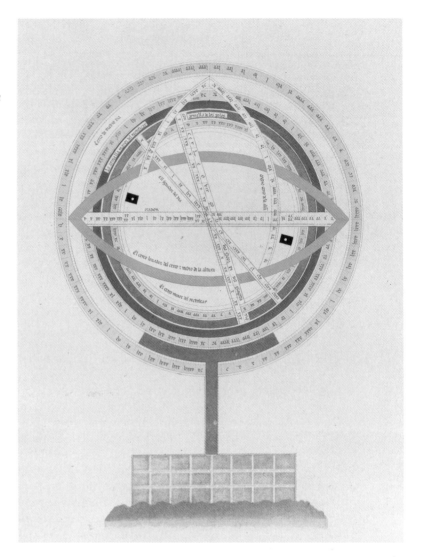

the position of the planet on the epicycle. This adjustment is furnished by the so-called table of equations shown in Figure 8.3.

Notice that this table contains not just one column of corrections but five, so the matter is clearly a little more complicated. Since the earth is not located at the center of the epicycle's orbit, even if the epicycle were moving around uniformly it would not appear that way to us. Sometimes, for example, the epicycle will appear closer and thus seem to be moving faster. Another column of corrections takes care of this detail.

But if the epicycle varies its distance from earth, then a given motion around the epicycle will have a larger effect when the epicycle is closer than when it is farther. This situation is a little more subtle, but Ptolemy, in one of his greatest mathematical tricks, found how to cover all the possible cases by multiplying together two numbers from two different columns. (He slightly complicated matters by using sometimes one pair of columns and sometimes another, so that three – rather than two – more columns were required.)

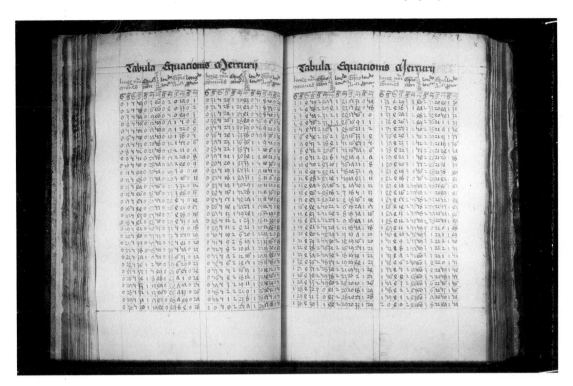

Figure 8.3 This manuscript of the Alfonsine 'equation' tables for Mercury was copied in Vienna in the fifteenth century. Reproduced by permission of Houghton Library, Harvard University.

Two centuries after Alfonso, a Polish student at the University of Cracow bought a newly printed set of the *Alfonsine Tables*, and he used them often, as the stains visible on his copy indicate. His name was Nicholas Copernicus, and two decades later he came up with a radical new sun-centered arrangement for the planets. At the end of his first small tract proposing this heliocentric arrangement (the so-called *Commentariolus*), Copernicus says, 'Behold, the entire ballet of the planets in only 34 circles.'

To nineteenth-century ears this pronouncement sounded as if Copernicus had enormously simplified things. Thus began the rumor that Ptolemy had required 80 circles in his system, and that Alfonso had used even more, piling epicycles upon epicycles. And then a famous remark attributed to Alfonso apparently gained great currency, to the effect that if he had been present at the creation, he could have given the Good Lord some hints about simplification!

If we pause to think about the structure of the *Alfonsine Tables*, we realize that the addition of a single epicycle on an epicycle would add many more columns to the tables. Not only would there be a column to correct for the new epicycle, but the tables would have to take into account whether the smaller epicycle was on the near side or the far side of the bigger one, and in turn whether the bigger one is near or far from the earth. Even Yehuda ben Moses and Rabbi Zag would have had difficulty calculating all these extra columns, not to mention what would have happened if there had been an epicycle on an epicycle on an epicycle!

In fact, what these scholars produced was even purer Ptolemy than the eleventh-century *Toledan Tables* on which the *Alfonsine Tables* were based – the models stayed exactly the same and the parameters were closer to Ptolemy's original numbers. Alfonso's astronomers had done a good, if not highly original, job. These tables were the basis of virtually all astronomical almanacs and ephemerides until the middle of the sixteenth century, and Copernicus' revisions were not all that much better for the mundane task of predicting planetary positions.

Alfonso's name was surely known to every university student through the sixteenth century – astronomy was a required subject. I doubt that he ever made the famous quip concerning some suggestions to the Creator. But I have no doubt at all that he got his money's worth as a royal patron of astronomy!

Notes and references

This chapter is based on a talk given by the author at an Anniversary Symposium for Alfonso the Wise held at Harvard University on 17 November 1984. It appears at greater length and with detailed citations in Francisco Márquez-Villanueva and Carlos Alberto Vega (eds.), *Alfonso X of Castile, The Learned King (1221–1284)* (Cambridge, Massachusetts, 1990), pp. 30–45. For a modern version of the tables, consult Emmanuel Poulle, *Les Tables Alphonsines avec les Canons de Jean de Saxe*, Paris, 1984. Since this chapter was written, Poulle has argued rather convincingly that what we know as the *Alfonsine Tables* originated in Paris around 1327 and were named in honor of the Castilian monarch, but otherwise had little to do with his astronomers, see 'The Alfonsine Tables and Alfonso X of Castile,' *Journal for the History of Astronomy*, **19** (1988), 97–113.

9 *From Aristarchus to Copernicus*

Two of the giants of ancient Greek thought, the astronomer Aristarchus and the mathematician Archimedes, lived in the third century BC. Aristarchus was born around 310 BC on the island of Samos, where the illustrious Pythagoras had lived some two centuries earlier. The astronomer left his homeland early on, and it is not known whether he ever returned. Undoubtedly he headed for Alexandria, the intellectual capital of the Hellenistic world that was renowned for its great library.

Archimedes was born about 30 years after Aristarchus, in Syracuse on Sicily, then part of 'Magna Graecia' (greater Greece). He too left his home island and headed for Alexandria. It therefore seems entirely possible that these two powerful intellects met in Greek-speaking Egypt.

Among the many treatises written by Archimedes is the *Sand-Reckoner*, an attempt to calculate how many grains of sand the universe could hold. Actually, his book is not so much a serious attempt to gauge the size of the world as it is a *tour de force* in the manipulation of large numbers.

In Archimedes' day, the Greeks used their alphabet to represent numbers. Hence, alpha (α) was one, beta (β) two, and so on through theta (θ) for nine. If you know the modern Greek alphabet, you will quickly realize that this scheme is one letter short. The Greeks employed the now-obsolete letter digamma (ς), borrowed from the Phoenicians, to represent the number six.

Then they used the letters iota (ι) through pi (π) plus another special symbol 'koppa' (ς) to represent the decades 10 through 90. The letters from rho (ϱ) to omega (ω), plus another special symbol with the relatively modern name 'sampi' (⅄), represented 100 through 900. Numbers between 1000 and 9000 were created by reusing the first nine letters and preceding them with a subscript accent. Thus, 1000 was written as ,α. The number 10 000 was a myriad, designated by M; thus 11 999 would have been written as M ,α $\text{⅄}\varsigma\theta$. Finally, utilizing alphabetic letters written above the M, they could, in a rather clumsy fashion, generate even larger numbers if necessary.

The limitations of this system challenged Archimedes. 'There are some,' he said, 'who think that the numbers of the sands is infinite, and some who, without regarding it as infinite, yet think that no number has been named that is great enough to exceed its magnitude.' And so he proceeded to find such a number by calculating how many tiny poppy seeds would be required to fill the universe. Just to cover all bets, at each step in the calculation he arbitrarily increased the size of things just to make sure his result would really exceed any conceivable 'useful' number. In the end he got something like 10^{63} poppy seeds. Of course, this was expressed more clumsily in combinations of Greek letters.

Although it is beyond the scope of this article to show all of Archimedes' individual steps, several deserve attention. For example, he mentions that the circumference of the earth is 300 000 stadia. During his lifetime his Alexandrian friend Eratosthenes calculated, from the noon altitudes of the sun at Alexandria and at Syene, that the circumference was 252 000 stadia. Quite possibly Archimedes knew of this measurement, but for his purposes such a detail was irrelevant. He proceeded to adopt not 300 000 but 3 000 000 stadia, just to make sure his result would come out sufficiently large!

The next step was more difficult because information about the size of the planetary system was sketchy at best. Aristarchus had written a treatise 'On the Sizes and Distances of the Sun and Moon,' but despite the title he had not actually worked out the distances. Archimedes noted that Aristarchus placed the sun 18–20 times farther from us than the moon, and so he adopted a value of 30 on the same principle of excess as before. (Little did he realize that the correct factor is not $1\frac{1}{2}$ but 20 times larger!)

He also knew from reports of solar eclipses that the moon subtends the same angle as the sun; therefore, if the sun were 30 times farther away than the moon, its diameter would necessarily be 30 times larger. This information, however, is inadequate to link the distance of the moon or sun with any terrestrial measure. (It was not until nearly a century later than Hipparchus determined a reasonable value for the distance to the moon, about 60 earth radii).

What could Archimedes do? To resolve the problem, he made an interesting assumption, namely, that the moon's diameter was not larger than the earth's. This meant that the sun was not more than 30 times larger than the earth. Knowing the apparent diameter of the sun, which he measured as something larger than $\frac{1}{1000}$ of a complete circle, he had an admirable little problem in geometry to calculate that the sun was not more than 10 000 earth diameters distant.

So far so good — a splendid exercise in geometry coupled with wildly speculative astronomy. (Perhaps this is not so different from some of our modern cosmological or life-on-other-worlds exercises!) But how was Archimedes to get the distance to the fixed stars? And how could he be sure his number was big enough?

Here is where the most memorable part of the *Sand-Reckoner* comes in. We can well imagine Archimedes and Aristarchus sitting under an awning in the Alexandrian marketplace, speculating on this very problem. Archimedes reports:

Aristarchus of Samos brought out *graphai* consisting of certain hypotheses, wherein it appears, as a consequence of the assumptions made, that the universe is many times greater than the 'universe' just mentioned. His hypotheses are that the fixed stars and the sun remain unmoved, that the earth revolves about the sun in the circumference of a circle, the sun lying in the middle of the orbit, and that the sphere of the fixed stars, situated about the same center as the sun, is so great that the circle in which he supposes the earth to revolve bears such a proportion to the distance of the fixed stars as the center of a sphere to its surface.

This single passage is our principal witness from antiquity attesting that Aristarchus had proposed a heliocentric cosmology! Were these

Figure 9.1 The ancient Alexandrian marketplace was perhaps where Aristarchus and his younger colleague Archimedes sat to discuss mathematics in the third century BC.

graphai a now-lost book on the subject, as Sir Thomas Heath thought? Or were they 'drawings or delineations,' as the standard Liddell-Scott *Greek-English Lexicon* gives as the first definition? Or was it just a verbal description with explanatory drawings, as Giovanni Schiaparelli supposed – perhaps just a splendid speculation tossed out during a vigorous discussion between the Alexandrian mathematicians? Of course, these are now unanswerable questions, though I personally suspect that Sir Thomas was engaging in wishful thinking about his hero.

In any event, Archimedes promptly proceeded to criticize Aristarchus on the mathematical grounds that, 'since the center of the sphere has no magnitude, we cannot conceive it to bear any ratio whatever to the surface of the sphere.' In other words, the Aristarchan picture would place the stars an infinite distance away. In essence Aristarchus was saying that the stars had to be so far away that there would be no measurable parallax, but Archimedes wanted something more denumerable.

Hence, he arbitrarily supposed that the distance of the fixed stars is to the earth–sun distance what the latter is to the size of the earth itself. By this device he could calculate, in terms of the earth's diameter, a number surely greater than the bounds of our universe. This, in turn, could be related to a stadium and to the width of a

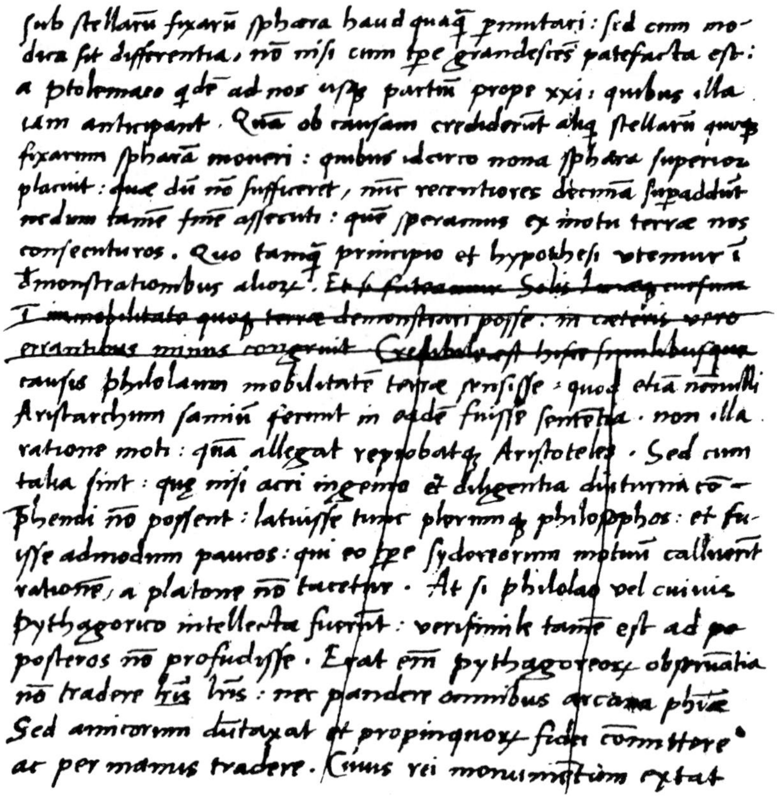

Figure 9.2 As described in the text, this is the famous passage in Copernicus' manuscript for his De Revolutionibus *where the section about Aristarchus was crossed out shortly before publication.*

poppy seed ('less than $\frac{1}{40}$ fingerbreadth,' with one stadium being less than 10 000 fingerbreadths). Incidentally, had Archimedes really been convinced by Aristarchus' heliocentric speculations, would he not have used the size of the sun rather than that of the earth in this last grand ratio?

More important, why did the Greeks, after evolving a heliocentric hypothesis, allow it to fall into neglect almost immediately? Thus asks William Stahl in the perceptive entry on Aristarchus in the *Dictionary of Scientific Biography*. "The common attitude of deploring the 'abandonment' of the heliocentric theory as a 'retrogressive step' appears to be unwarranted when it is realized that the theory, however bold and ingenious it is to be regarded, never attracted much attention in antiquity," he writes. "Aristarchus' system was the culmination of speculations about the physical nature of the universe that began with the Ionian philosophers of the sixth century, and it belongs to an age that was passing away."

What followed was an astronomy based increasingly on ingenious measurements and observations, one far less dependent on bold but unprovable speculations. Within two generations after the tragic death of Archimedes (at the hand of a Roman soldier in the sack of Syracuse), Hipparchus was born. The Bithynian astronomer carefully collected and arranged Babylonian observations as well as his own; among other accomplishments, he discovered the precession of the equinoxes and derived the remarkably accurate distance to the moon mentioned above.

The problem of the scale of the universe continued to perplex astronomers, however, and by the time of Ptolemy a clever line of reasoning (though very minimally rooted in observation) yielded an acceptably pleasing solution. Ptolemy supposed that the epicyclic mechanism for Mercury would begin at the outermost edge of the lunar machinery, that the spheres for Venus would begin precisely where those of Mercury left off, and so on.

This 'plenum universe' was conceptually very tidy, with no leftover holes, and it brilliantly matched up with the numbers: Given 64 earth radii for the greatest distance of the moon, the mean distance of the sun came out as 1210 earth radii. Hence, the distance of the sun was about 20 times the mean distance of the moon, just as Aristarchus had determined. Never mind that both Aristarchus' and Hipparchus' methods for getting the sun's distance were founded on observational quicksand – astronomers found the plenum universe quite believable because it was so compellingly elegant.

Fourteen centuries after Ptolemy, even Copernicus chose his numbers to get a solar distance of roughly 1200 earth radii. However, because his system was linked together in an entirely different way – by using the earth–sun distance as the common measure – the Copernican system was actually about half as large insofar as planetary distances are concerned. It is interesting to compare the two:

Distances in earth radii

	Geocentric (Ptolemy)	Heliocentric (Copernicus)
Moon	33 –64	(60)
Mercury	64 –166	433
Venus	166 –1079[a]	827
Earth–Sun	1160[a]–1260	1142
Mars	1260 –8820	1740
Jupiter	8820 –14187	5940
Saturn	14187 –19865	10860
Fixed stars	20000	'vast'

[a] This fit was considered convincingly good!

The essential difference was in the distance to the fixed stars, which for Copernicus, like Aristarchus, had to be far enough away to show no annual parallax. Copernicus gives no number, saying only, 'So vast, without any question, is the divine handiwork of the Almighty Creator.' If we suppose that the unaided human eye can resolve one arc minute, then the distance of the fixed stars would have to be greater than 7000 Copernican earth–sun units or 7 500 000 earth radii for the annual parallax to be invisible. This is about 14 times smaller than Archimedes' estimate of 100 000 000.

Our rapid romp through cosmic dimensions to the year 1543 has left one further question hanging: Did Copernicus get the idea for the heliocentric system from Aristarchus, 'the Copernicus of Antiquity'?

Certainly Copernicus eventually knew of the Samian's sun-centered speculations, for he somewhat vaguely attributed the idea to Aristarchus in the manuscript of his *De Revolutionibus*, where he stated, 'Philolaus believed in the mobility of the Earth, and some even say that Aristarchus of Samos was of that opinion.' But before publication, this section of the book was revised and partially recast into the preface, with the result that Aristarchus' name got omitted, though probably quite inadvertently.

Toward the end of his life, when Copernicus was writing his book, he knew at least something about Aristarchus' ideas. Where did he get his information? The *Sand-Reckoner* itself was not published until 1544, the year after Copernicus' death. Copernicus himself reports, 'I first found in Cicero that Hicetas supposed the earth to move. Later I also discovered in Plutarch that certain others were of this opinion.' In *De Revolutionibus* Copernicus quotes from *Opinions of the Philosophers*, a work then attributed to Plutarch, and a few pages before the sentences he quoted there is a passage that reads: 'Ought the earth . . . be understood to have been devised not as confined and at rest, but as turning and whirling about in the way set forth later by Aristarchus and Seleucus, by the former only as a hypothesis, but by Seleucus beyond that as a statement of fact?'

My guess is that Copernicus first read this not-too-enlightening account in the first Latin edition of the pseudo-Plutarch work (printed in 1516), a copy of which was subsequently bound with other works for the cathedral library where he was a canon. Since Copernicus had already defended the heliocentric idea before 1514, it seems most likely that he first heard of Aristarchus' priority only after he had already begun independently to work out the details of the theory.

Notes and references

A partial translation of the *Sand-Reckoner* is found in T. L. Heath, *The Works of Archimedes* (Cambridge, 1897), reprinted in *Great Books of the Western World*, Vol. 11 (Chicago, 1952). The passage quoted here is from Heath's slightly different translation in his *Aristarchus of Samos* (Oxford, 1913). The geocentric distances in the table in the text are drawn from Albert Van Helden's book *Measuring the Universe* (Chicago, 1985). A thorough discussion of all ancient references to the Aristarchan heliocentric hypothesis, with the conclusion that all of them probably derive from Archimedes' *Sand-Reckoner*, is given by Byron Emerson Wall, 'Anatomy of a Precursor, the Historiography of Aristarchos of Samos,' *Studies in the History and Philosophy of Science* 6 (1975), 201–28. My article, 'Did Copernicus Owe a Debt to Aristarchus?' is available in *Journal for the History of Astronomy* 16 (1985), 36–42.

10 The Great Copernicus Chase

When my astrophysicist friends ask me why I no longer compute model stellar atmospheres, I tell them that I'm the victim of anniversaries. In 1971 the astronomer Johannes Kepler had his 400th birthday, closely followed by the Copernican quinquecentennial in 1973. As an astronomer with strong historical interests, I knew I would be expected to have an opinion on these men, but I hoped, by taking at least a small look at some original sources, to avoid the time-honored clichés about their lives and discoveries.

Kepler was an enthusiast, a man who wrote a 5-foot shelf of books and who left scores of surviving letters and unexamined manuscripts. In contrast, Copernicus was essentially a one-book author. No personal letters survive, and precious few working manuscripts. To study Copernicus means studying his *De Revolutionibus Orbium Coelestium*, a formidably technical treatise published in 1543, just in time to reach the Polish astronomer on his deathbed.

But did anyone really read Copernicus's magnum opus? He wrote in an age when merely being able to understand Ptolemy's *Almagest* was considered the highest achievement of an astronomer. Copernicus's own treatise, heavily dependent on Ptolemy's 1400-year-old work, was no easier to read. It is entirely possible that more people alive today have read through *De Revolutionibus* than in the entire sixteenth century. Or so my fellow Copernicanist, Jerome Ravetz, and I decided in a bull session one night in York in 1970.

Two days after that discussion, I found myself at the Royal Observatory in Edinburgh, and there I took the occasion to examine its first edition Copernicus. To my astonishment, the book was carefully annotated from beginning to end. If there were so few knowledgeable readers in the sixteenth century, then by no decent probability should I have found so quickly a copy with errors perceptively marked to the very last page. As I weighed the possibilities, it occurred to me that, by an improbable chance, I might just have in hand a copy from the small group of Lutheran scholars who were responsible for getting the book printed.

There was, first, young Georg Joachim Rheticus, the 22-year-old radical professor who went off to northern Poland to persuade the aging and reluctant Copernicus to publish his work. I call him a radical because anyone who adopted heliocentrism in those days had to be something of a revolutionary, but it was more than that. Neither Joachim nor Rheticus (meaning 'from Rhaetia') was his real family name. He was christened Iserin, but when he was still a teenager, his father was charged with sorcery and beheaded, and the lad was forced to adopt a new name. Later on, after he was driven from his post as astronomy professor at Leipzig following a drunken homosexual incident, he left astronomy and became a medical doctor. As a doctor he adopted Paracelsianism, which in medicine was just as radical as heliocentrism was in astronomy. If a little more biographi-

Figure 10.1 This sixteenth-century oil painting, which hangs in the Town Hall of Torun, Poland, may be a copy from a self-portrait sketched by Copernicus.

cal data were available, Rheticus would make a prime subject for a psychobiography.

In any event, the well-annotated *De Revolutionibus* I had found in Edinburgh had not been worked over by Rheticus. I might have pinned down the annotator more promptly had I seen, on folio 96, a Latin note citing 'Our Joachim . . .,' clearly a reference by one of his near-colleagues. As it was, I soon found the owner's initials stamped among the decorations on the cover: E R S, which I eventually realized stood for Erasmus Reinhold Salveldensis – that is, Erasmus Reinhold from Saalfeld. Reinhold was the senior professor of mathematics and astronomy at Wittenberg; at the time of his Polish trip, Rheticus was the junior professor there. Unlike Rheticus, Reinhold was a solid, conservatively oriented man who steadily worked his way up the academic ladder, finally becoming rector, a much admired teacher, and the most influential astronomer in the Lutheran world. As for Copernicus's treatise, he ignored the new-fangled heliocentric cosmology while immersing himself in such safe technical questions as the motion of the moon (which in any system went around the earth) and the slow precessional motion of the starry firmament. That he held such an attitude is quite precisely revealed by the places where he did and where he did not annotate his *De Revolutionibus*.

The significance of having available the annotated *De Revolutionibus* of the foremost astronomy teacher of the mid-sixteenth century was not lost on me, for I realized that such annotations provided an ideal way to find out how teachers and students of that era approached Copernicus's astronomy. Fortunately, the mid-1500s were still close enough to the tradition of the Middle Ages, when marginal glossing of manuscripts was a common practice; also fortunately for me, the *De Revolutionibus* was printed with generous margins, despite the fact that paper was the major expense for publishers. And so, with the discovery of Reinhold's *De Revolutionibus*, the Great Copernicus Chase began.

In 1970 no one had much of an idea about how many copies of the *De Revolutionibus* had been printed, or how many survived. An obviously incomplete list of 70 copies had been published in 1943, and that provided a starting point, especially for the German libraries. When Rheticus brought the manuscript of Copernicus' book back to Germany, the obvious printer was Johannes Petreius in Nuremberg. In the 1540s Petreius had become the leading publisher of scientific texts, and his press had both the facilities and the distribution network to cope with a major technical treatise.

It has always seemed anomalous to me that Nuremberg, where the *De Revolutionibus* was originally printed, no longer has a copy from the first edition. Although still a cultural mecca, Nuremberg has long since yielded the intellectual centrum to Munich, the Bavarian capital. Munich still has two first editions in the Staatsbibliothek, but two other copies at the university perished in the air raid of July 1944 – a tragedy at least partly the fault of the librarian who I am told refused to protect the books on the grounds that such pessimism would be unpatriotic. In Frankfurt copies were also lost, but elsewhere in both West and East Germany more than three dozen copies survive. Germany is the only place in Europe where first editions outnumber the second, still reflecting the original saturation from the Nuremberg printing.

The 1943 list of 70 first editions provided a particularly strong compendium for Germany, but the task of ferreting out copies elsewhere in Europe and America required other detective techniques: letters of inquiry, consultation with dealers, advertising in bibliophilic journals. In Oxford and in Cambridge there existed central lists for the libraries – something which is not the case for London, so it took several years before I realized that London outstripped either of the traditional university towns. Who would have guessed that Dr Williams's Library, rich in theology, would also hold an *editio princeps* of the *De Revolutionibus*? Or that the Victoria and Albert Museum would hold another?

Finding the copy at the Victoria and Albert was something of a victory for dogged, systematic sleuthing. In an effort to ferret out as many copies as possible, I went through the Book Auction Records from their inception in 1888, and under 1897 I found the sale of a Copernicus in a Grolier binding. Jean Grolier, an illustrious French bibliophile of the sixteenth century, was noted for the distinguished book bindings in his library, which are nowadays great collector's items. My eyes jumped to attention when I saw the citation in the auction index, for I realized that I had never seen a Copernicus in a Grolier binding. An examination of the original Sotheby's auction catalog cast more light on the sale. 'The bindings,' read the description, 'though quite modern, are historically correct – they are, in fact, modern imitations of very high quality.' What Sotheby's delicately called imitations, most modern collectors have unhesitatingly called forgeries. The sale included fakes of a wide variety of distinguished libraries, all cleverly made by the French forger Hagué. Eventually my investigation traced the fake Grolier copy to the Victoria and Albert, which had purchased it for their decorative arts collection. And there I found it complete with genuine but entirely irrelevant old bookplates that Hagué had captured from who knows where.

Because the intellectual capital of France is so obviously Paris, it has come as something of a surprise for French savants to realize that there are even more copies of the *De Revolutionibus* in the provinces than in Paris itself, despite the fact that there are more Copernicuses in Paris than in any other city – twelve first editions and as many seconds by my latest count. The key to the French provinces was nevertheless held by the Bibliothèque Nationale, which contains printed catalogs from most of the outlying libraries. My systematic search in the Bibliothèque Nationale turned up numerous locations, and simultaneously scholars at the Centre Alexandre Koyré in Paris queried the provincial libraries. Eventually, in a series of field trips, I visited virtually every French library known to hold a copy of the first or second edition. Many of the provincial libraries have remarkable collections of early books, generally taken over from monasteries or Jesuit colleges at the time of the French Revolution. When I congratulated the librarian at Verdun on owning a Copernicus, she said,

'C'est rien. You should see our manuscripts!' And she promptly led me into the manuscript room and handed me the most exquisite illuminated manuscript that I have ever examined.

But owning a Copernicus *is* something, as far as printed books are concerned, and I have little doubt that the first edition Copernicus in some of these libraries represents the most valuable book in their collection, a fact probably unknown to the majority of the librarians. I must admit that I was startled in one of the libraries in southern France. Their copy had a particularly distinguished provenance from one of the sixteenth-century Pléiades, and I wanted to photograph its title page. The match-box sized library was literally too small to provide an area sufficiently bright, and after some negotiation it was decided that I should take the book onto the front steps facing the town square and use the sunlight there.

The price of the *De Revolutionibus* has climbed steeply in recent years, upward of $40 000, and will no doubt climb even higher now that the Japanese have entered the bidding. (Three of the last ten copies to enter the market have gone to universities in Japan.) The price for the first edition was considerably enhanced in 1974 following the Sotheby's auction of the first part of the scientific collection formed by Harrison Horblit. There, in a psychological configuration extremely favorable to the seller, his copy was pushed to $110 000, then the highest price ever paid for a printed work in the history of science. The Horblit copy was and is the most important example in private hands. In the absence of a copy autographed by Copernicus himself, the next best 'association copy' is a presentation from Rheticus, Copernicus's only direct disciple, and such is the copy that set the record-breaking price.

To me, this pricey copy has a feature that is much more interesting than Rheticus's signature. On the flyleaf is a long Greek poem written in 1543 by Joachim Camerarius, the leading professor at Leipzig, and inscribed into the book by Camerarius himself. 'What is this book?' asks the stranger in the poetic dialogue. 'A new one, with all kinds of good things in it,' replies the philosopher. 'O Zeus! How great a wonder do I see! The earth whirls everywhere in aethereal space,' the stranger cries out. 'But,' warns the philosopher, 'do not merely wonder, nor condemn a good thing as the ignorant do before they understand, but examine and ponder all these things.'

Such laudatory poems were commonplace in scholarly books of the sixteenth century, and their absence in the *De Revolutionibus* is conspicuous. Could Camerarius's poem have been commissioned by Rheticus for the book? Rheticus, lured by a particularly attractive salary, had left the Nuremberg printery before the front matter was finished, leaving the final part of the proofreading in the hands of Andreas Osiander, a local and learned preacher. To Rheticus's consternation, the final work contained no introductory poems, but an anonymous foreword from Osiander that cautioned the reader to treat the new cosmology as hypothetical, 'not necessarily true nor even probable.' In annoyance and anger Rheticus crossed off Osiander's foreword with a red crayon before sending the book to his friend Andreas Aurifaber, then dean at Wittenberg. Was Rheticus

Figure 10.2 Jerzy Dobrzycki examines the first edition De Revolutionibus *in which Rheticus inscribed a presentation to the Dean at Wittenberg, while its owner, Harrison Horblit, looks on. Since that time in the 1970s the book has been twice auctioned, most recently for $400 000.*

disturbed by the philosophy, or mostly just cross that Camerarius's poem had not been used? From this distant epoch, we can hardly even guess.

While Harrison Horblit still owned the Rheticus copy, I took Jerzy Dobrzycki, a colleague from Warsaw, down to Connecticut to see it. Professor Dobrzycki immediately realized that the Greek poem was the source of a Latin poem that Kepler had written into his own copy of the *De Revolutionibus*. Kepler's copy and the poem had long been known to scholars. I have seen and photographed it in Leipzig a couple of times. Kepler signed the poem 'I K,' but until Dobrzycki made the connection, no one realized that the initials stood not only for Iohannes Kepler but for the Greek iota–kappa of Ioachim Camerarius – a charming example of the wordplays that so delighted Kepler. (One of his anonymous books contains three consecutive anagrams of his name on the title page!)

The Rheticus and the Kepler copies both rate three stars in my census, but such extraordinary annotations are comparatively rare. More common are minor, rather uninspired, marginal tracks: catchwords that provide a running index. Usually such notes peter out after the first few chapters. Sometimes the remarks are more substantive, and occasionally even highly critical. Christopher Clavius, the Jesuit astronomer who engineered the Gregorian calendar reform, wrote beside a faulty trigonometric theorem, 'Here Copernicus is dreaming!' ('Hallucinatur hic Copernicus'). At least two annotators quoted the Sicilian astronomer Maurolycus, who had said that Copernicus 'deserved whips and lashes' for his unorthodox cos-

Figure 10.3 The title page of Horblit's Copernicus copy shows Rheticus' inscription to his friend Andrew Aurifaber, in 1543. The central paragraph played the role of the modern dust jacket blurb, urging its beholder to 'Buy, Read, Profit.'

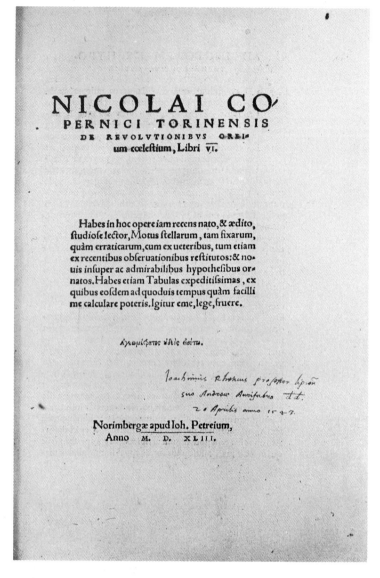

mology. In the richly annotated Yale copy, a long Latin critique finally breaks off into German: 'Der Himmel ist aber zum Narren worden, er musz gehn wie Copernicus will' ('The heavens would behave like a clown if they had to go as Copernicus wants'). However, the next owner added an ameliorating postscript; he noted that there was indeed a typographical error in the text, but it didn't matter because Copernicus actually used the correct number in subsequent calculations.

Despite the paucity of really well-annotated copies, there have been just enough to keep up the excitement of the chase. And gradually a major and unexpected result emerged from the survey: not only did several score books contain significant notes, but these notes ran in related families. Marginal clarifications and criticisms were passed from tutor to pupil, sometimes through several generations. For example, parts of Reinhold's comments have propagated into over a dozen specimens, and another independent set of notes appears in six more far-flung volumes. The critique in the Yale copy appears elsewhere thrice more, always untempered by the Yale postscript. These secondhand annotations reveal that even if Copernicus's revolutionary new doctrine failed to find a place in the regular university curriculum, a network of astronomy professors scrutinized the text and their protégés carefully copied out their remarks, setting the notes onto the margins of fresh copies of the book with a precision impossible by aural transmission alone. Clearly the students sat with the master book before them as they transcribed key words, data, diagrams, or whole paragraphs of elucidation. How wrong Ravetz and I were in supposing that only a handful of sixteenth-century readers were familiar with the book!

Even as the personal examinations drew to a close, well-annotated copies occasionally turned up. One of the last first editions for which information became available – from Debrecen in Hungary – turned out to have marginalia matching copies in Pisa, Soissons, and Ann Arbor. Another, from a specially unlocked safe in London, was filled with glosses apparently quite independent of any other copy, made by a sixteenth-century Spanish bishop. However, the unannotated copies of the *De Revolutionibus* are far more common and, apart from their bindings and state of preservation, they are as alike as peas in a pod and nearly as uninteresting. Often the circumstances of seeing a book are far more memorable than the book itself.

The visit to a secluded castle – one of three English buildings still surrounded by a water-filled moat – was all the more sweetly savored with the knowledge that no historian of science had been allowed to enter its precincts for several decades. At that time the Tate Gallery in London had mounted a two-million-dollar campaign to save two bucolic Stubbses for England, and so it was with double interest that we sat in the castle's drawing room and counted half a dozen splendid oils by George Stubbs on the walls.

In Budapest I was actually handed a first edition to take to my room overnight; I slept with the precious volume under my bed. And at the university in Vienna I was assured that the book couldn't be fetched until the next day – a day I did not have. There ensued a delicate maneuvering for position, and it quickly became clear that they were waiting for me to announce myself as a professor. Then the magic sprang into effect. Yes, I could have the books within an hour in the professional aula and, if I wanted, they would be available at any time till 8 p.m. 'We're not so efficient as American libraries,' a charming young assistant confided. 'Our rules go back to Franz Josef.' 'Never mind,' I replied, knowing full well that there are few university libraries in America where one can examine a *De Revolutionibus* at night.

In the professors' room that evening, all alone and unsupervised, I carefully made a rubbing of the blind-stamped binding of the Vienna Copernicus, and I mused on the fact that I hadn't been asked to produce a shred of evidence that I really was a professor. Had I shown them my Harvard dazzler letter, maybe they would have furnished tea and pastry as well!

I'm not sure if the Harvard dazzler is a uniquely Harvard phenomenon, but somehow I have trouble envisioning a Yale dazzler or a Princeton dazzler. The Harvard dazzler comes on an engraved letterhead of the president and fellows, with a gold seal bright enough to compete with the crown jewels. The corporation secretary certifies the bearer's professional credentials and expresses appreciation for any courtesies that the relevant libraries, museums, and consular officers can offer.

Only one and a half times was the dazzler letter really indispensable. I had always envisioned the Ecole Polytechnique as the MIT of France, the home of Pasteur and Gay-Lussac. In reality it is more like the US Military Academy, but considerably more formidable than West Point, where I had little difficulty in seeing *their* first edition Copernicus. The gate of the Ecole Polytechnique was barred by three officers behind wickets, who assured me that I needed permission from the commandant to see a book in the library. I followed their instructions to the next gate, where the porter simply sent me directly to the library. The librarian was horrified: were I to publish anything about their Copernicus the commandant was sure to find out and he would demand to know how I had got in without his permission. So, armed with my dazzler, I tracked back to the commandant's office, where a suitably dazzled officer promptly issued me a pass. I think the speed of this transaction left the librarian a bit chagrined. It must have been decades since anyone had asked to see the *De Revolutionibus*, and there was considerable scurrying before the volume was eventually produced for my inspection.

The other time when the dazzler seemed at least half needed was at the Bibliotheca Apostolica Vaticana. Casual visitors to Saint Peter's or to the Sistine Chapel are protected from the fact that Vatican City is a political enclave of its own, and fully as bureaucratic as Italy itself. To visit the library, a form and a visa are demanded at the gate to the City, and then a pass from the secretariat. The check-in procedure gives one a locker key, which subsequently entitles its holder to see three books per day. Special permission from the prefect allows more books to be examined, but the book fetchers will look affronted by anyone audacious enough to suppose he can read more than three volumes in a day! Special permission will also allow foreign scholars to work in the afternoon as well. Since that is a privilege accorded only male researchers, in the afternoon a professor is allowed to read the books without wearing his jacket.

Not only has the trail led to great institutional libraries such as the Vatican (which has three first editions, as do Turin, Princeton, Glasgow, Trinity College in Cambridge, and the Bibliothèque Nationale) and to small provincial libraries, but also to a score of private collections.

Fortunately for the census taker, American collectors are not particularly secretive, an attribute probably closely tied with the comparative honesty of our income tax structure. Nevertheless, tracking down these copies has required a lot of generous help from rare-book dealers, and has had its moments of tragicomedy. I was, for example, barred from a West Coast collection on the grounds that I came from

Massachusetts, the only state that had had the audacity not to cast its electoral votes for Mr Nixon. On another occasion, I was literally smuggled into the library of a collector who had grown senile, an act surreptitiously carried out while the nurse was taking the owner on his daily outing. But otherwise the collectors have all accepted my visits with good humor and splendid hospitality.

The list of memorable visits to private collections included one in Pittsburgh, where we met the man who brought neon advertising signs to the world. Henry Posner told us how, in his younger years, he had met the inventor of a new kind of light, the neon bulb, and he had asked the inventor what he intended to use it for. Well, said the inventor, it would be very good on an automobile to detect if the spark was active. 'It gives a very bright light doesn't it?' Posner observed, and shortly thereafter he installed the first outdoor neon sign for a large Parisian department store. Posner invested his ensuing fortune in rare books of all sorts, from Copernicus to Omar Khayyám, from one of two known copies of the first printing of the Bill of Rights to – so he assured us – a Gutenberg Bible kept in a safe in the cellar. By that time Posner was advanced in age and occasionally not too exact in his claims, so I mentally dismissed his Gutenberg Bible as a mild delusion of grandeur. A couple of years later I was chagrined to learn that he supposedly did have most of Gutenberg's Old Testament. Since then Posner has died and his Gutenberg pages have never come to light so perhaps my first hunch was correct.

It is comparatively easy to know where the Gutenberg Bibles are, because this is one of the few books for which a comprehensive census exists. Surprisingly, only three other major books have their locations listed with any serious claims to completeness: the first folio of Shakespeare (over one third of the copies are in the Folger Shakespeare Library in Washington), the elephant folios of Audubon's birds, and now the first two sixteenth-century editions of Copernicus's *De Revolutionibus*. As a subsidiary result of the Copernicus census, bibliographers will have extensive information on the provenances and movements of an important early scientific book.

For example, there are 47 first editions of the *De Revolutionibus* in America. The second edition, which was issued in Basel in 1566, presumably after the 1543 edition went out of print, is virtually as rare as the first, and 44 copies are currently recorded in this country. The first *De Revolutionibus* to come to the Americas, as far as anyone knows, arrived on a ship that docked in Veracruz in 1600. That such details were recorded in the bill of lading, and that the bill of lading survives, completely astonishes me. It is only sad that the book itself can no longer be found. Apparently the earliest surviving copy was the third edition, brought over at the request of James Logan in Philadelphia around 1700, and his second edition may have come as early as 1709. The earliest first edition is undoubtedly the one in the Boston Athenaeum, which arrived around 1825. Thomas Jefferson bought a second edition for the University of Virginia, and Harvard had one at an unknown time in the nineteenth century. The larger movement of these books to America did not take place until the twentieth century – a migration of such proportion that there are more first editions here than in any other country. There are also probably more copies in private hands in America than anywhere else, but the number is rapidly declining as the books are handed over to institutional libraries. Whether the tide has turned on the flow of Copernicus books to America is hard to say, but a first edition auctioned by Philadelphia's Franklin Institute (to help pay off its Bicentennial debt) recrossed the Atlantic to the Old World once again, perhaps a harbinger of the Gutenberg Bibles that have since returned to Mainz and to Stuttgart. More recently two more copies of the *De Revolutionibus*, from the Robert Honeyman collection, have followed suit – with the result that the United Kingdom has now almost nudged out the United States as the leading repository of the first edition, as it has been of the second.

Keeping track of the movements of the *De Revolutionibus* has proved quite an obstacle to the formation of a complete census, although most dealers and private owners, recognizing the uniqueness of the Copernican survey, have been exceptionally cooperative. After the distinguished Rheticus presentation copy was bought in 1974 by a combine of dealers in New York and London, such was the secrecy of its disposition that when the book eventually found a buyer (at a suitably augmented price), even the London owners did not know its destination. Nonetheless such things are hard to keep secret, and eventually the book world buzzed with rumors that the new owner lived in Cambridge, Massachusetts. I dropped the putative collector a note about my interests, but had no response for several months. Then one day the phone rang, and to my surprise the new owner was on the line telling me about his remarkable collection of great works in the history of ideas.

Meanwhile I often wondered about the earlier provenance of this distinguished copy. What had happened to it after its recipient, Andreas Aurifaber, had died in Danzig? There seemed to be no evidence of ownership between the sixteenth century and the early 1950s, when the volume had turned up on the London book market without an intervening pedigree. And all this leads, strangely enough, to the libraries of Italy.

At the end of the sixteenth century, the ownership pattern of the *De Revolutionibus* was quite different in Italy from that in Germany. There was no teaching tradition for the book in Italy, and few copies from the Nuremberg edition came down from Lutheran territory to the cisalpine regions. The 1566 Basel edition was apparently much easier for Italians to obtain and hence became widespread in the libraries of the Italian universities and religious orders. The evidence for this distribution has been preserved in an unexpected way: through the censorship of the book.

Early in the seventeenth century a brilliant but controversial physicist named Galileo Galilei began to argue that ultimate truth might be found not only in the Book of Scripture but in the Book of Nature – in other words, that cosmological theories might represent physical truth rather than mere hypotheses.

As a direct result of Galileo's polemics, Copernicus's *De Revolutionibus* was in 1616 placed on the *Index of Prohibited Books* 'until suitably corrected.' Such was the sensitivity of the affair that for this book, and for this book only, the changes were exactly specified. The corrections, duly announced in 1620, have much in common with the rewriting of California high school biology textbooks. The Inquisition ordered a dozen statements that sounded too positive to be replaced by weaker sentences confirming that the cosmology was meant only as a hypothesis, not as a fact or a physical law. Of the 30 copies of the second edition now in Italian libraries, nearly 60 percent have been censored. The number includes Galileo's own copy, which has the offending text only lightly canceled but with the corrections written in his own hand. Perhaps he corrected the book while under house arrest to demonstrate his good behavior. On the other hand, of the first editions only 14 percent have been censored. The conclusion follows naturally that the majority of the first editions now in Italy came after 1700, when heliocentrism was no longer a burning issue. Circumstantial evidence suggested that the Rheticus–Aurifaber copy participated in this southern migration.

According to its catalog, the Biblioteca Palatina in Parma should have a first edition, but when I visited that library a few years ago an assistant showed me an empty space on the shelf and through the fuzzy filter of the language barrier I understood that the copy couldn't be found. The empty slot suggested that the loss was recent, although in retrospect I realize that the shelves in question were laden with marvelously bound old books that would seldom, if ever, get moved. The catalog seemed to say something about a manuscript note on the preface attributing it to Camerarius, and I mentally filed this away in case I should ever find such a copy in another collection.

A year later I asked the librarian at Parma to send me a copy of the catalog description of the still-missing Copernicus, and then I promptly realized that the Latin did not say the *preface* was attributed to Camerarius but that a manuscript poem about the book by Camerarius was *prefixed* to the book. Precisely the description of the Rheticus copy! How the copy given by Rheticus to Aurifaber got to the Parma Palace Library was a mystery, but at least it would have been in keeping with the movement of first editions into Italy after the seventeenth century.

I was rather curious about all this, so I arranged to have a closer look at the copy. One Saturday morning the present owner brought it around to my observatory office, and there, armed with an ultraviolet lamp, we thoroughly scrutinized the volume, searching for evidence of erased inscriptions or library stamps. The ultraviolet light revealed an early and now faded signature, illegible even under the probing of the short wavelengths. However, absolutely no physical evidence emerged for an earlier ownership by the Italian library – no missing end leaves, no telltale remnants of bookplate paste, no trace of the Parma shelf mark. Nevertheless the circumstantial evidence provided by the old catalog description suggested that Parma had been its resting place for at least a century.

Now, as I prepare to reprint this essay from *The American Scholar*, I can add an unexpected postscript. Recently a European dealer contacted me, saying that he was on the trail of a first edition Copernicus

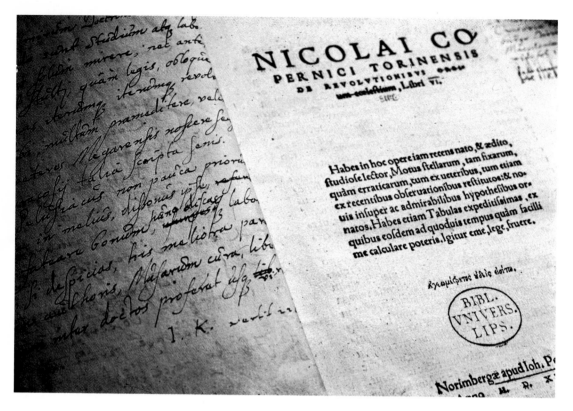

Figure 10.4 The first edition De Revolutionibus *at the Leipzig University Library was owned and annotated by Johannes Kepler. The flyleaf contains a prefatory poem translated by Kepler and signed with his initials.*

with some very interesting annotations. When I visited his shop, he showed me two xerox pages, which I instantly recognized as another copy of the manuscript poem, also in the hand Camerarius. And then he told me a curious story. He had been telephoned by an agent of a private library in Italy, who wished to sell the collection. The mystery caller did not give any names or addresses, but said that some information would come by post. Eventually a packet arrived, with no return address; it contained xeroxes of title pages of many old books, including a *De Revolutionibus*, and there were also the two additional sheets showing the Copernican poem. Since then, the agent has not been heard from. In light of this new information, it seems quite likely the missing Parma copy is not, after all, the one in Cambridge.

Whatever its previous homes, the present location of the Rheti-cus–Aurifaber copy is but another manifestation of the continual movement of rare books – a movement that puts books of English provenance into Poland, books of Polish provenance into Scotland, and books of Scottish provenance into England. This movement, although guaranteeing a built-in obsolescence for any census, at least has often the merit of making the books visible.

More troubling to any census maker is the thought of any number of copies sequestered in remote monasteries, in quiescent private collections, or even in major institutions. Recently I learned of a previously missed second edition at the University of Liverpool. When the curator of rare books there sent me its description, he remarked, 'I assume you are only interested in this edition, as we also have two of the first.' With that offhand comment, my census of the 1543 edition went from 243 to 245.

I am often asked how many copies were printed in the first place. Disappointingly little information survives concerning press runs in the fifteenth and sixteenth centuries. However, an interesting way to reconstruct empirically the original number of *De Revolutionibus* copies goes as follows: Make a list of possible sixteenth-century owners, and then check how many of their copies have actually been found. Since about half of the expected ownerships have turned up, this method suggests that at least half of the entire first edition is now accounted for. Hence, a printing of 400–500 copies seems likely.

Because Copernicus's book was quickly recognized as a classic, few copies would have been thrown away, and since it was formidably technical, not many copies would have worn out from overuse. But how do books disappear? In principle by fire and flood, but apart from those lost in modern warfare, cases are hard to document. Pirates threw at least one Copernicus into the Mediterranean because they were so outraged that the cargo they had just captured consisted of nothing but books. And it is quite possible that a few copies went up in smoke in the Great Fire of London, although specific evidence is still lacking.

For better or for worse, I am reconciled to producing an incomplete census, and I know that clever book dealers will keep turning up unrecorded copies. But there remains a bittersweet comfort: the Great Copernicus Chase will go on!

Notes and references

For further reading, see my chapter 'Copernicus's De Revolutionibus: An Example of Renaissance Scientific Printing' in *Print and Culture in the Renaissance*, edited by Gerald P. Tyson and Silvia S. Wagonheim (Newark, Delaware, 1986), reprinted in my *The Eye of Heaven: Ptolemy, Copernicus, Kepler* (New York, 1992).

11 The Tower of the Winds and the Gregorian calendar

Figure 11.1 The Tower of the Winds observatory rises above the Vatican Secret Archives, behind St Peter's Cathedral, in this recent photograph by Gordon Moyer. The rooms were the seat of the Vatican Observatory until 1906.

For over a decade I have been earnestly ferreting out all extant copies of Copernicus' magnum opus, to examine any marginal annotations for clues as to the way this book was studied in past centuries. One of the most exciting moments came on a visit to the Vatican Library in 1973, where I found a richly annotated *De Revolutionibus* with dated cosmological sketches that appeared to be in the hand of Tycho Brahe, the famed sixteen-century Danish astronomer. At this point I enlisted the aid of the late D. J. K. O'Connell, former director of the Vatican Observatory, and through his good offices the library agreed to produce photographs of critical pages within an hour.

To pass the time while we waited, Father O'Connell suggested that we visit the Tower of the Winds, a remarkable observatory built at the Vatican in the early 1580s. A narrow stairway led above the Vatican Secret Archives and up through the apartments of Christina. In 1655, after abdicating as Queen of Sweden, she had endorsed Catholicism and descended upon the timorous Pope Alexander VII.

We then entered what seemed to me a windowless, cubical room about 25 feet on a side – the strangest observatory I had ever seen. Each wall was decorated by one or more frescoes depicting the winds: to the west was the shipwreck of St Paul, to the south Christ calming the Sea of Galilee, and to the north, the allegory of the north wind.

The most curious feature was the orifice in the mouth of the south wind, an opening that allowed a spot of sunlight to fall on an inlaid marble meridional sundial line across the floor. Here, Father O'Connell related, the Jesuit astronomer Christopher Clavius had persuaded Pope Gregory XIII that the calendar was running ten days behind. Had the pope visited the Meridian Room around noon on 21 March 1582, he would have found the spot of sunlight crossing the line over 60 cm north of the equinoctial position marked with the symbol ♈ for Aries, since the Sun had actually passed the vernal equinox ten days earlier. Thus, the necessity of calendar reform would have been very evident.

As we descended from the tower, Father O'Connell remarked that for many years there had been a rumor that some part of the frescoes had been painted over when Christina arrived from Sweden to take up residence in the nearby apartments, but no one could remember just what had been eliminated. Possibly the diaphanous green drapery covering the north wind had been added to protect her sensibilities. Eventually, when a restoration was undertaken, something quite different was revealed: a Biblical motto paraphrased from *Jeremiah 1:14*, '*Ab Aquilone pandetur omne malum* (All bad things come from the north)'!

At the time of my first visit, the Meridian Room was filled with

Figure 11.2 The Meridian Room contains a series of frescoes including the Calming of the Sea of Galilee on the south wall, based on the account in Matthew 8:23–33. *Note the orifice in the mouth of the South Wind.*

scaffolding, and probably for that reason I didn't notice that the room was neither exactly cubical nor windowless. Recently I had occasion to see it again, when a small group of historians gathered at the Pontifical Academy of Sciences in celebration of the quadricentennial of Pope Gregory's calendar reform. The frescoes, now carefully restored, follow the ideas of the astronomer, instrument-maker, and cartographer Egnatio Danti (1536–86). He was reputedly brought to Rome to assist in the calendar reform, but a careful scrutiny of the dates involved (and presented at the symposium) now suggests that Danti and his Tower of the Winds had less to do with the reform than has been commonly believed.

Resetting the date of the vernal equinox to 21 March, and fixing a pattern of leap days to prevent the drift from repeating itself in the future, was actually only a small part of the problem troubling Gregory XIII's astronomical advisors. The real difficulty lay in the incompatible periods of the sun and moon, and therefore in fixing the holy days. Easter was traditionally the Sunday following the Jewish Passover, which depended on a lunar cycle. In the first centuries of the Christian era, the Hebrew lunar calendar was kept in rough step with the solar year by the occasional addition of a thirteenth lunar month. The Jewish authorities in Jerusalem (a committee appointed by the Sanhedrin) decided this matter by observing the state of vegetation near spring each year. Needless to say, this arrangement

Figure 11.3 The North
Wind in the Vatican fresco
on the north wall of the
Meridian Room includes at
the bottom a motto saying
'All bad things come from
the north.' The line was
painted out when Christina
arrived from Sweden.

proved increasingly unsatisfactory, especially when both the Christian and Jewish communities spread farther from Jerusalem.

Meanwhile, the Roman lunisolar calendar had been almost equally confused, with an extra month sometimes thrown in between 23 and 24 February. By the time of Julius Caesar the calendar year was about three months ahead of the seasons, and calendar reform was desperately needed. The Julian reform adopted a solar year of exactly $365\frac{1}{4}$ days with a pattern of leap days every fourth year. While this regularized the solar period, it did nothing for the religious holidays that depended on the moon. Various churchmen devised schemes for setting Easter by some recurring pattern, but none of these was universally adopted.

There had long been known, however, a rather close combination of lunar and solar periods, the Metonic Cycle: 12 years of 12 lunar months plus 7 years of 13 lunar months match sufficiently well so that a 19-year cycle can recur for several centuries without the years and lunar months getting seriously out of step. (The discrepancy is about an hour and a half in 19 Julian years.) Such a basis for Easter computation was introduced in Alexandria toward the end of the third century, and it became almost universally adopted in Christendom after the Council of Nicea in AD 325. (Around the same time the Jewish community independently adopted a 19-year reconciliation of the solar and lunar calendars.)

The use of a 19-year basis by no means settled all the questions of how, specifically, Easter was to be set, and these questions still plagued the makers of the Gregorian calendar. As Gordon Moyer has pointed out , the real achievement of Luigi Lilio, the man behind the Gregorian calendar reform, was not in his scheme of having 97 instead of 100 leap days every 400 years, but in his arrangement of the so-called system of epacts for finding the Paschal full moon, which is then followed by Easter Sunday.

Calendar reform had for many years been an earnest desire of the Catholic Church. In fact, the opinion of Copernicus had been specifically invited when the Fifth Lateran Council (1512–17) was in session. However, the matter remained undecided at that time because it was believed that the periods of the sun and moon were still insufficiently known. In fact, the numbers of the *Alfonsine Tables* were good enough for such calendrical purposes, and, had a reform been proposed around 1517, it might have been universally adopted throughout the western world. Unfortunately for the calendar, by the time of Gregory's reform in 1582, Christendom had been severely divided by the Protestant Reformation, and acceptance of the new scheme became a matter of religious and political contention.

In the Roman Catholic countries and their colonies the transfer to the Gregorian calendar generally took place in 1582, as planned. In practice this meant that Italy, Spain, Portugal and Poland dropped days in October, going from the 4th to the 15th. In France the bishopric of Strasbourg moved first, in November, and the rest of the country followed in December after an edict from Henry III. Belgium and the Catholic states of the Netherlands likewise switched from the Julian to Gregorian calendars at the end of 1592. The transalpine

Catholic regions adopted the reform in a wildly patchwork manner. Some parts of Austria moved to the new calendar in October 1583, and others in December of that year.

The Protestant regions of Germany were bitterly opposed to the new calendar. Michael Maestlin, the astronomy professor at Tübingen University, led the opposition on technical grounds and even edited a book of sermons against it. Protestant Germany and Denmark held out against the Gregorian calendar until the very end of the seventeenth century. A face-saving move then became possible in the work of the Jena astronomer Erhard Weigel. He proposed an 'improved calendar,' which matched the Gregorian scheme with respect to the dates, but which used astronomical reckoning according to Kepler's *Rudolphine Tables* rather than the cyclical system for establishing the Paschal full moon.

Germany finally abandoned the Julian pattern in February 1700, and Denmark also made the change then. This was a critical time, because in the Gregorian reckoning 1700 was not a leap year; hence the two systems moved out of phase by 11 days in March of that year. The capitulation of the remaining German principalities was apparently too much for the northern Swiss cantons, and they abandoned the Julian calendar at the beginning of 1701. Likewise, the Protestant states in the Netherlands made the changeover at this time.

The differences between the Gregorian and the 'improved calendar' adopted in the Protestant lands became apparent in 1704 when the date of Easter by the cyclical calculation fell on 23 March and by the astronomical calculation on 20 April. Again in 1724 the dates fell differently, a week apart. This situation was not straightened out until the time of Frederick the Great in 1775, when the astronomical calculations were abandoned as the basis for Easter.

The story of calendar reform in England is unusually long and well documented. After nearly two centuries of resistance, a turning point came on 25 February 1751, when Lord Chesterfield introduced into the House of Lords a bill entitled 'An Act for Regulating the Commencement of the Year [which was then observed in Britain on 25 March]; and for Correcting the Calendar now in Use.' The way for Chesterfield's bill had been paved by George Parker, the second Earl of Macclesfield, an amateur astronomer who had erected his own observatory at Shirburn Castle. In 1750 Lord Macclesfield presented a paper to the Royal Society on the solar and lunar years, which must have catalyzed the sentiments for calendar reform in England.

Philip Stanhope, the second Earl of Chesterfield, was a cosmopolitan man, now perhaps best known for the long series of letters to his son, and it is in these letters that we find a charming vignette on the passage of the calendar reform bill in Parliament. He wrote:

It was notorious, that the Julian calendar was erroneous, and had overcharged the solar year with eleven days. Pope Gregory the Thirteenth corrected this error; his reformed calendar was immediately received by all the Catholic powers of Europe, and afterward adopted by all the Protestant ones, except Russia, Sweden, and England. It was not, in my opinion, very honorable for England to remain in a gross and avowed error, especially in such

Figure 11.4 In England in 1752, September had only 19 days. This page from The Ladies Diary: or Woman's Almanack *shows the gap where 11 days were skipped between the 2nd and 14th.*

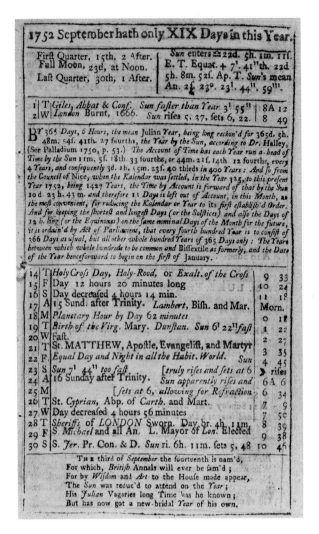

company.... I determined, therefore, to attempt the reformation; I consulted the best lawyers and the most skillful astronomers, and we cooked up a bill for that purpose. But then my difficulty began: I was to bring in this bill, which was necessarily composed of law jargon and astronomical calculations, to both [of] which I am an utter stranger. However, it was absolutely necessary to make the House of Lords think that I knew something of the matter; and also to make them believe that they knew something of it themselves, which they do not. ... I gave them, therefore, only an historical account of calendars ... amusing them now and then with little episodes; but I was particularly attentive to the choice of my words, to the harmony and roundness of my periods, to my elocution, to my action. Lord Macclesfield, who had the greatest share in forming the bill, and who is one of the greatest mathematicians and astronomers in Europe, spoke afterward with infinite knowledge, and all clearness that so intricate a matter would admit of: but as his words, his periods, and his utterance, were not near so good as mine, the preference was most unanimously, though most unjustly, given to me.

The bill provided that in 1752 the English calendar be brought into line with the Gregorian by dropping the days between 2 September and 14 September. The bill met with a mixture of militant Protestantism and resentment of change, and 'Give us back our Eleven Days' became a campaign slogan. Nevertheless, the change seems to have gone relatively painlessly both in Britain and in its overseas colonies.

The next stage in the story of calendar reform was something that, in retrospect, appears as a comical step backwards. The French revolutionaries, in their anticlerical zeal, adopted the 'calendar of reason' on 5 October 1793. They established a year of 12 months, each with 3 weeks of 10 days, plus 5 or 6 epagomenal days. The day from midnight to midnight was divided into 10 hours, and a hundredth part of one of these hours became a decimal minute.

The calendar actually began retroactively, starting on the autumnal equinox in 1792, and the months were renamed beginning then with

Autumn	Vendémiaire	Brumaire	Frimaire
Winter	Nivose	Pluviose	Ventose
Spring	Germinal	Floréal	Prairial
Summer	Messidor	Thermidor	Fructidor

All of these radical revisions and the novel nomenclature inspired a certain amount of ridicule, and English wit devised the following parody:

Autumn	wheezy	sneezy	freezy
Winter	slippy	drippy	nippy
Spring	showery	flowery	bowery
Summer	hoppy	croppy	poppy

Needless to say, the scheme met with considerable resistance, especially the idea of observing a day of rest only every tenth day. In 1801 Napoleon Bonaparte concluded an agreement with the pope to reinstitute the observance of Sunday and several of the Christian festivals including Easter and Christmas. Ultimately, France had the opportunity to adopt the Gregorian calendar a second time, going over on 11 Nivose XIV – on 1 January, 1806.

The civil acceptance of the Gregorian calendar in the rest of the world had to await the twentieth century, except for Alaska, which switched following its 1867 purchase by the United States from Russia, and for Japan, which accepted the system on 1 January, 1873. The calendar was introduced to China by Sun Yat Sen in 1912, but not until 1929 did it officially replace the Chinese calendar.

The general disruption associated with World War I brought the Gregorian calendar to Eastern Europe, including Bulgaria, Lithuania, Latvia, Estonia, Russia, Yugoslavia, and Romania. Calendar reform came last to Greece; the Greek adoption followed a congress of the Eastern Orthodox Church in Constantinople in May, 1923. It is fascinating to learn that this last calendar improves upon the year length of the Gregorian scheme, so that leap years will include 2000, 2400 (as in the Gregorian calendar), but also 2900 and 3300 instead of 2800 and 3200. Thus, all the dominions of the Eastern Orthodox Church will differ by a day from the rest of the world in the twenty-ninth century, provided tendentious civilizations can last that long!

Although other calendars (such as the Jewish and Islamic ones) have continued in use, and although occasional attempts at calendar reform have gained some notoriety since 1924, the entire world has been united in the secular use of the Gregorian calendar dates.

Notes and references

For a detailed description of the Meridian Room in the Vatican Tower of the Winds, see the article by John W. Stein, S. J., Specola Astronomica Vaticana *Miscellanea Astronomica*, **97** (1950).

Key sources for this chapter are the articles by Olaf Pederson and by me in *Gregorian Reform of the Calendar: Proceedings of the Vatican Conference to Commemorate its 400th Anniversary* (Vatican City, 1983) (G. V. Coyne, M. A. Hoskin and O. Pederson, editors), respectively 'The Ecclesiastical Calender and the Life of the Church,' pp. 17–74, and 'The Civil Reception of the Gregorian Calendar,' pp. 265–79. See also Gordon Moyer 'Luigi Lilio and the Gregorian Reform of the Calendar,' *Sky and Telescope*, **64** (1982), 418–19.

Incidentally, Robert Westman and I eventually discovered that the Copernicus book mentioned above was annotated not by Tycho Brahe, but by Paul Wittich – see our note in *Journal for the History of Astronomy*, **12** (1981), 53–4, and the reference following the next chapter.

12 Tycho Brahe and the Great Comet of 1577

In 1577 one of the most influential comets of all time blazed forth in the northern heavens. When first generally seen, in mid-November it rivaled Venus in brilliance and its tail stretched more than 20°, from Sagittarius into Capricornus. The comet moved swiftly across the sky into Pegasus, where it was last seen in January, 1578.

With the possible exceptions of the sun-grazing comet of 1680, whose motion helped Newton perceive the universality of gravitation, and the comet of 1682, whose elliptical orbit was first calculated by Edmond Halley, the Great Comet of 1577 exceeded all others in its impact on the development of astronomical ideas. As analyzed by Tycho Brahe, the outstanding observational astronomer of the sixteenth century, the comet of 1577 provided powerful evidence against the prevailing Aristotelian view of the cosmos.

According to this Greek philosopher, the central terrestrial sphere, full of birth, death, and mutability, stood in distinct contrast to the eternal and unchanging celestial spheres. All phenomena of change – rainbows, meteors, comets – belonged to the sublunar world. Tycho's claim, that the comet traveled in the heavenly realms beyond the moon, helped to demolish ancient authority and led to the acceptance of radical new ideas.

Tycho himself first noticed the comet on the evening of 13 November as he was trying to catch some fish for supper from one of the ponds on the Island of Hven. (He had been given the small island in the previous year as a fiefdom on which to build his observatory; eventually he also constructed a series of fishponds whose water fed his paper mill.) The great Danish observer promptly began precise measurements of the comet's position. Although the full panoply of instruments for which he was to become famous had not yet been constructed, he had available a quadrant with a 16-inch radius by means of which he could establish altitudes and azimuths. There was also a sextant – not the compact modern navigational device, but simply a graduated sixth of a circle – with which he could determine the comet's distance from various fixed stars.

Tycho was not the first to observe the comet, however. In his book on this subject, published ten years later in 1588, he noted that some seamen had seen the comet as early as 9 November, and eventually a report going back as early as 1 November came in from Peru. Quite by chance, in 1882 Tycho's biographer, J. L. E. Dreyer, found a series of early observations made from London beginning 2 November, all handwritten in an old printed comet tract that he had acquired in Copenhagen.

This little tract was but one of many that poured from the presses in 1577 and 1578. Comets traditionally had been seen as fearful omens, and the apparition of 1577 was no exception. Astrological themes ran

Figure 12.1 This little-known portrait of Tycho Brahe at age 50 hangs in the Gavno Castle in southern Denmark. Recorded as a copy, the painting shows the red-haired astronomer in 1596, the year before he left Hven. The name of the original artist is unknown.

EFFIGIES TYCHONIS BRAHE OTTONIDIS ÆTATIS SVÆ ANNO 50 Completo.

through virtually all of the tracts. The comet portended the wrath of God, and the threat of the Turks in the east furnished a frequent refrain.

A particularly fine collection of comet tracts, including the annotated one that Dreyer acquired, was brought together at the end of the last century by the Scottish Lord Crawford and is now at the Royal Observatory in Edinburgh. Most of the black-and-white illustrations accompanying this article come from the Crawford collection and are reproduced through the courtesy of the Astronomer Royal of Scotland.

In addition to the tracts, numerous comet broadsides had appeared – poster sheets printed on one side, something like handbills, which were forerunners of our modern newspapers. Early printers struck them off to record monster births, meteorological phenomena, natural calamities, and whatever else seemed marvelous and noteworthy to sixteenth-century readers. One of the best collections of these now rare ephemera was formed at the time by a Swiss clergy-

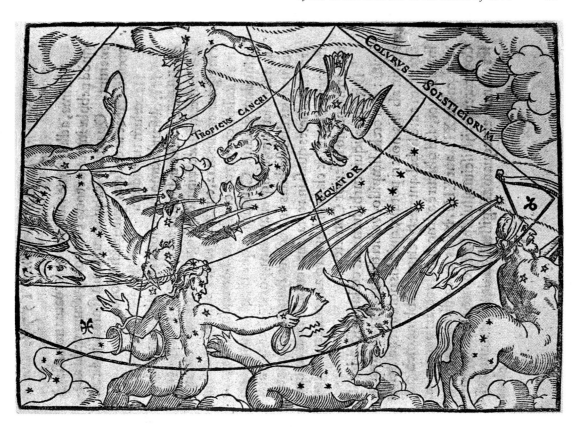

Figure 12.2 The Great Comet of 1577, after passing through perihelion on 27 October of that year, was discovered in the evening sky on 1 November and remained visible until 26 January 1578. This map showing its motion from the head of Sagittarius (right) to the forelegs of Pegasus (left) is from Descriptio cometae *by Thaddeus Hagecius, Prague, 1578. During early November the comet was bright enough to be seen through clouds, and on the 13th it could be seen before sunset.*

man, Johann Jakob Wik (1522–88), and about 400 sheets are presently preserved in the Wikiana collection at the Zentralbibliothek in Zurich. My favorite includes a group of well-dressed spectators, among whom an artist works by lantern light to sketch the nocturnal visitor. The illustrations seen here come from among approximately 20 Great Comet of 1577 broadsides in this rich collection, which has many of the small comet tracts as well.

From a modern point of view, the majority of tracts and broadsides are more akin to folklore than to science. Most of them, insofar as they were explicit, adopted the Aristotelian view that comets were phenomena of the earth's own atmosphere. Nevertheless, several publications from distinguished astronomers contained valuable observations. Chief among these was Michael Maestlin's 60-page treatise. Best known as Kepler's teacher at Tübingen, Maestlin argued that the comet lay beyond the moon; he used a Copernican scheme to show that the comet traveled in the region of Venus. He added some astrological remarks almost as an afterthought.

Cornelius Gemma, son of the Louvain astronomer Reiner Gemma Frisius, also concluded that the comet's distance exceeded the moon's, and he proposed that it moved in the region of Mercury. In contrast, the well-known Prague astronomer Thaddeus Hagecius declared that his own measurements placed the comet below the moon. Several years later, after Tycho had pointed out the errors in that analysis, Hagecius retreated and recognized the supralunar path of the comet.

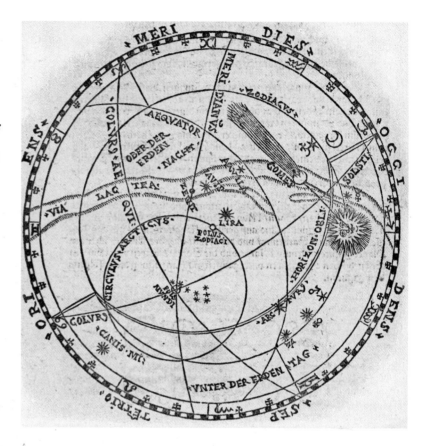

Figure 12.3 The comet is located in the Milky Way in this planisphere by Georg Busch, Erfurt, 1577. Its tail is pointing away from the sun, which is below the horizon (the narrow double line). The north celestial pole and the stars of Ursa Minor are at lower right, and Vega is the bright star near the center.

Unlike most of his colleagues, Tycho Brahe did not rush into print concerning the new comet. Instead, he bided his time, carefully accumulating the observations of other astronomers. His examination of the nova of 1572, observed almost exactly five years earlier, had established that the new star was situated beyond the moon, thereby dealing a heavy blow to the Aristotelians. Even at that time he had wondered in print whether comets might also lie beyond the moon, and the comet of 1577 afforded the opportunity he sought. By collecting data from observers elsewhere in Europe he was able to demonstrate more thoroughly than the others that the object appeared in the same position among the stars for everyone at any given time; this absence of parallax proved that the comet was supralunar.

Tycho's large treatise on the comet, singularly devoid of astrology, finally appeared in 1588. Entitled *De mundi aetherei recentioribus phenomenis* ('Concerning the quite recent phenomena of the aethereal region'), it offered systematic documentation about the celestial position of the comet. In addition it exhibited for the first time the so-called Tychonic system, his newly devised geocentric arrangement of the planets.

In the Tychonic system the earth remains fixed at the center. 'I am convinced,' wrote Tycho, 'that the earth occupies the center of the universe and is not whirled about with an annual motion as Copernicus wished.' In his system the moon, sun, and stars revolve around

Figure 12.4 In this picture printed at Prague by Peter Codicillus, the comet's tail stretches above the moon and Saturn as an artist draws it, aided by men holding his sketch block and a lantern. The heading reads: 'Concerning the fearful and wonderful comet that appeared in the sky on the Tuesday after Martinmass [12 November] of this year MDLXXVII.'

the earth, whereas the planets, revolving about the sun, are in turn carried around the earth as the sun moves. 'Thus a clear reason is furnished why the simple motion of the sun is necessarily involved in the motions of all five planets,' he wrote. 'Therefore the sun regulates the whole harmony of the planetary dance in order that all the celestial appearances may be subject to his rule as if he were Apollo in the midst of the Muses.'

However, Tycho's scheme requires the circular orbit of Mars around the sun to intersect the circular orbit of the sun around the earth. If the planets revolved in solid crystalline spheres, as was the general belief in Tycho's day, such an interpretation would be inconceivable. In adopting his new cosmology, Tycho was obliged to discard the crystalline celestial spheres, a truly radical step.

Had the solid crystal spheres existed, then the comet of 1577 would have smashed them – or so it is commonly said today in repetition of a statement once made by Kepler. Actually the development of Tycho's cosmology is more subtle than this, and it is worthwhile to look at the historical evidence a little more carefully.

Although Tycho was not the first to observe the comet of 1577, he was apparently the last. On 26 January 1578, a friend had asked to

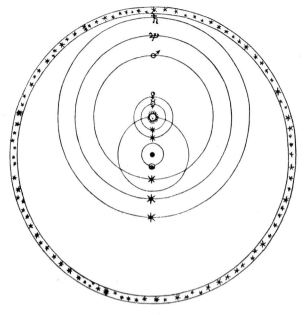

Figure 12.5 Tycho's drawing of his planetary system model has the earth at the center, around which move the moon and sun, with the other planets from Mercury to Saturn circling the sun. Around the whole is the sphere of fixed stars.

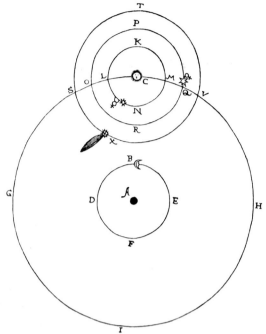

Figure 12.6 A second Tychonic diagram shows how he fitted the comet of 1577 into his system. Here A is the earth, B the moon, and C the sun; K, P and T are the orbits of Mercury, Venus, and the comet, respectively.

Figure 12.7 A typical prognostication, connecting the comet with two lunar eclipses (25 September 1577, and 16 September 1578) and with death and destruction by the Turks, is from a pamphlet by Andreas Celichius, Magdeburg, 1578; his illustration borrows heavily from a woodcut published in Nuremburg by Johannes Portantius a few months earlier.

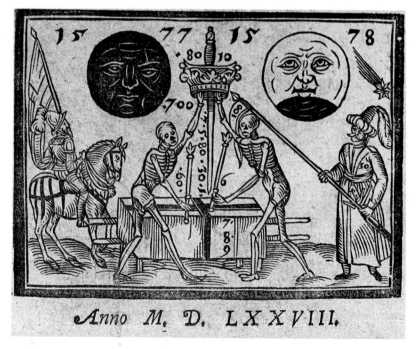

Figure 12.8 This dramatic view is from an anonymous German broadside printed in Augsburg by Valentin Schönigk. The text below states that the comet which appeared in November, 1577, was equal to those of 1531 and 1556, and that on 13 and 14 November it stood in Capricornus near Saturn. The broadsheet goes on to say that this comet foretells great heat and mortality during the coming dozen years.

be shown where the recent comet had last been seen, and when Tycho pointed out the place, he caught one final glimpse of the fading object. With the blazing comet out of sight, Tycho continued to think about its place in the universe. While rejecting the dizzying motions of the earth in the Copernican system, he nevertheless liked the idea of linking the other planets into orbits around the sun, and such an arrangement provided a place for the comet of 1577.

Copernicus accepted the traditional distance to the sun, about 1140 earth radii. However, when he moved the sphere of Venus from a position *between* the earth and sun, to a position *around* the sun, he suddenly had a lot of extra space in his system. Copernicus is silent on this point. Probably he believed the space was filled out with

Figure 12.9 One of several variants of the broadsheet in Figure 12.8, this one was also printed at Augsburg, by Bartholomew Räppeler.

Ein kurtze erinnerung / von dem Cometen / so auff den
12. tag Nouembris des 1 5 7 7. Jars zu Augspurg erschinen /
vnd erstmals gesehen worden.

additional transparent aether, but it certainly was not required for the planetary mechanisms.

Tycho was more explicit; in commenting on the 'enormously vast interval' in which the planets perform their wonderful motions, he remarked that 'we should assign the comet some particular place in the very wide space of the aethereal region in order to establish between which orbs it will direct its path.' It was precisely into the void between the moon and Venus that Tycho proposed to place the comet of 1577. By locating the comet in this zone, there was no danger that it would smash any of the crystalline spheres. Therefore the motion of the comet did not lead directly to Tycho's new cosmology nor to his dismissal of the solid spheres, although it certainly helped undermine the older Ptolemaic–Aristotelian construction of the heavens.

Not until 1583 did Tycho make the observations that eventually helped lead to his rejection of the solid spheres. These were measurements of Mars at its close approach. Actually, they offered him no fundamentally different information than he already had, but by drawing his attention once again to the arrangement of the planets, these studies pushed him further towards a new cosmology. Eventually another comet, in 1585, and correspondence with the German astronomer Christoph Rothmann, convinced Tycho that his own geo-heliocentric version was the most satisfactory arrangement for the sun, earth, and planets.

In any event, Tycho's discussion of the Great Comet of 1577 offered the vehicle for proposing a new cosmological system. Although his geocentric world picture is generally seen as a giant step backward, his radical rejection of the crystalline spheres opened the way for an innovative new physics. It was the search to find an alternative physics to replace the spheres as the machinery of the heavens that led Kepler to the elliptical orbits in a heliocentric system, and beyond him to Newton's universal gravitation. Although not directly responsible for this chain of events, the Great Comet of 1577 came at a propitious time for the advancement of astronomy, and in particular for Tycho Brahe's own intellectual development.

Notes and references

In the original 1977 version of this chapter, I falsely believed that a group of cosmological manuscripts in the Vatican Library represented Tycho's early halting steps towards his new cosmology. A few months after writing this *Sky and Telescope* article, Robert S. Westman and I discovered that the Vatican manuscripts were in fact written by a little-known sixteenth-century astronomer named Paul Wittich. Wittich visited Tycho in 1580, showed him the cosmological diagrams, and may well have had an important effect on Tycho's thinking. We have described this fascinating story in 'The Wittich Connection: Priority and Conflict in Late Sixteenth-Century Cosmology,' in the *Transactions of the American Philosophical Society* (1988). Some other subtleties in Tycho's path to his geo-heliocentric cosmology are described in Chapter 32 of this anthology.

13 *Galileo and the phases of Venus*

Few astronomers of the past have attracted so much sympathetic attention from the modern scientific community as Galileo Galilei. Born in 1564, the son of a pioneering Renaissance musician, he became the most renowned physicist since Archimedes. But he died in 1642 under house arrest, the sorry result of a series of events set in motion by his astonishing telescopic discoveries.

Today Galileo is still newsworthy, and the press has over the past several years carried a variety of stories ranging from Pope John Paul II's attempts to rehabilitate the seventeenth-century prisoner, to claims that Galileo was an unrepentant plagiarist. It is one of these latter episodes that I will examine here, and interested readers will be able to find a fuller account in the articles cited at the end, particularly the one by Stillman Drake.

Galileo was professor of mathematics in Padua when he first heard of a new invention that was being exhibited in nearby Venice. It was an optical tube (not yet named the telescope) that made distant objects appear closer. Galileo considered what arrangement of lenses might produce such an effect, and then he set to work. Fortunately, Venice had a flourishing glass industry – in fact the only one outside of Holland, where the telescope had been invented. Thus Galileo could obtain clear glass for the required lenses.

By November 1609, Galileo had pointed one of his new instruments at the moon. The view was surely nowhere near as satisfactory as with a modern pair of binoculars. Nevertheless, it enabled him to conjecture that the changing pattern of light and dark arose from the play of light and shadow on mountains and plains, and he could even estimate the height of the lunar features. He could distinguish at least one crater (Albategnius), and he speculated that more existed.

Then, in January, 1610, he turned his optical tube toward Jupiter and saw three, and then four, starlike companions. These, he brilliantly deduced, were satellites of the distant planet.

Quickly, Galileo wrote out an account of his discoveries and rushed it to the printer. He called his book *Sidereus Nuncius*, the 'Starry Messenger.' Like many books of that age, it was dedicated to a nobleman in the hope of some suitable honorarium. Actually, Galileo was playing for higher stakes: He was tired of tutoring students and yearned for a secure government position in Florence at the court of the Grand Duke Cosimo II de' Medici.

There were four Medici brothers, precisely matching the number of the new satellites, so Galileo considered naming the moons 'Medicean Stars.' However, fearing this might dilute the impact on Cosimo, the eldest and most powerful of the group, he settled on the name 'Cosmican Stars.' Barely had the printing begun, however, when the duke's secretary suggested that such an appellation sounded too much like 'cosmic stars' and hence the significance of the name would be lost. Galileo hastened back to the print shop and arranged

Figure 13.1 This portrait by Domenico Robusti, the son of Tintoretto, was painted around 1605–7 when Galileo was about 42 years old and living in Padua. It is probably the earliest existing original portrait of Galileo.

to have a small label pasted over the original text, changing it to read 'Medicean Stars.'

There were other changes to be made as well. He had meanwhile examined the Milky Way and nebulous patches recorded in Ptolemy's star catalog, such as Praesepe and the head of Orion. Each of these nebulosities gave way before the power of his telescope and stood revealed as a group of faint stars. Galileo added four more pages in the middle of his book, unnumbered and out of the regular arrangement, in order to include this latest discovery. The opening pages, printed last, refer to all these findings, and *MEDICEA SIDERA* stands forth on the title page like the lights on a theater marquee.

Galileo's astonishing discoveries and his publishing ploy had the desired effect. He got the job offer from the duke, and by September 1610, he had packed up and moved to Florence. He relished his new-found celebrity status and was determined to make as many further astronomical discoveries as his small telescope could possibly deliver.

However, as anyone who has bought an inexpensive spyglass for astronomical purposes soon learns, there are a limited number of sights available with such modest means. Mars is pretty disappoint-ing, and Saturn is confusing without good definition and fairly high

Figure 13.2 In the Ptolemaic system (left) Venus lies between the earth and the sun, so that it is never possible to see the fully illuminated face of Venus. In the Copernican system (right) Venus displays an entire set of phases like the moon.

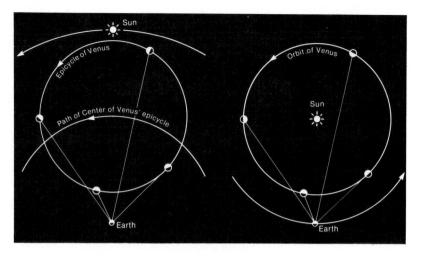

power. The sun can be interesting if one discovers how to observe such a bright object without destroying the eyes or the eyepiece in the process. And Venus is rather fine only for several weeks out of its 19-month synodic cycle.

We would naturally suppose that Galileo lost no time in looking at the other planets, but not all of them were well placed for observation in the winter of 1610. Mars was relatively inconspicuous and certainly uninteresting in the evening sky, Saturn was lost in the twilight, and Venus was rapidly disappearing in the morning dawn. By July, however, Saturn had reappeared, and Galileo was baffled by its strange appendages. (Not until the late 1650s would Huygens explain the ringed nature of this planet.)

Since Jupiter had four satellites and the earth one, Mars, which lay in between, was a good candidate for having two, and undoubtedly Galileo searched hard for them. In fact, when Galileo sent Kepler an anagram containing his concealed discovery of the curious 'companions' of Saturn, Kepler deciphered it as referring to the discovery of two Martian moons!

As it was, Galileo had nothing to say about Mars. Only Venus was left, and by late summer 'the mother of loves' was finally visible in the evening sky. Galileo must surely have checked this bright object to see if it had any companion. Finding none, he wrote in mid-November to Cosimo's brother, Giuliano de' Medici, ambassador in Prague and a pipeline for information to Kepler, that there was nothing going on 'around the other planets.'

There was, however, another key role for the Cytherian planet, Venus. In Ptolemy's ancient geocentric system, Venus always lies *between* the sun and the earth. In the new and controversial Copernican system, however, Venus circles *around* the sun. In principle the light curves corresponding to each system should differ, but since there are no good comparison objects for this brilliant planet, a naked-eye test is exceedingly difficult. With a telescope, however, it is a simple matter to check out the phases. In the Ptolemaic system Venus could never show a fully illuminated face, because it is never on the far side of the sun from the earth. Hence in that scheme it

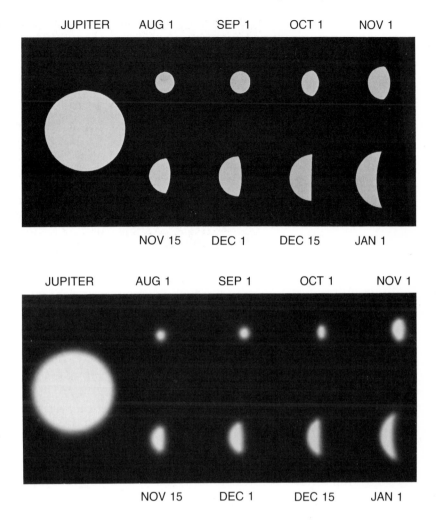

Figure 13.3 The appearance of Venus in 1610–11 compared with Jupiter. The out-of-focus panel below simulates the blurred appearance in a simple telescope like Galileo's.

shows only crescent phases. Quite the contrary is true in the Copernican system, where Venus will show a complete range of phrases from crescent to full.

And herein lies one of the recent controversies of Galilean studies. In October 1983, the eminent scholar of seventeenth-century science Richard S. Westfall spoke to the annual meeting of the History of Science Society on the relationship of Galileo and patronage, and particularly on Galileo's attempts to secure financial support from the Medici. Westfall noticed, as had a few others before him, a remarkable coincidence of timing between a letter that Galileo received and a mysterious anagram sent to Giuliano de' Medici in Prague.

On 5 December 1610, Galileo's former student Benedetto Castelli had written to say that telescopic observations of Venus could help distinguish between the Ptolemaic and Copernican systems. A few days later – approximately the time it took for him to get the letter – Galileo sent to Giuliano de' Medici an anagram containing the key information about Venus' phases, but not until New Year's Day of 1611 did Galileo decipher it for him.

The rearranged anagram read '*Cynthiae figuras aemulatur mater amorum,'* that is, 'The mother of loves imitates the shapes of Cynthia.' This meant that Venus, like our moon (Cynthia), has a complete set of phases. The observations showed decisively that Venus went around the sun (as in the Copernican system) rather than lying always between us and the sun (as in the Ptolemaic arrangement). Soon thereafter, Galileo wrote to Castelli that he had been observing Venus for three months.

Noticing the very short time between the dispatch of Castelli's letter to Galileo and the latter's anagram to Giuliano, Westfall argued that Galileo stole the idea of the significance of Cytherian phases from Castelli and that he had not really examined Venus before he had received Castelli's December letter. According to Westfall, Galileo, in order to guarantee his priority for a possible major new discovery, sent the cryptic anagram to Giuliano before he knew the appearance of Venus. Thus, Westfall claimed, Galileo had lied in telling Castelli that he had been observing the Cytherian planet for a number of months.

Even before Westfall's lecture, I had noticed that Venus, in a telescope such as Galileo's, was small and uninteresting in the summer of 1610, and not until late October did any conspicuous changes of phase occur. When Galileo sent the anagram, Venus was almost exactly half illuminated, and only at the end of December, when he unscrambled the anagram, did it show a distinguishable 'horned' or crescent appearance.

With the increased interest in this topic, I undertook to examine these phases more exactly. When the apparent sizes and phases of Venus are plotted for selected dates in 1610 and placed alongside the corresponding image of Jupiter, it immediately becomes apparent that Venus must have been rather uninspiring in a small telescope from August until early November 1610. This is all the more evident when a set of slightly out-of-focus images are used to simulate the performance of Galileo's telescope (see Figure 13.3).

By September 1610, Venus, the brightest celestial object apart from the sun or moon, was well placed for observation in the western evening sky. Surely Galileo must have examined it as soon as he had unpacked his telescopes after the move from Padua. It is exceedingly difficult to imagine otherwise. As the figure shows, however, hardly enough could have been seen to warrant even a sketch. A natural conclusion might have been that tiny Venus was shining by its own light.

Only in the days after mid-November did Venus become a truly exciting object. It reached dichotomy (half of the visible disk illuminated) at just about the time he wrote to Giuliano with the famous scrambled message, and the 'horned' appearance of the Cytherian disk did not become obvious until two weeks later, when Galileo deciphered the anagram.

Thus an examination of the actual appearance of Venus in 1610 makes plausible not only Galileo's actions but also his late-December claim that he had been observing the planet for three months. There was simply not enough observational data to be on secure grounds before 11 December, when Galileo sent out the anagram. Hence, I

Figure 13.4 The title page of Galileo's Starry Messenger, *wherein he advertises himself as a Florentine teaching in Padua who has recently used a 'perspicillium' (not yet named a telescope) to examine the moon, fixed stars, and the Milky Way and to discover the four 'Medicean Stars' moving around Jupiter with 'amazing velocity.' His publishing ploy worked, and Galileo got the job he sought at the Medici court in Florence.*

SIDEREVS
NVNCIVS
MAGNA, LONGEQVE ADMIRABILIA
Spectacula pandens, suspiciendaque proponens
vnicuique, præsertim verò
PHILOSOPHIS, *atq; ASTRONOMIS, quæ à*
GALILEO GALILEO
PATRITIO FLORENTINO
Patauini Gymnasij Publico Mathematico
PERSPICILLI
*Nuper à se reperti beneficio sunt obseruata in LVNÆ FACIE, FIXIS IN-
NVMERIS, LACTEO CIRCVLO, STELLIS NEBVLOSIS,
Apprime verò in*
QVATVOR PLANETIS '
Circa IOVIS Stellam disparibus interuallis, atque periodis, celeri-
tate mirabili circumuolutis; quos, nemini in hanc vsque
diem cognitos, nouissimè Author depræ-
hendit primus; atque
MEDICEA SIDERA
NVNCVPANDOS DECREVIT.

VENETIIS, Apud Thomam Baglionum. M DC X.
Superiorum Permissu, & Priuilegio.

think there is no reason to believe that Galileo stole the idea from Castelli or lied when he said he had been observing Venus for three months, that is, since the beginning of October, 1610.

More likely, Galileo was keeping a systematic watch on Venus. He knew that the round disklike appearance was incompatible with the Ptolemaic arrangement if Venus shone by reflected light, but until the phases began to appear he could not rule out the possibility that Venus shone by its own light or lay always beyond the sun. Meanwhile, he must have become increasingly nervous that his potential discovery would be scooped by another observer.

If Galileo had actually received Castelli's letter by the time he sent out the anagram (and that is by no means certain), then the letter might simply have precipitated the move to protect his priority. As Drake has pointed out, it was an unusual challenge to reassemble the anagram into a new text. The encrypted form of the anagram (*Haec immatura a me iam frustra leguntur o y*) means, 'These premature from me are at present deceptively gathered together,' as if, indeed, Galileo had been forced to act before he was quite ready.

As events unfolded, Galileo had no competitors in the telescopic examination of Venus. Not only was the observational discovery his, but he took the lead in exploring its consequences. Galileo realized that the Cytherian phases provided the first observational disproof of the time-honored Ptolemaic system. This, in turn, furnished him with a powerful defense for the new Copernican cosmology.

Soon Galileo was arguing that the heliocentric system could be reconciled with scriptural statements, and this line of attack promptly got him into hot water with the Catholic theologians. The Inquisition placed Copernicus' treatise on the *Index of Prohibited Books* in 1616 and ordered Galileo not to teach the new cosmology. When a new pope, Urban VIII, took office in 1623, the Florentine astronomer finally gained permission to write about the great world systems of Ptolemy and Copernicus, but his *Dialogo* was so lopsidedly enthusiastic for the heliocentric view that the attempt backfired and he ended his days as a prisoner of the Inquisition.

Had Galileo not been so fiesty or combative he surely would have not lived his last nine years under house arrest, nor would he now be under suspicion of plagiarism from some quarters. His great *Dialogo* would not have been half as interesting, nor would Galileo himself have been such a memorable figure in the annals of science!

Notes and references

A more detailed account of Galileo and the phases of Venus will be found in a group of articles by Stillman Drake, William Peters, and me in the October, 1984, issue of the *Journal for the History of Astronomy*. Richard S. Westfall's differing interpretation of these events appears in an article entitled 'Science and Patronage: Galileo and the Telescope,' *Isis*, **76** (1985), 11–30. English translations of the *Sidereus Nuncius* and related pieces are given by Stillman Drake in *Discoveries and Opinions of Galileo* (Garden City, New York, 1957). For the dates of Galileo's early observations see Ewan A. Whitaker, 'Galileo's lunar observations,' *Journal for the History of Astronomy*, **9** (1978), 155–69.

14 The Galileo affair

Galileo's difficulties with the Roman Catholic church, which ultimately led to his trial and humiliation, have often been described as a confrontation between empirical science and blind dogmatism. Notwithstanding his abjuration, Galileo clearly believed in the truth of the heliocentric Copernican system. Today, with the sun-centered arrangement of the planets firmly established, it is easy to see Galileo as right and the church as wrong. In Galileo's time, however, the issues were by no means obvious or clear-cut.

Galileo defended the Copernican system by a series of ingenious arguments, many of them based on his new telescopic observations. From a modern point of view Galileo's defense seems immediately compelling, but when he presented his ideas, there was as yet no observational proof of the new cosmology, and even he remarked that he could not admire enough those who had adopted the heliocentric system in spite of the evidence of their senses. By the standards of his time his reasoning was not only contrary to traditional church doctrine but also flawed in its logic. Indeed, I would contend that Galileo was breaking the accepted rules of science, but by doing so he created new rules that have been accepted ever since.

The outcome of the Galileo affair, in which the church won the battle but lost the war, had important historical consequences, most notably a shift of scientific enterprise northward into Protestant countries. Some 350 years later, at a time when some individuals are again asserting a religious claim on cosmology, Galileo's experience still has much to say about the practice and the philosophy of science. What was at issue was both the truth of nature and the nature of truth.

To understand the Galileo affair it is necessary to know something about the introduction of the Copernican cosmology some decades earlier. In 1543, when Copernicus' magnum opus, *De Revolutionibus Orbium Coelestium (On the Revolutions of the Heavenly Spheres)* was finally published, there was not a single item of unambiguous observational evidence in its favor. Copernicus' achievement had been in the mind's eye. What he had noted was that by rearranging the planetary orbs so that the sun was near their center a wonderful regularity emerged. The fastest planet, Mercury, had the orbit closest to the sun; the slowest planet, Saturn, was at the outside, and the planets in between also were placed in the order of their periods. Furthermore, the scheme gave a natural explanation to several previously unrelated observational facts, such as the geometry of each planet's retrograde arc (the segment of the orbit in which the planet seems to reverse direction across the sky). This explanatory power came at a high cost, however: it threw the earth into a dizzying flight around the sun, and the earth somehow had to bring the moon along with it. In the framework of the accepted Aristotelian physics the entire scheme was ridiculous. ''Tis all in pieces, all

Figure 14.1 Cosmological dispute is represented in the frontispiece of Galileo's Dialogue concerning the Two Chief World Systems, *printed in Florence in 1632. The three figures are Aristotle (left), Ptolemy (middle), who carries a model of nested geocentric spheres, and Nicolaus Copernicus (right), who bears an emblem of his own heliocentric theory. In discussions with Pope Urban VIII, Galileo had agreed to write a neutral account of the Ptolemaic and the Copernican systems, but the* Dialogue *was far from impartial. Galileo's advocacy of the heliocentric cosmology led to his prosecution by the Congregation of the Inquisition. The banner carries the dedication of the* Dialogue *to Ferdinand II de' Medici, the Grand Duke of Tuscany. 'Linceo' identified Galileo as a member of the Academy of Lynxes, a scientific society formed in 1603.*

coherence gone,' John Donne was to lament a few generations later. And coherency is cherished above all else in science; it is the touchstone by which crank theories can be rejected.

In order to convey the viewpoint of the astronomical community about 50 years after the publication of *De Revolutionibus*, I should like to describe an imaginary congress of the International Astronomical Union in 1592. The vice-president, Christoph Clavius of Rome, has risen to praise the remarks of the president, Tycho Brahe of Denmark. Tycho has lately introduced still another cosmological system, in which the planets orbit the sun but the sun itself and the accompanying planets are in orbit around a motionless earth. Clavius remarks that Tycho's system beautifully preserves the relations found by Copernicus in the harmonious spacing of the planets and gives a fully natural explanation of retrograde motions, just as Copernicus' hypothesis does. As Tycho himself has said, the Copernican arrangement nowhere offends the principles of mathematics, but it gives to the earth – this lazy, sluggish body, unfit for motion – a movement as swift as that of the ethereal planets. The Tychonic system brilliantly saves physics by keeping the earth at rest, and it is consistent with the scriptures, such as Psalm 104: 'O Lord my God . . . who laid the foundations of the earth, that it should not be removed for ever.'

An informal poll I have taken among the delegates indicates a somewhat mixed reaction: about half accept Tycho's view, but the rest say the choice of systems does not matter since all such geometric schemes are only hypothetical anyway. Some of those who adopt the latter attitude cite the anonymous preface to Copernicus' book: 'Beware if you expect truth from astronomy lest you leave this field a greater fool than when you entered.' Fewer than 10 percent agree with the Sicilian astronomer Franciscus Maurolycus that Copernicus deserved whips and lashes. The great majority find the Copernican tables preferable for calculating planetary positions, but that does not require a commitment to the heliocentric cosmology because the tables are set up independent of any particular arrangement of the planets.

Although gracious amity prevails between the Jesuit Clavius and the Lutheran Tycho, international tensions are evident. Michael Maestlin of Tübingen is vociferous in his criticism of Clavius' new calendar. Maestlin is also nettled that his graduate student Johannes Kepler, who has come along on a young astronomer's grant, does not agree that the Gregorian calendar is the work of the devil. As for the 27-year-old Galileo Galilei, an untenured mathematics professor at Pisa, no one at the congress has heard of him.

Nearly four centuries later millions of people who could not identify Tycho, Clavius, Maestlin or Kepler know the name of Galileo. One reason for Galileo's prominence is the importance of his contributions to both physics and astronomy, but his trial at the hands of the Inquisition has surely added to his fame. In the quaint words of the nineteenth-century physicist David Brewster, Galileo became a 'martyr of science.' Now, 350 years after his trial and abjuration, the Vatican itself has moved to reopen his case.

Galileo's ordeal is often referred to as a trial for heresy. Strictly speaking the Copernican system was never officially declared heretical, nor was Galileo condemned for heresy. The judge, who was out to get Galileo, raised a charge of 'a vehement suspicion of heresy.' He also found that an unofficial panel of theologians had agreed the Copernican system ought to be considered heresy, but this opinion never became the official position of the church. To understand these points it is necessary to examine both the historical circumstances of the trial and what was at stake philosophically.

At the end of the sixteenth century there was still no compelling reason to accept the Copernican doctrine as a physical picture of the universe. Astronomers were all well aware of its general idea, yet few believed it described the real world. There was widespread agreement that truth resided not in astronomy but in the Bible. Since the Book of Scripture had been literally dictated by God, it had a unique status.

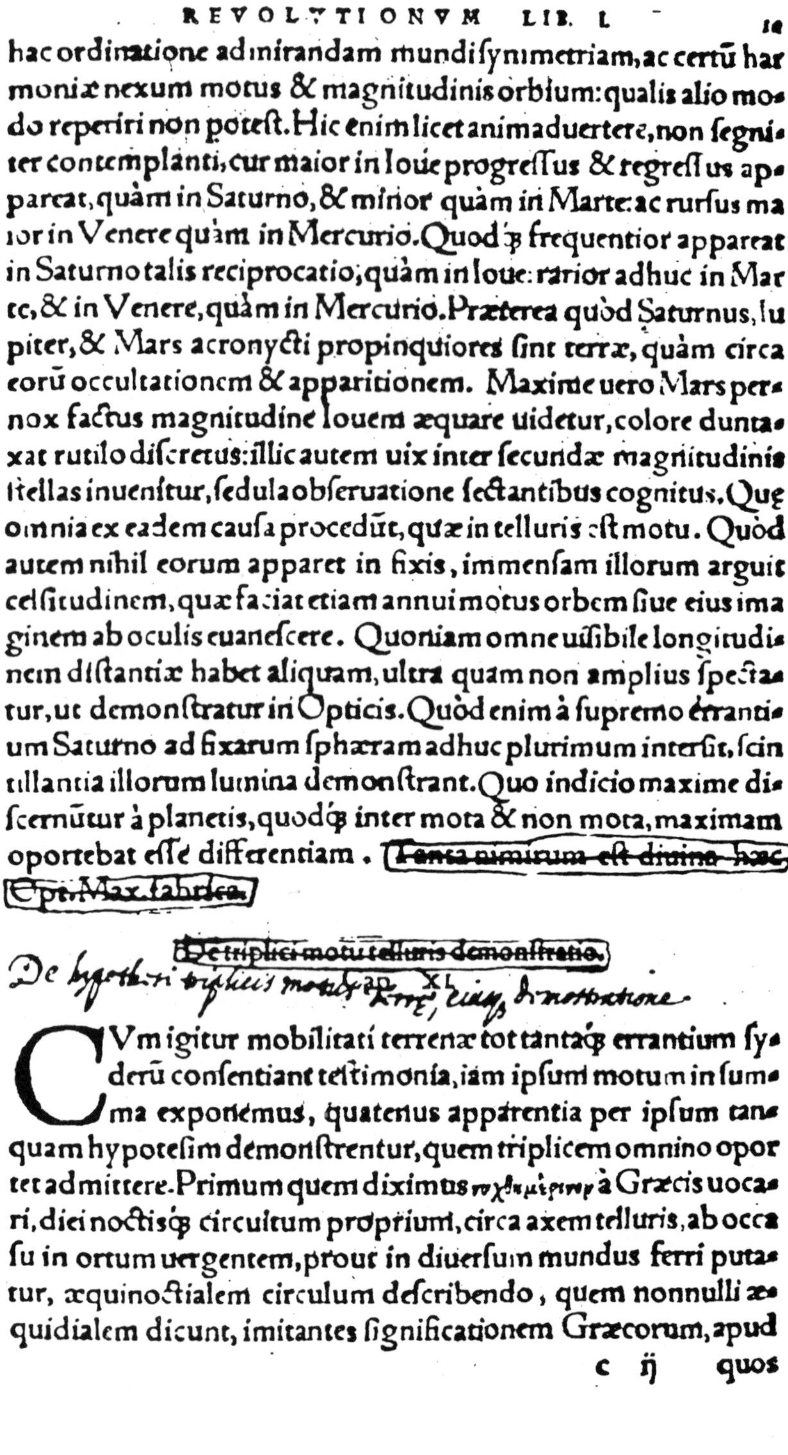

Figure 14.2 Censored passages of Copernicus' De Revolutionibus Orbium Coelestium were altered to make the heliocentric cosmology acceptable to the church by making it strictly hypothetical. The emendations shown are in Galileo's copy and were entered in his own hand. Following the instructions of the Holy Congregation of the Index, Galileo struck out the last sentence of Chapter 10, which had read: 'So vast, without any question, is the divine handiwork of the most excellent Almighty.' He also changed the title of the next chapter from 'On the Explication of the Threefold Motion of the Earth' to 'On the Hypothesis of the Threefold Motion of the Earth and its Explication.' The decision to censor De Revolutionibus rather than ban it was made in 1616, partly at the urging of Maffeo Cardinal Barberini, who later became Pope Urban VIII. At the same time Galileo was warned against speaking too forcefully in favor of the Copernican cosmological system, although he was not officially enjoined against teaching it.

hac ordinatione admirandam mundi symmetriam, ac certum har
moniae nexum motus & magnitudinis orbium: qualis alio mo
do reperiri non potest. Hic enim licet animaduertere, non segni
ter contemplanti, cur maior in Ioue progressus & regressus ap
pareat, quàm in Saturno, & minor quàm in Marte: ac rursus ma
ior in Venere quàm in Mercurio. Quod si frequentior appareat
in Saturno talis reciprocatio, quàm in Ioue: rarior adhuc in Mar
te, & in Venere, quàm in Mercurio. Praeterea quòd Saturnus, Iu
piter, & Mars acronycti propinquiores sint terrae, quàm circa
eorum occultationem & apparitionem. Maxime uero Mars per
nox factus magnitudine Iouem aequare uidetur, colore dunta
xat rutilo discretus: illic autem uix inter secundae magnitudinis
stellas inuenitur, sedula obseruatione sectantibus cognitus. Que
omnia ex eadem causa procedunt, quae in telluris est motu. Quòd
autem nihil eorum apparet in fixis, immensam illorum arguit
celsitudinem, quae faciat etiam annui motus orbem siue eius ima
ginem ab oculis euanescere. Quoniam omne uisibile longitudi
nem distantiae habet aliquam, ultra quam non amplius specta
tur, ut demonstratur in Opticis. Quòd enim à supremo erranti
um Saturno ad fixarum sphaeram adhuc plurimum intersit, scin
tillantia illorum lumina demonstrant. Quo indicio maxime di
scernitur à planetis, quodque inter mota & non mota, maximam
oportebat esse differentiam . [Tanta nimirum est diuino hac
Opt. Max. fabrica.]

[De triplici motu telluris demonstratio.]
De hypotesi triplicis motus...

CVm igitur mobilitati terrenae tot tantaeque errantium sy
derum consentiant testimonia, iam ipsum motum in sum
ma exponemus, quatenus apparentia per ipsum tan
quam hypotesim demonstrentur, quem triplicem omnino opor
tet admittere. Primum quem diximus νυχθημερινον à Graecis uoca
ri, diei noctisque circuitum proprium, circa axem telluris, ab occa
su in ortum uergentem, prout in diuersum mundus ferri puta
tur, aequinoctialem circulum describendo, quem nonnulli ae
quidialem dicunt, imitantes significationem Graecorum, apud
c ij   quos

Even Galileo accepted this doctrine without hesitation. He did not necessarily agree, however, that the intellectual road to truth lay solely within the territory of the theologians. The Book of Scripture could be ambiguous, he argued, whereas God's Book of Nature could be probed and tested. He conceded that the Bible had its place, but he

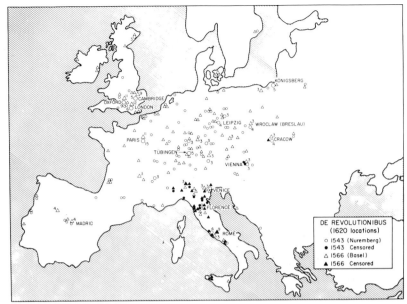

Figure 14.3 Inscriptions now found in copies of Copernicus' book allow us to deduce where about half of them were located in Galileo's day, as shown here. The solid symbols show the censored copies, which were concentrated almost exclusively in Italy.

also believed the Bible told how to go to heaven, not how the heavens go.

How *do* the heavens go, and how is their motion revealed by the Book of Nature? A flippant answer might be: by observing with the telescope. For Galileo the telescope had an enormous psychological impact. For years he had been at best a timid or even an indifferent Copernican, and he had taught his students in Pisa and later in Padua the standard arguments for a fixed, central Earth. Then, in the fall of 1609, with an optical tube of his own making, a perspicillum as he called it, he turned his attention to the heavens and was staggered by what he saw. Within a few months his book reporting on his observations was off the presses: *Sidereus Nuncius*, or *The Starry Messenger*. It told of mountains on the moon and of stars and satellites unknown to the ancients. The moon was earthlike and not the ethereal globe of pure crystal imagined by his predecessors. The Milky Way was revealed to be the confluence of innumerable stars. Most unexpected of all, Jupiter was circled by four companions, Galileo craftily called them the Medicean stars, with hopes of a government-supported position in Tuscany at the court of ·Grand Duke Cosimo II de' Medici.

Galileo's observations with the telescope must have shaken his complacency, but his account in *The Starry Messenger* gives no unambiguous evidence that he espoused the Copernican system. The book had only just been printed, however, when he made another remarkable finding: the phases of Venus, which in a stroke falsified the Ptolemaic system.

Venus had been too close to the sun to observe when Galileo was making his astonishing discoveries at the end of 1609 and the beginning of 1610. Sometime in the autumn a former student, Benedetto Castelli, remarked to Galileo that in the Copernican system Venus should show the entire range of phases, from a dark disk through crescent and gibbous forms to a fully illuminated disk. In the Ptolemaic system, on the other hand, the epicycle of Venus is locked between the earth and the sun, and Venus therefore has only crescent phases; it never passes behind the sun for full illumination.

Not until October did Galileo train his perspicillum on Venus, which was then in its distant gibbous phase. By early December, when the planet had waned to a miniature half moon, he put forth his discovery in an anagram: '*Haec immatura a me iam frustra leguntur o y*' ('These premature from me are at present deceptively gathered together.' The letters '*o y*' are part of the original sentence but did not fit into the anagram.) He undoubtedly chose this veiled form of announcement to give himself time to be sure of his finding; after all, Venus might lie always beyond the sun, in which case it would go back into a gibbous phase. By this stratagem Galileo also guarded his priority; since Castelli had mentioned the possibility in the first place, others might have been on the verge of making the same discovery.

Galileo was a scrambling social climber. His discoveries had gained him a new post as mathematician to the Medici and had brought him the fame he greatly relished. Fame in turn brought power of a kind, perhaps the power to persuade the entire Catholic hierarchy to adopt the Copernican system. At least Galileo was egotistical enough to expect that it would.

In Galileo's rush to assert a claim of priority he was sometimes more aggressive than might seem prudent. He got into a squabble with the Jesuit Christoph Scheiner when each asserted he had been the first to observe sunspots. Moreover, Scheiner preferred to believe the sun was unblemished and the spots were intervening clouds. Galileo proved otherwise, with rather little charity to Scheiner. Giorgio de Santillana, in his book *The Crime of Galileo*, hints darkly that Scheiner never forgot and years later led the Jesuits in a vendetta. It is true Scheiner was in Rome at the time of Galileo's trial, but there is no evidence he had anything to do with those machinations. Nevertheless, the story might make a play along the lines of *Amadeus*, the Broadway production about the supposed poisoning of Mozart.

By New Year's Day of 1611, just after Venus had rounded its western elongation, the crescent phase began to emerge and Galileo unscrambled his anagram for Kepler. It read, '*Cynthiae figuras aemulatur mater amorum.*' ('The mother of loves imitates the shapes of Cynthia'), or in other words, Venus goes through the same series of phases as the moon. When Galileo realized that the observed phases of Venus are incompatible with the Ptolemaic arrangement, he could hardly fail to notice that the Book of Nature was indeed saying something about how the heavens go. With the Ptolemaic scheme eliminated Galileo threw his support behind the Copernican system, ignoring the Tychonic plan.

At a breakfast with Cosimo de' Medici and his mother, the Dowager Grand Duchess Christina, the question of the reality of the Jovian satellites came under discussion. Galileo himself was not present but Castelli was there. Through Galileo's influence Castelli had just become professor of mathematics at Pisa, and he entered into a spirited discussion with Christina on the issue of whether there is any

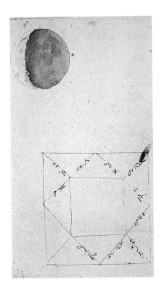

Figure 14.4 Earthlike features on the surface of the moon were among the observations cited by Galileo in support of the heliocentric theory. The presence of mountains, craters and other 'blemishes' indicated that the heavenly bodies are not fundamentally different from the earth. It therefore became reasonable to suppose the earth is a planet and not a fixed sphere with a quite different status. The drawing was made by Galileo after he constructed an astronomical telescope in 1609. On the same page is the start of a horoscope he cast for Cosimo II de' Medici.

conflict between the Bible and the heliocentric theory. As a direct result of that debate Galileo was challenged to defend his view that the Book of Scripture raises no insuperable objections to the Copernican system. Galileo wrote a cogent analysis, including the splendid epigram about the Bible's teaching how to go to heaven, not how the heavens go. (Actually Galileo had borrowed the saying from Caesar Cardinal Baronius, the librarian of the Vatican.)

It was one thing to argue that the heliocentric arrangement is compatible with the Book of Scripture and quite another to prove that the Book of Nature speaks unmistakably in favor of Copernicus. To understand this part of the controversy it is necessary to keep in mind the two forms of Aristotelian logic: induction and deduction.

Induction is the process of drawing general conclusions from particular instances; it is, I think, the basic process whereby learning takes place. Consider the reproduction of birds: chickens lay eggs, robins lay eggs, ostriches lay eggs and so on, and thus we generalize that all birds reproduce by laying eggs. We have not proved this conclusion, however, since there is always the possibility that a counterexample will be found. For this reason inductive reasoning, as all the scholastic philosophers of Galileo's time knew, cannot lead to indubitable truth.

Deduction is another matter. Given true premises, a conclusion reached by valid deduction must be rigorously true. Consider this syllogism:

(A) If it is raining, the streets are wet.
(B) It is raining.
(C) Therefore the streets are wet.

    Now consider the converse:

(A) If it is raining, the streets are wet.
(B) The streets are wet.
(C) Therefore it is raining.

To students of logic this procedure of confirming the consequent was a well-known fallacy. After all, the street could be wet for other reasons: the winter snow could be melting, the street-cleaning department might be out in force or the Lippizaner horses might have been on parade.

How does this logical analysis apply to Galileo's defense of Copernicanism? Consider this syllogism:

(A) If the planetary system is heliocentric, Venus will show phases.
(B) The system is heliocentric.
(C) Therefore Venus will show phases.

True enough, but this was not the form of Galileo's argument. He had exchanged the second premise and the conclusion:

(A) If the planetary system is heliocentric, Venus will show phases.
(B) Venus shows phases.
(C) Therefore the planetary system is heliocentric.

Clearly Galileo had committed an elementary blunder of logic, and even Kepler criticized him for it. There might well be other explanations for the observed phases of Venus; indeed, the Tychonic system also predicted them.

When Galileo's '*Letter to Christina*' was circulated in Rome in 1616, it elicited the following response from Roberto Cardinal Bellarmino, the leading Catholic theologian of the day, who wrote to another Copernican, Father Paolo Antonio Foscarini:

I have gladly read the letter in Italian and the essay in Latin that Your Reverence has sent me, and I thank you for both, confessing that they are filled with ingenuity and learning. But since you ask for my opinion, I shall give it to you briefly, as you have little time for reading and I for writing.

First, I say that it appears to me that Your Reverence and Signor Galileo did prudently to content yourselves with speaking hypothetically and not positively, as I have always believed Copernicus did. For to say that assuming the Earth moves and the Sun stands still saves all the appearances better than eccentrics and epicycles is to speak well. This has no danger in it, and it suffices for mathematicians.

But to wish to affirm that the sun is really fixed in the center of the heavens and that the earth is situated in the third sphere and revolves very swiftly around the sun is a very dangerous thing, not only by irritating all the theologians and scholastic philosophers, but also by injuring our holy faith and making the sacred Scripture false. For Your Reverence has indeed demonstrated many ways of expounding the Bible, but you have not applied them specifically, and doubtless you would have had a great deal of difficulty if you had tried to explain all the passages that you yourself have cited. . . .

Further, I say that if there were a true demonstration that the sun is in the center of the universe and that the sun does not go around the earth but the earth goes around the sun, then it would be necessary to be careful in explaining the Scriptures that seemed contrary, and we should rather have to say that we do not understand them than to say that something is false. But I do not think there is any such demonstration, since none has been shown to me. To demonstrate that the appearances are saved by assuming the sun at the center and the earth in the heavens is not the same thing as to demonstrate that in fact the sun is in the center and the earth in the heavens. I believe that the first demonstration may exist, but I have very grave doubts about the second.

(The translation is abridged from one done by Stillman Drake.)

Galileo knew he could not logically establish the Copernican system by deduction, but the situation was not quite that simple. The Copernican system not only predicted the phases of Venus but also, as a model, explained many other things. If the earth was a planet, the other planets might well be earthlike, and so indeed the moon turned out to be when he examined it with his telescope. The Copernican system arranged the planets naturally by period; similarly, when the telescope revealed the satellites of Jupiter, they were found to be arranged sequentially by period, as in a miniature solar system.

Galileo's process of reasoning was similar to induction but more sophisticated. It was, in an embryonic state, what is now called the hypothetico-deductive method: the testing of a hypothetical model, which attains ever more convincing likelihood as it passes each test successfully. Today it is not the word 'truth' but the word 'model' that continually decorates the pages of scientific journals.

As far as the theologians were concerned, the Copernican system was not really the issue. I can hardly emphasize this point enough. The battleground was the method itself, the route to sure knowledge of the world, the question of whether the Book of Nature could in any way rival the inerrant Book of Scripture as an avenue to truth. In the opinion of Cardinal Bellarmino and the other Catholic theologians, Galileo's procedures were essentially inductive and therefore potentially fallacious. Such contingent arguments were insufficient to force a reinterpretation of scripture that might erode the concept of the inerrancy of Holy Writ.

To be quite sure of avoiding confusion in the popular mind (particularly because issues of interpretation were central in the ongoing battle with the Protestants) the church officials found it prudent to condemn the Copernican teaching. The first step was to seek a theological opinion on two separate propositions: the immobility of the sun and the mobility of the earth. The report, which was essentially an internal memorandum, said the immobility of the sun was foolish and formally heretical because it violated the literal meaning of the Scriptures, but the mobility of the earth was merely erroneous. The question then was what to do about the report. Two actions were planned: to rein in Galileo and to put *De Revolutionibus* on the *Index of Prohibited Books*.

The latter measure, however, entailed certain practical difficulties. Copernicus' book was considered an important contribution to the reform of astronomy, on which the calendar and the accurate determination of the date of Easter depended. Accordingly the Holy Congregation of the Index decided not to proscribe the book but instead to expurgate and emend it.

From these deliberations a bit of gossip has survived in the diary of Giovanfrancesco Buonamici, a diplomatic secretary from Galileo's province of Tuscany. Buonamici wrote that 'Pope Paul V was of the opinion to declare Copernicus contrary to the faith; but Cardinals Bonifacio Caetani and Maffeo Barberini withstood the Pope openly and checked him with the good reasons they gave.' The two cardinals were central figures in the cosmological controversy. Barberini was later to have an even larger role in the story of Galileo's life, and Caetani drafted the opinion recommending censorship of *De Revolutionibus*.

Caetani's opinion declared that the Copernican teaching was false and opposed to Scripture, but not that it was heretical. This may seem to be a distinction without a difference, but it was certainly not so in the seventeenth century. The instructions for the censorship read:

If certain of Copernicus' passages on the motion of the earth are not hypothetical, make them hypothetical; then they will not be against either the truth or the Holy Writ. On the contrary, in a certain sense they will be in agreement with them, on account of the false nature of suppositions, which the study of astronomy is accustomed to use as its special right.

We, Roberto Cardinal Bellarmino, having heard that it is calumniously reported that Signor Galileo Galilei has in our hand abjured and has also been punished with salutary penance, and being requested to state the truth as to this, declare that the said Signor Galileo has not abjured, either in our hand, or the hand of any other person here in Rome, or anywhere else, so far as we know, any opinion or doctrine held by him; neither has any salutary penance been imposed on him; but that only the declaration made by the Holy Father and published by the Sacred Congregation of the Index has been notified to him, wherein it is set forth that the doctrine attributed to Copernicus, that the Earth moves around the Sun and that the Sun is stationary in the center of the world and does not move from east to west, is contrary to the Holy Scriptures and therefore cannot be defended or held. In witness whereof we have written and subscribed these presents with our hand this twenty-sixth day of May, 1616.

Even as the Holy Congregation of the Index was moving against Copernicus' book, Galileo was in Rome aggressively lobbying on behalf of the heliocentric system. It seems he was convinced he could single-handedly sway the Catholic leaders to his view. Indeed, he had powerful friends in Rome who were sympathetic to his ideas, even among the churchmen, but the conservative forces were also strong, and they included Pope Paul V.

While Galileo was in Rome the other part of the pope's response to the theological opinion was put into action. Galileo was to be called before Cardinal Bellarmino and cautioned against speaking out too forcefully on behalf of the Copernican system. The pope told Bellarmino that if Galileo proved intractable, he was to be ordered to keep quiet. To be sure the pope's wishes were enforced the interview was conducted in the presence of two Dominican friars, members of the order charged with administering the Inquisition.

As it turned out, Galileo was cooperative in accepting Bellarmino's warning. After the conference, however, rumors began to circulate in Rome that Galileo had been officially enjoined against teaching the Copernican doctrine. Galileo was naturally disturbed by the rumors, and he sought and received a letter from Bellarmino saying that no such thing had happened. It read in part:

> We, Roberto Cardinal Bellarmino, having heard that it is calumniously reported that Signor Galileo Galilei has in our hand abjured and has also been punished . . . declare that the said Signor Galileo has not abjured . . . any opinion or doctrine held by him; neither has any salutary penance been imposed by him; but that only the declaration made by the Holy Father and published by the Sacred Congregation of the Index has been notified to him, wherein it is set forth that the doctrine attributed to Copernicus . . . is contrary to the Holy Scriptures and therefore cannot be defended or held.
> (The translation is by de Santillana.)

Thus for the time being Galileo was silenced. For seven years he remained in Florence and complied with Cardinal Bellarmino's advice. He was as feisty as ever, but he reserved his scrappiness for other subjects, such as the comets of 1618. In his book on the comets (*Il Saggiatore.* or *The Assayer*) he avoided discussing the Copernican system, but he included so many interesting remarks on the nature of science that the book is sometimes called his scientific manifesto. He stated, in Italian:

> Philosophy is written in this grand book, the universe, which stands continually open to our gaze. But the book cannot be understood unless one first learns to comprehend the language and read the letters in which it is composed. It is written in the language of mathematics, and its characters are triangles, circles and other geometric figures. . . . Without these one wanders about in a dark labyrinth.

The printing of *Il Saggiatore* was not finished when news arrived that cheered all liberal Catholics. The newly elected pope, who had taken the name Urban VIII, was Maffeo Barberini, one of the cardinals who had intervened to prevent the proscription of *De Revolutionibus*. Barberini was also a friend of the arts and a fellow member with Galileo of the small Academy of Lynxes, one of the earliest scientific societies. The delighted Lynxes had just enough time to change the title page on Galileo's book so that it could be dedicated to the new pontiff. Before a year had passed Galileo was in Rome for a series of papal audiences. Urban assured him that *Il Saggiatore* had been read to him, to his great pleasure. Galileo hinted he would like to write more, in particular a book on the relative merits of the Copernican and the Ptolemaic systems, but his enemies prevented him.

From what is known of the two men it is possible to speculate on how the conversation went. 'Nonsense,' the pope may have responded. 'I helped to keep this from becoming heresy before, and I can protect you now. But remember, your account should be neutral, since you have no physical proof of the Copernican system.'

'Ah,' replied Galileo, 'but I do. I believe the tides are the proof of a moving earth, and I propose to call my book *On the Flux and Reflux of the Sea.*'

'No,' said Urban, 'that won't do at all. That title would give too much prominence to what you take to be a physical proof, but God could have created the tides in any way he liked, and not necessarily by moving the earth.' Note that Urban's argument was the same as Bellarmino's: even if a moving earth would produce tides, the observation of tides does not necessarily imply the movement of the earth. The case is particularly ironic, because Galileo's physical argument based on the tides was quite wrong. (He attributed them to the daily change in velocity that results from the earth's compound motion of rotation and revolution.)

Galileo was elated to have the gag order removed by the highest possible authority, and he returned to Florence to work on his book. He adopted the popular form of a dialog, just as his father, a distinguished musician, had done in writing a *Dialogue on Ancient and Modern Music*. Galileo's three speakers are Simplicio, a traditionalist, named after a sixth-century commentator on Aristotle; Salviati, who most often speaks for Galileo himself, and Sagredo, an open-minded man of the world who asks intelligent questions and is generally persuaded by Salviati's reasoning.

The arguments marshaled on behalf of the Copernican system include the phases of Venus, the harmony of the arrangement of the planets and the existence of the tides. The work could hardly be considered neutral, but it ends with the pope's argument in the following words spoken by Simplicio:

I confess that your hypothesis on the flux and reflux of the sea is far more ingenious than any of those I have ever heard; still, I esteem it neither true nor conclusive, but, keeping always in mind a most solid doctrine I once received from a most eminent person, I know that if you were asked whether God in his infinite power and wisdom might confer upon the element of water the reciprocal motion in any other way, both of you would answer that he could, and in many ways, some beyond the reach of our intellect.
(The translation is based on one by Drake.)

The passage seems quite innocuous, and yet it is singularly inappropriate as the closing argument of the preceding four days of dialog. Throughout the work Galileo had attempted to show that reasoning from the Book of Nature can, at the very least, establish that one world view is far likelier than another. This has surely been the method of science ever since. Indeed, one might argue (as Alfred

North Whitehead did) that since an omnipotent Creator could have made the world in any way he liked, it is all the more incumbent on scientists to discover which way God chose to make it.

When the *Dialogue* appeared, Galileo's enemies were outraged, and they quickly persuaded the pope that the book was heavily weighted in favor of the Copernican system. Furthermore, they convinced the pope that he had been made to look a fool by having his argument given to Simplicio, whose very name suggested 'simpleton.' The pope, agreeing that Galileo had gone too far, unleashed the Inquisition.

There were two stumbling blocks to prosecution: Copernicus' doctrine had never been publicly declared heretical, and the *Dialogue* had received a license from the censors. From the Vatican Archives, however, the Inquisitors produced a fascinating document: a report of the 1616 meeting between Galileo and Bellarmino. The report stated that an official injunction had indeed been served on Galileo, and that the astronomer had promised not to teach or defend the Copernican doctrine in any way. The pope was furious; it appeared Galileo not only had made him into a fool but also had deceived him about the outcome of the proceedings of 1616.

In February, 1633, Galileo was ordered to Rome, and he was told to come immediately in spite of the rigors of winter travel for a man of almost 70. Before a tribunal of ten cardinals he was accused of disobedience. The archival evidence, however, was quite irregular: the document was neither signed nor notarized, as such an injunction should have been. Bellarmino had died, and so it was difficult to clarify the status of the document. Hence the Inquisitors, without revealing the source of their accusations, tried to get Galileo to admit that he had been served an injunction, which would have established the legitimacy of the earlier document. Ultimately Galileo played his trump card. Having been alerted by his friends and their spies, he knew that the Inquisition was looking into his 1616 visit to Rome, and so he had brought a copy of Bellarmino's letter. Galileo's unexpected move threw the Inquisition into disarray, and the cardinals decided to recess.

It was a duel of wits, and Galileo had outwitted the pope. Nevertheless, all the secular power remained in the hands of the church, and the pope could not afford the embarrassment of bringing Galileo to Rome for naught. Even Galileo could appreciate this, and so some plea bargaining ensued. It could all be settled out of court: Galileo would confess that he had gone too far, would repent and then would be sent home and enjoined to avoid writing about cosmology.

One can imagine Galileo's shock on 16 June 1633, when he found that the agreement had been overruled and the following sentence was entered in the Book of Decrees: 'Galileo Galilei . . . is to be interrogated concerning the accusation, even threatened with torture, and if he sustains it, proceeding to an abjuration of the vehement [suspicion of heresy] before the full Congregation of the Holy Office, sentenced to imprisonment. . . .' He was also forbidden to write further on the mobility of the earth, and the *Dialogue* was banned.

On the next page the results of the interrogation are recorded. In Italian are Galileo's words: 'I do not hold and have not held this opinion of Copernicus since the command was intimated to me that I must abandon it.' Then he was again told to speak the truth under the threat of torture. He responded: 'I am here to submit, and I have not held this opinion since the decision was pronounced, as I have stated.' Finally, there is a notation that nothing further could be done, and this time the document is properly signed in Galileo's hand.

Galileo was sent back to his house at Arcetri, outside Florence, where he remained under house arrest until his death in 1642. Partly as a consequence of his persecution, the center of creative science moved northward to the Protestant countries notably the Netherlands and England.

I am fascinated by the choices the Vatican confronts today in reopening Galileo's case. In the first place, it would do no good to announce that the Copernican doctrine should never have been declared heretical, since strictly speaking it never was. Second, Galileo was tried not so much for heresy as for disobeying orders, and it seems clear beyond question that he ignored the earlier decree of the Index when he published his *Dialogue*.

Where there is room for maneuver, it seems to me, is in accepting Galileo's arguments about the reconciliation of science and Scripture. The truth of the Bible, for those who wish to affirm it without rejecting the findings of science, must not be found in a literal six days of Creation, in the sun standing still for the battle of Gibeon or in a physically real star of Bethlehem. I quote Galileo, as he quoted Cardinal Baronius: 'The Bible teaches how to go to heaven, not how the heavens go.' Such a judgment, it seems to me, would confirm what has long since been accepted by both Catholic and Protestant theologians. It would also speak to the current controversy over Darwinian evolution and the so-called Creation science.

I had a sense of *déjà vu* when the Creationists in California tried to have evolution presented in biology textbooks as a mere hypothesis. This was precisely the tactic the Inquisition adopted with Copernicus' book: they made it acceptable by making it appear hypothetical. I expect the Creationists will have about as much success as the Holy Congregation of the Index did. Of course, Galileo believed the Copernican system could be defended as physically real, and not simply as a hypothetical geometric arrangement. It is an irony of history that Galileo's own methods of scientific argument were instrumental in showing that what passes for truth in science is only the likely or the probable; truth can never be final and never absolute. What makes science so fascinating is the task of pushing ever closer to the unattainable goal of complete knowledge.

It is this process that the poet Robinson Jeffers had in mind when he wrote: 'The mathematicians and the physics men have their mythology; they work alongside the truth, never touching it; their equations are false but the things *work*.' The mathematicians and the physicists cannot really claim truth, but they have certainly sorted a lot of things that do not work, and they are building a wondrously

Figure 14.6 Book of Decrees of the Congregation of the Inquisition records the sentencing of Galileo in 1633. The proceedings against him had come to a halt after he had produced Cardinal Bellarmino's letter of 1616. Thereafter an agreement had been reached: Galileo would repent and would promise to write no more on cosmology. The agreement was overruled, however, and he was forced to submit to the humiliating ritual of abjuration, followed by house arrest, the banning of the Dialogue *and a prohibition of further writing on the Copernican system.*

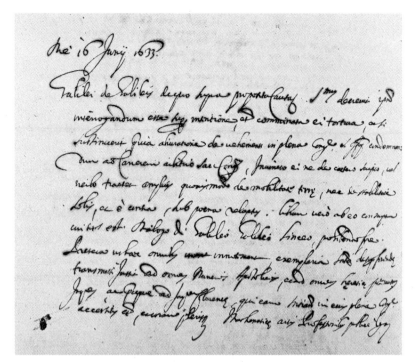

> *16 June 1633*
>
> *Galileo Galilei, for the above reasons, as decreed by his Holiness, is to be interrogated concerning the accusation, even threatened with torture, and if he sustains it, proceeding to an abjuration of the vehement [suspicion of heresy] before the full Congregation of the Holy Office, sentenced to imprisonment at the pleasure of the Holy Congregation, ordered, in either writing or speaking, not to treat further in any way either the mobility of the Earth or the stability of the Sun; or otherwise he will suffer the punishment of relapse. The book actually written by him, whose title is* Dialogo di Galileo Galilei Linceo, *is to be prohibited. Furthermore, that these things may be known by all, he ordered that copies of the foregoing sentence shall be sent to all Apostolic Nuncios, to all Inquisitors against heretical depravity, and especially the Inquisitor of Florence who shall publicly read the sentence to his whole congregation and even in the presence of as many of those who teach mathematics as he can summon together.*

coherent picture of the universe. The Copernican system is surely a part of that coherency. A universe billions of years old and evolving is also part of that coherency. Galileo made a noble effort to convey such a picture of beauty and rational coherency to his public. Scientists today would honor him by helping their own public to understand better not only the majesty and the beauty of the modern scientific picture of the universe but also the process of hypothesizing and testing by which that view is achieved.

Figure 14.7 Galileo's abjuration appears in the Book of Decrees following his sentencing. He retired to his house at Arcetri outside Florence, where he was confined until his death in 1642.

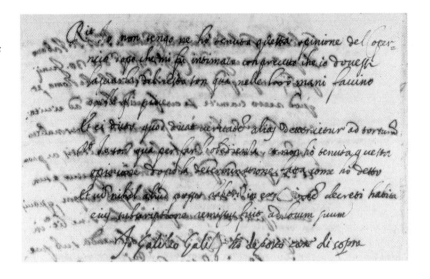

'*I do not hold and have not held this opinion of Copernicus since the command was intimated to me that I must abandon it; for the rest, I am here in your hands – do with me what you please.*'

Being once more bidden to speak the truth, otherwise recourse would be had to torture:

'*I am here to submit, and I have not held this opinion since the decision was pronounced, as I have stated.*'

And since nothing further could be done in execution of the decree, his signature was obtained, and he was sent back to his place.

'*I, Galileo Galilei, testify as above.*'

Post Script

My *Scientific American* article generated a continuing stream of responses; Pierre Conway, OP, reminded me that I could have carried the account of the procedures against Galileo one day further. I ended the story on 21 June 1633, the last day on which the panel of cardinals received testimony, and neglected to mention the final sentencing and extended abjuration on 22 June 1633.

In a sense, Napoleon is responsible for this lapse. In 1810 the Vatican Archives were hauled to France as part of the aggrandizement of the Bibliothèque Nationale. The records of the Galileo trial were an especially important desideratum for the Napoleonic officials. When the Archives were returned in 1814, many documents were missing, including this important group. Eventually only one of the three record books turned up – the slim volume containing the letters from Bellarmine and Galileo's notarization of the testimony, all illustrated in my article. Thus the Vatican Archives today has no seventeenth-century record of the actual sentencing. A contemporary Italian translation exists in the Modena State Archives (although my request for a copy had gone unanswered); the Latin version was published in full by the Jesuit J. B. Riccioli in his 1651 *Almagestum novum*, but in writing my article I had temporarily forgotten this source as well as the English translation in Giorgio de Santillana's *The Crime of Galileo*.

In the sentencing, the Inquisition accused Galileo of a 'vehement suspicion of heresy' and in the loyalty oath that Galileo was forced to read, he said in part,

after it had been notified to me that [this] doctrine was contrary to Holy Scripture, I wrote and printed a book in which I discuss this new doctrine already condemned and adduce arguments of great cogency in its favor without presenting any solution of these. Hence I have been pronounced by the Holy Office to be vehemently suspected of heresy, that is to say, of having held and believed that the sun is the center of the world and immovable and that the earth is not the center and moves.

Therefore, desiring to remove from the minds of your Eminences, and of all Catholic Christians, this vehement suspicion justly conceived against me, with sincere heart and unfeigned faith I abjure, curse, and detest the aforesaid errors and heresies and generally every other error and sect whatsoever contrary to the Holy Church, and I swear that in future I will never again say or assert, verbally or in writing, anything that might furnish occasion for a similar suspicion regarding me; but, should I know any heretic or person suspected of heresy, I will denounce him to this Holy Office. . . .

From this moment onward the heliocentric doctrine could have been interpreted as *de facto* heresy by virtue of this legal ruling, despite the fact that no official decree had been issued; in this sense Galileo's trial could be considered a trial for heresy even though up until 22 June 1633 no public declaration had been made to that effect. Nevertheless, when Riccioli summed up the arguments 18 years after the event, he wrote that the doctrine that the earth moves 'had been condemned as heresy, *or at least as erroneous* (my italics),' thereby expressing the ambiguity; meanwhile, in editions of the *Index of Prohibited Books* published in Spain, Copernicus' book was explicity permitted. In 1757 the Holy Congregation finally ruled that the decree prohibiting books teaching the immobility of the sun and mobility of the earth (i.e. the decree of 5 March 1616) should be omitted from the *Index*, but not until 1835 was a new edition issued that actually omitted the banned books by Copernicus, Kepler, and Galileo.

My conclusions therefore stand, that, at least up until Galileo's sentencing, the heliocentric doctrine had not been declared heretical, that he was tried more for disobeying orders than for heresy, and that the Vatican today has a curiously restricted set of options in exonerating Galileo.

Notes and references

Discussions with historians of science Joseph Clark, SJ, and Ernan McMullan as well as reading Giorgio de Santillana's *The Crime of Galileo* (Chicago, 1955), stimulated my thinking on this topic. Vatican astronomers George V. Coyne, SJ, and Martin McCarthy, SJ, were very helpful in providing some of the illustrations for this chapter. For further reading see Stillman Drake, *Discoveries and Opinions of Galileo* (Garden City, 1957) and Ludovico Geymonat, *Galileo Galilei* (New York, 1965), see also my 'The Censorship of Copernicus' *De Revolutionibus*,' *Annali dell'Istituto e Museo de Storia della Scienze di Firenze*, **4** (1981), 45–61.

15 *Johannes Kepler and the* Rudolphine Tables

Of the many books by Johannes Kepler, the *Rudolphine Tables* of 1627 is at first sight the least inspiring. A tall, slender volume, the *Tabulae Rudolphinae* is half filled with numerical tables, half with Latin explanations – scarcely exciting reading. Yet the tables are extraordinarily important, for they document in a unique way Kepler's great contributions to astronomy.

When he became an astronomer in the closing years of the sixteenth century, Kepler found a science in which planetary predictions typically erred by several degrees on the sky; the legacy of the *Rudolphine Tables* was a prediction scheme nearly 50 times more accurate. For Kepler these tables were the proof of the pudding, the substantiation of his laws of planetary motion. He called them 'my chief astronomical work.'

Johannes Kepler was born on 27 December 1571. Two events from his childhood left a lasting impression: in 1577 his mother showed him the Great Comet, and later his father let him watch a lunar eclipse. In 1589 Kepler entered the University of Tübingen, where the senate was soon to note that he had 'such a superior and magnificent mind that something special may be expected of him.'

Yet Kepler himself wrote that nothing indicated to him a special bent for astronomy. Hence he was surprised and distressed when, midway through his last year as a theology student, he was summoned to Graz, far away in southern Austria, to become an astronomy teacher and the provincial mathematician.

In Graz, Kepler's active and ever-speculative mind soon hit upon what he believed to be the secret key to the construction of the universe. He knew that there were five regular polyhedra, that is, solid figures each with faces all the same kind of regular polygon. (The cube is such a solid.) By inscribing and circumscribing these figures with spheres (all nested in the proper order), he found that the positions of the spheres closely approximated the spacings of the planets. Since there are five and only five regular polyhedra, Kepler thought that he had explained the number of planets in the solar system. In 1596 he published his scheme in his *Mysterium Cosmographicum*, 'The Cosmographic Secret.'

Before dismissing the idea as the work of a crank, we must remember the revolutionary context in which it was proposed. The *Mysterium Cosmographicum* was the first unabashedly Copernican treatise since *De Revolutionibus* itself, for without a sun-centered universe the entire rationale of the book would have collapsed.

Kepler also realized that, although in Copernicus' system the sun was near the center, it played no physical role. But he argued that the sun's centrality was essential, for the sun itself must supply the driving force to keep the planets in motion. This idea, which appears in

Figure 15.1 The frontispiece illustration of Johannes Kepler's epoch-making 'Rudolphine Tables' (1627), a book which made possible planetary predictions of previously unsurpassed accuracy. It depicts an allegorical Temple of Urania, in which such famous astronomers as Copernicus and Tycho Brahe are at work.

the latter part of the book, establishes Kepler as the first scientist to demand physical explanations for celestial phenomena. Although the principal idea of the *Mysterium Cosmographicum* was erroneous, never in history has a book so wrong been so germinal in directing the future course of science.

Kepler sent copies of his remarkable book to various scholars, including the most famous astronomer of the day, Tycho Brahe. Although unwilling to accept all these strange arguments, the Danish astronomer immediately recognized the author's genius, and invited Kepler to visit him. However, the long journey was out of the question for the impecunious young man. Thus, wrote Kepler, 'I ascribe it to Divine Providence that Tycho came to Bohemia.'

Tycho, fearing the loss of royal support by the King of Denmark, had in the meantime resolved to join the court of Rudolf II in

Figure 15.2 In Weil der Stadt (near Stuttgart in southwestern Germany), where Kepler was born, a seated statue honors the great astronomer. This view is from a color transparency taken by the author in 1966.

Prague. Emperor Rudolf was a moody, eccentric man whose twin loves were the occult and his collection of curiosities. He was more than willing to support a distinguished astronomer whose accurate planetary positions could make horoscopes more accurate. Tycho arrived at Prague in 1599, and Kepler, forced out of Graz by religious controversy, joined him there the following January. To Kepler was assigned the analysis of the observations of Mars.

The encounter between the young German theoretician and the famous Danish observer turned the course of astronomy. Otherwise, Tycho's observations would have remained largely unexploited, and without them Kepler would never have found the true key to planetary motions. Precisely how much Kepler learned from the imperious Tycho himself can never be established. In any event, within two years Tycho had died and his full set of observations, although claimed by his heirs, fell into Kepler's hands.

The young man's speculative drive was tempered but not suppressed as he wrestled with the unique observational legacy. Kepler's eagerness to publish seemed unbounded. We can see from the still-extant manuscript notes that he was aleady outlining the chapter headings for his *Commentary on Mars* long before he knew that its orbit was anything but a combination of circles. In fact, he had written 58 chapters of his *Astronomia Nova* in virtually their final form before he even discovered Mars moved in an ellipse. Fortunately, several other studies, including an analysis of the supernova of 1604, delayed the completion of this great book until he had fully established the elliptical orbit and law of areas for Mars. The *Astronomia Nova* was finally published in 1609. Its break with the traditional requirement of circular motions for the planets made it truly 'the new astronomy.' Seldom has a book been better titled.

At Tycho's death, Kepler received not only his observations but also his title of Imperial Mathematician. The chief duty of this post was to complete the great set of planetary tables originally envisioned by Tycho, to be based on his incomparable set of precisely determined positions and intended to provide much improved predictions. Kepler's analysis of Mars could be considered the first step in this great undertaking.

The discovery of the first two planetary laws had created an entirely new foundation for calculating the *Rudolphine Tables*. But since Kepler had derived these laws solely from the observations of Mars, he now had to demonstrate their validity for the other planets. Mercury created special difficulties, and the complex motion of the moon caused a great deal of trouble. Among the papers dating from Kepler's stay in Prague are hundreds of sheets of calculations that are apparently preparatory work for the *Tables*. Yet more than two decades were to pass before the work was published, and the emperor who had commissioned it was long dead.

The astronomer's efforts to work on the tables were continually diluted by the time lost in trying to collect his salary, and by a variety of astronomical events. The most important of these was Galileo's application of the telescope to the heavens, causing Kepler to take time out to write his *Dissertatio cum Nuncio Sidereo* – 'Conversation

(a)

TABULÆ
RUDOLPHINÆ,
QUIBUS ASTRONOMICÆ SCIENTIÆ, TEMPO-
rum longinquitate collapsæ RESTAURATIO *continetur;*
A Phœnice illo Astronomorum
TYCHONE
Ex Illustri & Generosa BRAHEORUM in Regno DANIÆ
familiâ oriundo Equite,
PRIMUM ANIMO CONCEPTA ET DESTINATA ANNO
CHRISTI MDLXIV: EXINDE OBSERVATIONIBUS SIDERUM ACCURA-
TISSIMIS, POST ANNUM PRÆCIPUE MDLXXII, QUO SIDUS IN CASSIOPEIÆ
CONSTELLATIONE NOVUM EFFULSIT, SERIO AFFECTATAI VARIISQUE OPERIBIS, CUM ME-
chanicis, tùm librariis, impensâ patrimonio amplissimo, accedentibus etiam subsidiis FRIDERICI II. DANIÆ
REGIS, regali magnificentia dignis, tractâ per annos XXV. potissimùm in Insulâ freti SUNDICI HUEN-
NA, & arce URANIBURGO, in hos usus à fundamentis extructâ:
TANDEM TRADUCTA IN GERMANIAM, INQUE AULAM ET
Nomen RUDOLPHI IMP. anno M D IIC.

TABULAS IPSAS, JAM ET NUNCUPATAS, ET AFFECTAS, SED
MORTE AUTHORIS SUI ANNO MDCI DESERTAS,
JUSSU ET STIPENDIIS FRETUS TRIUM IMPPP.
RUDOLPHI, MATTHIAE, FERDINANDI,
ANNITENTIBUS HÆREDIBUS BRAHEANIS, EX FUNDAMENTIS
Observationum relictarum; ad exemplum ferè partium iam extructarum; continuâ multorum annorum
speculationibus & computationibus, primùm PRAGÆ Bohemorum continuavit; deinde LINCII,
superioris Austriæ Metropoli, subsidijs etiam Ill. Provincialium adjutus, perfecit, absolvit,
adq; causarum & calculi perenni formulam traducit
IOANNES KEPLERUS.
TYCHONI primùm à RUDOLPHO II. Imp. adjunctus calculi minister; indeq,
trium ordine Imppp, Mathematicus:
Qui idem de speciali mandato FERDINANDI II. IMP.
petentibus instantibusq; Hæredibus,
Opus hoc ad usus præsentium & posteritatis, typis, numericis proprijs, cæteris
& prælo JONÆ SAURII, Reip. Ulmanæ Typographi, in publicum
extulit, & Typographicis operis ULMÆ curator adfuit.

Cum Privilegiis IMP. & Regum Rerumq; publ. vivo TYCHONI ejusq; Hæredibus,
& speciali Imperatorio, ipsi KEPLERO concesso, ad anno XXX.
ANNO M DC XXVII.

(b)

TABULÆ
RUDOLPHINÆ,
QUIBUS ASTRONOMICÆ SCIENTIÆ, TEMPO-
rum longinquitate collapsæ RESTAURATIO *continetur;*
A Phœnice illo Astronomorum
TYCHONE,
Ex Illustri & Generosa BRAHEORUM in Regno Daniæ
familiâ oriundo Equite,
PRIMUM ANIMO CONCEPTA ET DESTINA-
TA ANNO CHRISTI MDLXIV: EXINDE OBSERVATIONIBUS
SIDERUM ACCURATISSIMIS, POST ANNUM PRÆCIPUE MDLXXII.
QUO SIDUS IN CASSIOPEIÆ CONSTELLATIONE NOVUM EFFULSIT - SERIO AFFECTATA; VARIIS-
que operibus, cùm mechanicis, tùm librariis, impensâ patrimonio amplissimo, accedentibus etiam subsidiis FRE-
DERICI II. DANIÆ REGIS, regali magnificentia dignis, tractâ per annos XXV. potissimùm in Insula
freti SUNDICI HUENNA, & arce URANIBURGO, in hos usus à fun-
damentis extructâ:
TANDEM TRADUCTA IN GERMANIAM, INQUE AULAM ET
Nomen RUDOLPHI IMP. anno MDIIC.

TABULAS IPSAS, JAM ET NUNCUPATAS, ET AFFECTAS, SED
MORTE AUTHORIS SUI ANNO MDCI. DESERTAS,
JUSSU ET STIPENDIIS FRETUS TRIUM IMPPP.
RUDOLPHI, MATTHIÆ, FERDINANDI,
ANNITENTIBUS HÆREDIBUS BRAHEANIS; EX FUNDAMENTIS OB-
servationum relictarum; ad exemplum ferè partium jam exstructarum; continuâ multorum annorum spe-
culationibus, & computationibus, primùm PRAGÆ *Bohemorum continuavit; deinde* LINCII,
Superioris Austriæ Metropoli, subsidiis etiam Ill. Provincialium adjutus, perfecit, ab-
solvit; adq; causarum & calculi perennis formulam traducit.
IOANNES KEPLERUS,
TYCHONI primùm à RUDOLPHO II. Imp. adjunctus calculi minister; indeq;
Trium ordine Imppp. Mathematicus:
Qui idem de speciali mandato FERDINANDI II. IMP.
petentibus instantibusq; Hæredibus,
Opus hoc ad usus præsentium & posteritatis, typis, numericis propriis, cæteris, & prælo
JONÆ SAURII, Reip. Ulmanæ Typograph; in publicum extulit, & Typographicis operis ULM *curator affuit.*

Cum Privilegiis IMP. & RegumRerumq; publ. vivo TYCHONI ejusq; Hæredibus,
& speciali Imperatorio, ipsi KEPLERO concesso, ad Annos XXX;
ANNO M. DC. XXVII.

Figure 15.3 (a) The second and (b) the third stages of the title page of the Rudolphine Tables, the last exhibiting Kepler's well-defined esthetic taste. He had neglected to show Tycho's heirs the title page before it was printed, and they objected to wording implying that he had to correct his mentor's data. The second printing, (a), was hastily prepared at Prague, but Kepler found the typography so bad he had the third one, (b), printed at Ulm, presumably at his expense. Among other things, Kepler indicates near the end of the long title that he personally owned the type for the numbers used in the tables.

with the Sidereal Messenger' of Galileo. Kepler's enthusiastic reception of the Italian astronomer's discoveries and his confirmation of the new Jovian satellites served as strong witness for the new findings. Galileo wrote to him: 'I thank you because you were the first one, and practically the only one, to have complete faith in my assertions.'

In 1611 the political situation in Prague took an abrupt turn, ending Kepler's exhilarating atmosphere of intellectual freedom. The gathering storm of the Counter-Reformation reached the capital, and brought about the abdication of Rudolf II. As warfare and bloodshed surged around him, Kepler sought refuge in Linz, where he was appointed provincial mathematician and a teacher in the district school.

The school was even smaller than the one at Graz, but his principal work was to be the *Rudolphine Tables*. The Linz authorities charged him first of all to 'complete the astronomical tables in honor of the Emperor and the worshipful Austrian House, for the profit of . . . the entire land as well as also for his own fame and praise.'

After Rudolf's death in 1612, his successor Matthias confirmed Kepler as a court mathematician and agreed to his new residence away from Prague. But Kepler realized that as long as the *Rudolphine*

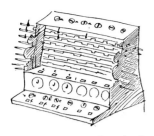

Figure 15.4 To aid Kepler in his arduous calculations, Wilhelm Schickard of Tübingen University invented a pioneer calculating machine, shown here in a sketch from a letter of Schickard to Kepler. Unfortunately it was destroyed in a workshop fire before it could be used. Numbers were entered by turning the series of dials near the bottom. The upper part is a set of Napier's rods for multiplication.

Tables were unfinished, he would be tied to Linz. Thus the work on the tables became part of his fate.

Several other great works took form in Linz. Foremost among them were the *Epitome of Copernican Astronomy*, which outlined the theoretical basis for Kepler's planetary tables, and the *Harmonice Mundi*, a great cosmological sequel to the *Mysterium Cosmographicum*. In 1618, Kepler discovered what has come to be called his third or harmonic law of planetary motions; it was published next year in the *Harmonice Mundi*. He continued to grapple with the difficulties of the moon's orbit, achieving success in 1620. This theory was included in the final installments of the *Epitome*, and in the preface Kepler urged his readers to use that work until the *Rudolphine Tables* came out.

Nevertheless, from all sides impatient reminders urged Kepler onward with the tables. 'Don't sentence me completely to the tread-mill of mathematical calculations – leave me time for philosophical speculations, my sole delight,' responded Kepler. But he added, 'I am as eager for the publication as Germany is for peace.'

Meanwhile, another innovation completely altered Kepler's original plan for the form of the tables – 'a happy calamity,' as he called it. In 1617 Kepler first saw John Napier's epoch-making work on logarithms and was deeply impressed by it, recognizing how this new invention would simplify the time-consuming computations of astronomy. However, not content to adopt the new aid as he found it, during the winter of 1621–2 Kepler composed his own book on the subject.

He exploited the new logarithms to solve two problems introduced for the first time by the novel form of the *Rudolphine Tables*. The first arises in the solution of what is now called Kepler's equation. For a planet moving in an ellipse, under Kepler's law of areas, there is no elementary way to find explicitly the position angle corresponding to a given time. However, the converse is easily calculated. Therefore he solved his equation for a set of uniformly spaced angles, which determine a set of nonuniformly spaced times. Kepler tabulated the *logarithms* of these intervals as a convenient means for interpolating to the desired times.

The second important use of logarithms arises from the thoroughly heliocentric nature of the book. In previous planetary tables, the motions of the sun and planets were combined into a single procedure. In the *Rudolphine Tables* we must find separately the heliocentric positions of the earth and planet in question. To find the *geocentric* position of the planet, these two positions must be combined – essentially a problem of vector addition. Kepler facilitated this maneuver by tabulating the logarithms of the radius vectors of earth and planet, and by providing a convenient double-entry table for combining them.

Kepler's tables, unlike the modern *Nautical Almanac*, do not show the daily positions of the sun, moon, and planets. Instead, they contain general tables from which it is possible to work out a planet's position for any time in the past or future. Besides the planetary data, Kepler included tables of logarithms, a catalogue of 1000 stars, and a list of the geographical longitudes and latitudes of a large number of

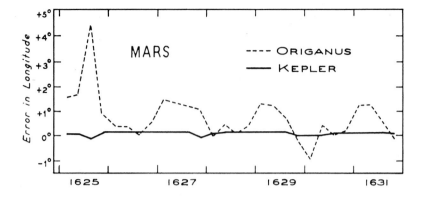

Figure 15.5 A modern test compares predictions of the geocentric longitudes of Mars during the years 1625–31, made with the Rudolphine Tables *of Kepler (solid line) and the* Brandenburg Ephemerides *of David Origanus (dashes). The latter man, who lived from 1558 to 1628, followed the* Prutenic Tables *of Copernicus. The great superiority of Kepler's predictions is obvious.*

cities. Approximately half the volume is made up of instructions for the use of the tables and numerical examples of their use.

At long last, in 1624, Kepler completed his *Rudolphine Tables* in their new logarithmic form. The printing of the tables was very difficult because of wartime conditions. Linz offered no really suitable press, and circumstances in the city became increasingly unpleasant. Kepler's lodgings were located on the city wall, and eventually he was obliged to open his house for the soldiers guarding the ramparts. To a correspondent he described the noise and smells of battle, but pointed out that he found solace in continuing his calculations. Finally, however, the printing was transferred to Ulm.

Financing the book's publication caused the astronomer much worry and trouble. A long trip to the imperial court in Vienna won some concessions for him, including an agreement to tax the city of Nuremberg for the arrears in his yearly allowance. Yet a visit to Nuremberg in 1625 was in vain, and he eventually decided to finance the work from his own pocket. Even this was not a simple operation, for Tycho's heirs claimed both a share in the profits and also censorship rights.

His preface to the tables explains that the long delay in publication was partly from these circumstances, but also from 'the novelty of my discoveries and the unexpected transfer of the whole of astronomy from fictitious circles to natural causes.' These, he said, required deep searching, and were especially difficult to explain or calculate, since no one had ever attempted anything of the kind before.

Because the *Rudolphine Tables* was in many ways the long-awaited climax to Kepler's entire scholarly output, he proposed that this book, alone among his works, should have an appropriate frontispiece. His Tübingen friend Wilhelm Schickard prepared a wash sketch of the proposed engraving. It showed the Temple of Urania, modeled after the foyer of Tycho's observatory, Stjerneborg, on the Danish island of Hven. On the ceiling of the temple was shown the geocentric Tychonic system, and within the building stood a series of notable astronomers, including Tycho himself.

Kepler submitted the sketch to Tycho's heirs, who at once objected. Their illustrious ancestor should be depicted more formally, they said, with his long ermine robe and the elephant medal around his

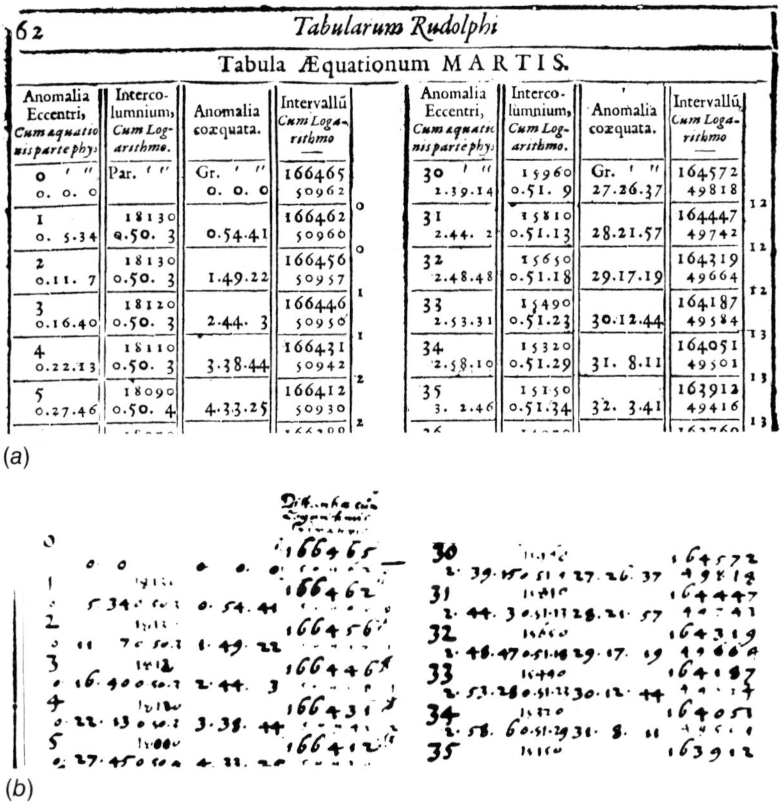

Figure 15.6 (a) A small portion of one of the six pages in the Rudolphine Tables *that provide the data for predicting the position of Mars on any date since 4000 BC. These particular numbers would be used in converting the planet's mean anomaly to eccentric anomaly, that is, in the solution of Kepler's equation. (b) The corresponding part of the manuscript in Kepler's handwriting.*

62				*Tabularum Rudolphi*			
Tabula Æquationum MARTIS.							
Anomalia Eccentris *Cum aquatio nis parte phy.*	Interco-lumnium, *Cum Log-arithmo.*	Anomalia coæquata.	Intervallū *Cum Loga-rithmo*	Anomalia Eccentri, *Cum aquatio nis parte phy.*	Interco-lumnium, *Cum Log-arithmo.*	Anomalia coæquata.	Intervallū, *Cum Loga-rithmo*
o ' '' 0. 0. 0	Par. ' ''	Gr. ' '' 0. 0. 0	166465 50962	30 ' '' 1.39.14	15960 0.51. 9	Gr. ' '' 27.26.37	164572 49818
1 0. 5.34	18130 0.50. 3	0.54.41	166462 50960	31 2.44. 2	15810 0.51.13	28.21.57	164447 49742
2 0.11. 7	18130 0.50. 3	1.49.22	166456 50957	32 2.48.48	15650 0.51.18	29.17.19	164319 49664
3 0.16.40	18120 0.50. 3	2.44. 3	166446 50956	33 2.53.31	15490 0.51.23	30.12.44	164187 49584
4 0.22.13	18110 0.50. 3	3.38.44	166431 50942	34 2.58.10	15320 0.51.29	31. 8.11	164051 49501
5 0.27.46	18090 0.50. 4	4.33.25	166412 50930	35 3. 1.46	15150 0.51.34	32. 3.41	163912 49416

(a)

(b)

neck, which he always wore while observing. (The medal was the highest award of the Danish monarchy.)

The frontispiece in its final form (see Figure 15.1) was a great elaboration over the trial sketch. Ten of the temple's 12 zodiacal columns are visible; their rich variety depicts the increasing elegance of astronomy. Those at the back, merely rough-hewn logs, represent the most ancient traces of the science. Nearby stands a Chaldean observer who, for want of an instrument, uses his fingers to measure the angular separations of the stars. The Greek astronomers Hipparchus and Ptolemy appear beside brick columns adorned with instruments of the times. Copernicus, seated beside an Ionian column, engages in spirited conversation with Tycho, who stands in his magnificence by an elaborate and splendid Corinthian pillar. The Danish astronomer points to the Tychonic system inscribed on the ceiling, and his Latin comment to Copernicus may be loosely translated, 'How about that?'

Surrounding the dome of the temple appear six goddesses, each recalling an important idea of Kepler's. To the far right stands Magnetica, with her lodestone and compass, reminding us of the magnetic forces that Kepler believed to control the planets. Next is Stathmica, goddess of the law of the lever and balance; the sun at the fulcrum reminds us that this is a form of Kepler's law of areas. The third divinity is Geometria, with her mathematical compass, square, and tablet bearing the Keplerian ellipse. The next figure is Logarith-

*Figure 15.7 The original
design drawn by Schickard
for the frontispiece of the
Rudolphine Tables was
much simpler than the
printed version and gave less
emphasis to Tycho Brahe.*

mica, who holds in her hands rods in the ratio of one to two, while
emblazoned on her halo is the natural logarithm of $\frac{1}{2}$. The fifth goddess holds a telescope, and the sixth a globe with its shadow, reminding us of Kepler's two books on optics: *Dioptrice* (1611) and
Astronomiae Pars Optica (1604).

   Above the temple hovers the imperial eagle, generously dropping
golden coins from his beak for the support of astronomy. A few of the
coins even manage to fall to the foundations of the temple where, in
the left panel, we see Kepler himself, calculating by candlelight. And
here is a marvelous personal touch, for prominently on the worktable
sits a replica of the dome of the Temple of Urania. In his subtle and
uncensored fashion, Kepler reminds us that although Tycho may
have built the most splendid column, the Temple of Urania would
never have been finished without Kepler himself laboring far into the
night!

Figure 15.8 The Kepler monument at Regensburg, Germany, photographed by the author in 1971. The unknown site of Kepler's grave is nearby.

Kepler personally oversaw the printing and worked almost daily with the typesetters. One thousand copies were printed – an enormous edition for a seventeenth-century science book. (Sixty years later, Newton's *Principia* was published in an edition of about 300 copies.) As a result, the *Rudolphine Tables* is by far the most common of all Kepler's books, though examples today generally command a price of nearly $1000.

Kepler's tables enabled him to predict the transit of Mercury over the disk of the sun on 7 November 1631, a phenomenon never previously observed. He did not live to see the prediction fulfilled, for he died on 15 November 1630, at Regensburg, where he had gone on yet another effort to collect his back salary. However, this transit of Mercury was actually witnessed by the astronomer Pierre Gassendi at Paris.

In a letter to Kepler's old friend Schickard, Gassendi wrote: 'But Apollo, acquainted with [Mercury's] knavish tricks from his infancy, would not allow him to pass altogether unnoticed. To be brief, I have been more fortunate than those hunters after Mercury who have sought the cunning god in the sun. I found him out and saw him where no one else had hitherto seen him.'

In 1665, J. B. Riccioli related that the tables of Ptolemy, Copernicus, and Longomontanus each erred by about 5° in predicting this transit, whereas Kepler missed by less than 10′ of arc! The overwhelming evidence of this crucial experiment convincingly established the power of the *Rudolphine Tables*. In this almost forgotten way, the geocentric world view was broken and the heliocentric system, together with Kepler's laws of planetary motion, triumphed.

Notes and References

The late Franz Hammer, editor of the *Johannes Kepler Gesammelte Werke*, first alerted me to the interesting story of the frontispiece. Following his lead, I researched the story and iconography of the engraving, and this article remains practically the only account in English. For more details on Kepler and an extensive bibliography, see my biographical entry in *Dictionary of Scientific Biography*, Vol. 7 (New York, 1973). A highly readable account of Kepler is in *The Watershed* (Garden City, 1960), an independently printed part of Arthur Koestler's *Sleepwalkers*; harder going but more authoritative is Max Caspar's *Kepler* (New York, 1959), translated and edited by C. Doris Hellman. Since 1971, when this article was written, the price of a copy of the *Rudolphine Tables* has greatly increased.

16 An astrolabe from Lahore

It was about 25 years ago that a student from Afghanistan turned up at Harvard Observatory with a curious old brass instrument $\frac{1}{4}$-inch thick and about 4 inches in diameter. I immediately recognized it as an astrolabe. I had seen scores of them in museums but had only a dim idea about how they were constructed or how they were supposed to work. But I did know that most astrolabes turning up nowadays are fakes.

Upon examining this astrolabe, I realized that the main body of the instrument was hollowed out, providing a nest for a series of circular plates, and that the upper plate was mostly cut away, leaving a delicate filigree structure. Every surface, except the underside of the cutaway plate, was filled with Arabic inscriptions, tables, or graphs.

With the help of a friend I deciphered a maker's name and date on the back of the instrument: Diya' ad-Din Muhammad, great grandson of the 'royal astrolabist of Lahore,' AH 1073. The Islamic date is with respect to the Hegira (Muhammad's flight from Mecca to Medina in AD 622) and could be converted to the Gregorian calendar by noting that the Islamic year is three-percent shorter than a solar year, since it is a strict period of 12 lunations with no intercalary months. Hence the Gregorian date came out as AD 1662 or $1663 = 622 + (0.97 \times 1073)$. During the lifetime of Diya' ad-Din the Mogul court in Lahore had reached its zenith, but the powerful Shah Jehan was already dead and had been entombed in the Taj Mahal five years before this astrolabe was made.

After my preliminary investigation, an opportunity to buy the instrument came along. I was a little uneasy, because it is difficult to find a genuine, unrecorded astrolabe. Still, the metalwork seemed old – I had had some experience with this, having lived in the Middle East for a few years – and the calibrations seemed accurately made. Thus, I took the plunge and have never regretted it. Since than I have had numerous opportunities to inspect fake astrolabes (which will be described in Chapter 17), and experience has indicated that the accuracy of the scales and graphs is one of the best criteria for determining whether or not a particular instrument is genuine.

After acquiring the astrolabe, a little research quickly showed that there is a vast literature on these devices; some of the best and most accessible material is listed at the end of this chapter. I promptly learned that the body of the astrolabe is called the *mater* (mother) or, in Arabic, the *umm*. The uppermost, cutaway plate is a rudimentary star chart, called the *rete* (net) or, in the case of an Arabic instrument, the *ankabut* (spider). The net is a framework that carries a number of small labeled pointers representing the stars; these pointers form the star chart. This chart is pivoted about the north celestial pole at its center, so that diurnal motion can be shown by rotating this disk. Underneath, on the plate that shows through whenever the star chart has been cut away, is inscribed a circle representing the

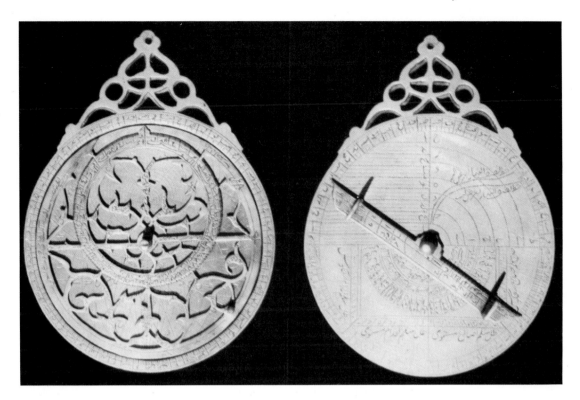

Figure 16.1 The author's astrolabe was made by Diya' ad-Din in Lahore in 1662 or 1663.

horizon. Hence, an astrolabe is somewhat like a metal version of an ordinary planisphere, except that the map is on top and the horizon below.

An astrolabe differs further from an ordinary planisphere in that it uses a stereographic projection, in which all circles on the celestial sphere are represented by circles on the plane of the instrument. The astrolabe maker must pay for this convenience by sacrificing the southern sky rather far from the equator; in a stereographic north polar projection, the southern regions become greatly expanded in scale and the instrument would either have to become uncomfortably large or undesirably squashed in the northern regions to include the sky much below declination $-23\frac{1}{2}°$, the limit of the ecliptic. In an ordinary planisphere, by contrast, the southern regions are compressed in a north–south direction and thereby distorted, so that the entire dome of the sky can be contained in an oval aperture.

In making such a map on a metal grid, several considerations govern the choice of stars. The astrolabist would not want too many of them, especially in the compressed polar regions. Further, stars would be chosen as symmetrically as the heavens allowed, both for the beauty of the design and for the balance of the rete when the instrument is suspended upright. Finally, the entire ecliptic would be included so that the sun and planets could be plotted; this means including an eccentrically placed circle tangent to the outer edge (the Tropic of Capricorn) and also tangent to a smaller inner circle in the north (the Tropic of Cancer).

The number of stars actually marked on an astrolabe varies from

Star names on the Diya' ad-Din astrolabe

Star	Name	Arabic Astrolabe Name	Meaning
1. α And	Alpheratz	*surrat [al-faras]*	navel [of the horse]
2. β Cas	Caph	*kaff al-khadīb*	hand of the colored one
3. ι Cet		*dhanab al-qītus sh[amālī]*	northern tail of the whale
4. β Cet	Deneb Kaitos	*dhanab al-qītus*	tail of the whale
5. β And	Mirach	*batn al-ḥūt*	belly of the fish
6. π Cet		*ṣadr al-qītus*	breast of the whale
7. γ Cet[a]		*fam al-qītus*	mouth of the whale
8. β Per	Algol	*[al]-ghūl*	the ghoul
9. γ Eri	Zaurak	*tālī masāf al-nahr*	the last one in this section of the river
10. α Tau	Aldebaran	*[al]-dabarān*	follower [of the Pleiades]
11. β Ori	Rigel	*rijl jawzā, yusrā*	left leg of Orion
12. α Aur	Capella	*ʿayyūq*	[traditional name, meaning unknown]
13. γ Ori	Bellatrix	*rijl[b] jawzā*	leg[b] of Orion
14. $\varkappa$ Ori	Saiph	*rijl jawzā, yamīnī*	right leg of Orion
15. α Ori	Betelgeuse	*yad jawzā*	arm of Orion
16. α CMa	Sirius	*shiʿrā yāmāiyya*	the southern Sirius
17. α CMi	Procyon	*shiʿrā shāmiyya*	the northern Sirius
18. α Hya	Alphard	*[al]-fard [al]-shujāʿ*	the one standing alone in the water serpent
19. ε Leo		*ra's [al]-asad*	head of the lion
20. α Leo	Regulus	*qalb al-asad*	heart of the lion
21. α Crt	Alkes	*qāʿidat al-bāṭiya*	base of the bowl
22. α UMa	Dubhe	*dhahr [=ẓahr] dubb*	back of the bear
23. γ Crv	Gienah	*janāḥ al-ghurāb*	wing of the raven
24. ζ UMa	Mizar	*ʿ[i]nāq*	embracing (?)
25. α Vir	Spica	*simāk aʿzal*	the unarmed simak
26. α Boo	Arcturus	*simāk rāmiḥ*	simak armed with a lance
27. α Lib	Zubenelgenubi	*kiffa, janūbī*	southern balance pan
28. α CrB	Alphecca	*nayyir [al]-fakka*	lucida of the broken dish
29. α Ser	Unukalhai	*ʿunuq al-ḥayya*	neck of the serpent
30. α Sco	Antares	*qalb al-ʿaqrab*	heart of the scorpion
31. ζ Oph		*rukbat ḥawwā*	knee of the serpent bearer
32. α Her	Rasalgethi	*ra's al-jāthī*	head of the kneeler
33. α Oph	Rasalhague	*ra's [al]-ḥawwā*	head of the serpent bearer
34. ν Oph		*yad ḥawwā*	arm of the serpent bearer
35. α Lyr	Vega	*nasr wāqiʿ*	the swooping eagle
36. β Cyg	Albireo	*minqār dajāja*	beak of the hen
37. α Aql	Altair	*[al]-nasr [al]-ṭā'ir*	the flying eagle
38. ε Del		*dulfīn*	dolphin
39. α Cyg	Deneb	*dhanab [al-dajāja]*	tail [of the hen]
40. δ Cap	Deneb Algedi	*dhanab al-jady*	tail of the goat
41. ε Peg	Enif	*fam faras*	mouth of the horse
42. δ Aqr	Skat	*sāq sākib al-mā'*	shin of the water pourer
43. β Peg	Menkib[c]	*mankib faras*	shoulder of the horse

[a] It is curious that Diya' ad-Din plots γ Ceti rather than the brighter α Ceti (Menkar).

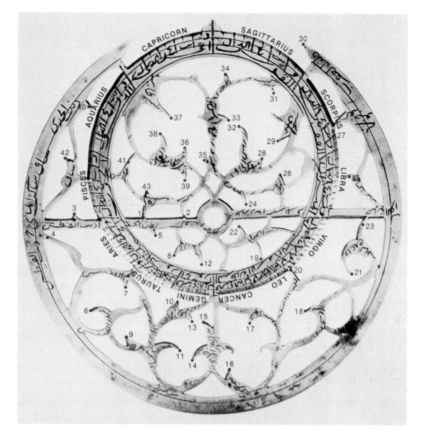

Figure 16.2 An enlargement of the star chart, reproduced here slightly larger than actual size. The off-centered circle represents the ecliptic; numbered star pointers correspond to entries in the table.

about a dozen to around 50. The Diya' ad-Din instrument has 43, which is unusually large for an astrolabe with a diameter of only about four inches. The Arabic names describe the part of the constellation figure in which they are found, such as *ras* for 'head' or *deneb* for 'tail'. Frequently, these are simply translations from Ptolemy's star catalogue. The list of star names on Diya' ad-Din's astrolabe is shown in the table. Many of the names are familiar, although the spelling in the strict transliteration is a little different: *rijl* for Rigel, *al-fakka* for Alphecca, *dhanab al-jady* for Deneb Algedi. But a Greek name like Antares or a Latin name like Regulus comes out quite differently: *qalb al-ᶜaqrab* and *qalb al-asad* for, respectively, 'heart of the scorpion' and 'heart of the lion.'

An important reason for restricting the number of stars on the rete is to allow a substantial view of the underlying plate. On the latter the horizon is inscribed *for a particular latitude,* and herein lies one of the most ingenious features of the instrument. By providing multiple plates, inscribed both back and front, the astrolabe maker could produce an instrument good for a wide variety of locations. The Diya'

---

[b] Here Diya' ad-Din should have written *yad al-jawzā al-yusrā,* left arm of Orion.

[c] Today Menkib is more commonly known as Scheat.

For brevity in the limited space on his instrument, Diya' ad-Din generally omits the article *al.* and I have followed him strictly except to indicate with brackets a few instances where the article should go in order to make the anglicized form of the name more obvious. I have translated *yad* as 'arm' rather than the more common 'hand' because of the placement of these stars both by Ptolemy and by as-Sufī.

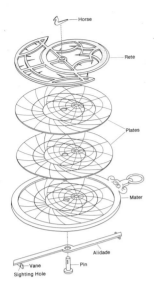

Figure 16.3 An exploded view of an astrolabe shows how its various parts nest together. The wedge that secures the pin traditionally resembles a horse's head.

Figure 16.4 The inscription of the maker's name is found just below the pivot on the back of the astrolabe.

ad-Din astrolabe has plates inscribed for latitudes 22°, 25°; 27°, 29°; 32°, 36°; and 18°. The eighth side has a combination chart for the equator and for ecliptic coordinates. I was puzzled about the distribution of latitudes for the plates until I deciphered the geographical gazetteer hidden in the *umm* underneath all the plates. In the list of 48 cities in the Islamic world, none were between latitudes 29° and 32°.

The underlying plates contain a great deal more than just the horizon circle. Within the latter is a whole series of circles representing lines of equal altitude around the sky and culminating in the zenith point overhead. These circles are called *almucantars*, which like 'zenith' is an Arabic word. Incidentally, an astrolabe with concentric almucantar circles is sure to be fake. In a genuine instrument, each successive almucantar circle must be slightly offset from the one before it, as shown in Figure 16.5.

Besides the almucantars, these plates contain lines for equal and unequal hours. The unequal hours divide the night into 12 equal parts regardless of season – hence, a night hour in the winter is considerably longer than a night hour in summer, and vice versa for the day hours. On the astrolabe, the unequal hours are shown by sweeping arcs that are symmetrical on the eastern and western halves of the plate. In contrast, the equal hours (used in modern timekeeping) always have the same duration, so more nighttime hours occur in winter than in summer.

For Diya' ad-Din the day began at sunset, so then the sun was on the western horizon, the first hour of night started, whether in equal or unequal hours. The line for the beginning of the second equal hour is nearly the same as the line for the beginning of the second unequal hour, but for each successive hour the lines differ more and more. On Diya' ad-Din's instrument the equal hours are shown by dotted arcs. In the stereographic projection each of these arcs is part of a large circle.

The back of Diya' ad-Din's astrolabe is typical of an Indian instrument and to some extent representative of astrolabes in general. For example, an astrolabe must have a movable sighting ruler, called an *alidade*; this Arabic word is used for both Arabic and Latin instruments. Furthermore, nearly all astrolabes have a pair of shadow squares in the lower half of the back. The shadow square, in conjuction with the alidade, can be used to measure the heights of buildings or to design sundials. Its horizontal part gives tangents, which are continued in the arc near the edge of Diya' ad-Din's astrolabe; the vertical part gives cotangents. The right half of the shadow square is divided into twelve parts, the left into seven, a traditional scheme that gives no particularly intentional relationship between the two sides.

In the upper left quadrant, Arabic astrolabes generally have a graph that may be used to find sines or cosines. In those from Lahore the pattern often consists of horizontal lines, in some submultiple of 60, whereas others and particularly those from Persia usually have a grid of both horizontal and vertical lines. In the upper right quadrant, Diya' ad-Din's astrolabe has a graph showing the maximum seasonal altitude of the sun for latitudes 27° and 32°.

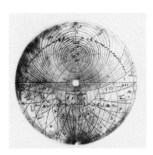

*Figure 16.5 The plate of 27°
with the nonconcentric
almucantar circles.*

Within the shadow squares of this instrument are two circular series of inscriptions, the inner giving the zodiacal signs and the outer the corresponding lunar mansions, 28 altogether. And inside this is one of the most interesting inscriptions of all, the maker's name and pedigree. It reads, 'The work of Diya' ad-Din Muhammad, son of Qa'im Muhammad, son of Mulla 'Isa, son of Shaikh Allah-dad, the royal astrolabist of Lahore.' For over a century this family worked at the Mogul capital of Lahore, and astrolabes or globes survive from each member as well as from Diya' ad-Din's uncle, Muhammad Muqim, and his cousin Hamid. Finally, within this signature is the Hegira date, 1073. Incidently, the date is the only set of digits on the astrolabe actually written in Arabic numbers – not the Arabic numbers familiar in the West, but the form used in the Arab world today. All of the other numbers are in a system called *abjab*, in which letters of the Arabic alphabet substitute for the digits.

What was the astrolabe actually used for? In principle it could be used to measure the altitude of the sun or a star, and by turning the star chart so that the object would fall on the right almucantar (matching the observed altitude), the time of day or night could be established. For a Muslim, this could be useful for establishing prayer times. Furthermore, treatises on the astrolabe frequently describe how the instrument could be used for elementary surveying or for finding the heights of buildings. To carry out such observations accurately would require a careful design of the alidade, which should be fully balanced so that it easily rests in any position when the astrolabe is held upright. In making his alidade, Diya' ad-Din followed the weak, unbalanced design of his forefathers. Had these astrolabes been employed regularly for observations, the design of the alidade probably would have evolved into something more accurate and stable. Hence, I am inclined to believe that this device served most commonly as a mathematical calculator, or perhaps merely to add an air to authenticity to an astrologer's salon!

Notes and references

The best general reviews are J. D. North, 'The Astrolabe,' *Scientific American*, **230** (January, 1974), 96–106, with excellent illustrations, and W. Hartner, 'The Principle and Use of the Astrolabe,' in A. U. Pope (editor), *A Survey of Persian Art*, Vol. 3, (Oxford, 1939), pp. 2530–54, reprinted in Hartner's *Oriens-Occidens* (Hildesheim, 1968), pp. 287–311. An excellent guide and inexpensive kit is the National Maritime Museum's *The Planispheric Astrolabe* and *Make-it-yourself Astrolabe* (Greenwich, 1979).

The single most comprehensive modern compendium is R. T. Gunther, *The Astrolabes of the World* (Oxford, 1932, reprint London, 1976), but it is not entirely reliable – see my review in the *Journal for the History of Astronomy*, **9** (1978) 69–70. The work contains (Vol. 1, 192) a list of stars on an astrolabe made by Diya' ad-Din's uncle, Muhammad Muqim, which is useful but unfortunately marred by numerous errors of transcription and identification. A more accurate list from another astrolabe by his uncle, containing only 27 stars, appears in an article by C. A. Crommelin in *Orientalia Neerlandica* (Leiden, 1948), pp. 240–56. A list of 41 stars, not completely identified, from a very similar Diya' ad-Din astrolabe in the State Library at Rampur, India, has been given by Pt. Padmakara Dube in *The Journal of the United Provinces Historical Society*, **4** (1928), 1–11.

In making up the star list on page 134 I have been generously guided by A. I. Sabra, professor of the history of Arabic science at Harvard University, and I have found Paul Kunitzsch's *Arabische Sternnamen in Europa* (Wiesbaden, 1959) to be an essential reference.

17 Fake astrolabes

Chicago's Adler Planetarium possesses the finest collection of historic astronomical instruments in this hemisphere, including one of the world's six largest holdings of astrolabes. The collection also contains some exceedingly curious specimens that defy the ordinary rules of astrolabe making.

It was at Adler, a quarter of a century ago, that I first learned about some of the pitfalls awaiting would-be astrolabe collectors. After I had admired the magnificent examples on display, an assistant offered to let me see the overflow in the storeroom. The morning spent with those instruments turned out to be a most educational experience. As I scrutinized one astrolabe after another, it gradually dawned on me that something was odd about each one of them.

I knew that an astrolabe consisted of a series of movable plates nested in a brass matrix or 'mother,' and that the uppermost of the plates, or rete, was a grillwork star chart. But a few of the Adler instruments had elaborately foliated retes with completely symmetrical left and right halves (see Figure 17.1) – obviously impossible for a real map of the starry sky. Other instruments had very peculiar graphs on their back sides or had crudely calibrated circumferences instead of the accurate protractors normally used with the alidade for measuring altitudes. Eventually I reported back to the assistant with the opinion that I had been turned loose in one of the world's largest reserves of fake astrolabes. 'Congratulations!' he said. 'You've passed the test.'

In retrospect, not all the instruments may have been forgeries. In some cases, such as the one illustrated in Figure 17.1, the 'mother' is a perfectly fine old specimen, but a lost rete has been replaced by a strange, nonfunctioning brass decoration. In other cases, the calibrations were made with high accuracy, but some of the details were idiosyncratic; these may well have been made by early astronomers according to their own ingenuity rather than following a somewhat stereotyped tradition. Ultimately, what gives away a fake astrolabe are clumsy calibrations or incorrectly drawn circles.

In owning several bogus specimens, Adler is in good company. Some of the imposters, while useless as scientific instruments, exhibit beautiful calligraphy and decorations, and therefore make their way into some of the best art museums. Examples include Boston's Museum of Fine Arts, the Freer Gallery in Washington, and the Detroit Institute of Arts.

Some years ago a student brought me photographs of a pretty astrolabe in the art museum at Oberlin College, but at the same time he remarked that certain features seemed rather troublesome. Not only was the rete completely symmetrical with no star names marked, but a detailed examination showed that the zodiacal signs marked around the ecliptic were 180° out of phase, with Cancer farthest south and Capricorn farthest north! The instrument bore a

Figure 17.1 A spurious and symmetrical replacement rete (star chart) has been added to this genuine ᶜAbd al-A'imma astrolabe at the Adler Planetarium in Chicago. Note the accuracy of the calibrations on the rim of the genuine 'mother' compared to the degenerate instrument on the next page.

famous maker's name, ᶜAbd al-A'imma, a Persian astrolabist who worked in Isfahan around 1700.

A quick survey showed that, while a number of his astrolabes also had striking peculiarities, many others were beautiful and highly accurate instruments. Did the Isfahan astrolabist have incompetent apprentices, or was he just a calligrapher who got things right only when some collaborator designed the astrolabe for him? I teamed up with two Arabists, David King and George Saliba, to take a systematic look at ᶜAbd al-A'imma's workmanship.

What we discovered was that the reputation of the genuine ᶜAbd al-A'imma was deservedly high — his instruments are among the most beautiful Islamic astrolabes known — which made him an obvious target for later forgers. (I say 'forger' rather than imitator or facsimile-maker because the astronomically degenerate instruments almost always carried dates as well as ᶜAbd al-A'imma's name, clearly done by someone who wanted to make the forgeries look old.) These fakes give themselves away by errors in their astronomical constructions. Frequently, the scales around the edge are subdivided into some crazy number of parts, so that 5° intervals might be broken into seven, eight, nine, or eleven parts, for example.

Other errors are a little more subtle, and measurement might be required to detect them. Astrolabes are based on the stereographic projection, and as seen in Figure 17.3, the ecliptic projects onto the rete as a very specific off-centered circle that is tangent both to the Tropic of Capricorn and, 180° away, to the Tropic of Cancer. The

*Figure 17.2 This ⁶Abd al-
A'imma forgery, now in the
Smithsonian's Freer Gallery
of Art in Washington, DC, is
a sophisticated fake. The use
of a bird to denote the star
Vega (the 'soaring eagle') is
typical of fake astrolabes.*

distance from the central pivot to the Tropic of Capricorn must
always be 2.33 times the distance from the central pivot to the Tropic
of Cancer. On the ⁶Abd al-A'imma forgeries the ecliptic circles are
generally too small, so that the ratios run from 2.6 to 3.5.

Figure 17.2 shows a typical ⁶Abd al-A'mma forgery. It is immediately evident that the ecliptic does not lie tangent to the Tropic of
Cancer as marked on the plate below the rete. Those who can read
Arabic will note with amusement that the zodiac runs the wrong way
around the ecliptic circle. The stars, however, are not reversed, yet
the rete's deliberate but erroneous symmetry forces their positions to
be only approximate. (This forgery is rather superior since at least it
has star names!) Around the border of the rete is a decorative band
with some nonsensical Arabic calligraphy – a feature common to
these forgeries and a detail that has led us to conclude that the forger
was probably a Persian who knew little or no Arabic.

Another sensitive test of accuracy, which requires no measurement, is also shown in the illustration of the stereographic projection.
The ecliptic *must* intersect the equator on a straight line that passes
through the central pivot, marking the celestial pole. The plates offer
another quick test of an astrolabe's accuracy and authenticity. The
second diagram in Figure 17.3 shows why the almucantar circles
each have slightly different centers. Any astrolabe with concentric
almucantars is just plain wrong and can't be genuine. Furthermore,

Figure 17.3 (a) Stereographic projection causes circles on the celestial sphere to project as circles on the plane of an astrolabe. (b) The representation of almucantars (circles of equal altitude) for an observer situated 50° in latitude north of the equator.

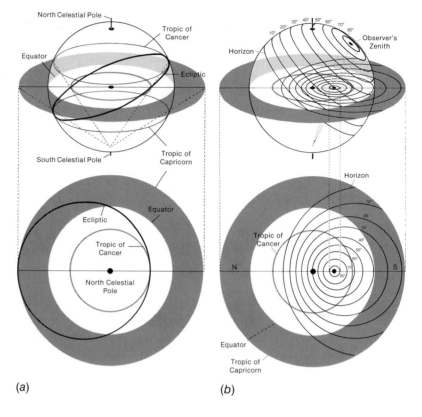

(a)   (b)

Figure 17.4 An astrolabe with completely inaccurate calibrations and small diagrams depicting signs of the zodiac, two traits that are typical of modern fakes.

on each plate the equator, horizon, and east–west line must all intersect at the same point.

It is difficult to say just how old the ʿAbd al-Aʾimma forgeries are; several have been in museums since the 1920s. As early as 1850 the few remaining astrolabes 'were sought by the greatest in the kingdom and were easily sold for 50 ducats, so much did they love to have one in their sight, although many could understand not one iota of it.' Clearly, this was a situation made to order for clever metalworkers who did not really understand the scientific principles of the older instruments they were imitating.

Even today the making of imitation astrolabes goes on apace. A friend gave me the small astrolabe shown in Figure 17.4. Its back side provides a fast clue for spotting it as a fake. The calibration lines in the border of the shadow square must be radial to the pivot, but this one has only crude scratches in a nearly parallel pattern. Another penchant of modern fakers is also obvious: the use of pictorial symbols for the zodiacal signs. Apparently such vignettes have strong appeal for tourists to the Middle East, so they are often found on 'astrolabes' brought back by visitors to Israel or the Islamic countries. Two or three genuinely old astrolabes with zodiacal pictures might exist, but surely not more.

Several years ago a local fine arts dealer phoned to report that he had just bought a whole boxful of astrolabes. Several of them had these pictorial devices, and one, he proudly announced, was in Hebrew. 'That's a bad sign,' I told him, 'because the number of

Figure 17.5 This Hebrew 'astrolabe' typifies brass devices made nowadays for the tourist trade in the Middle East. Notice the practically concentric almucantars beneath the heavily foliated rete.

genuine old Hebrew astrolabes can be counted on the fingers of one hand. But do bring them over.'

I could see the 'astrolabes' in the box even before he crossed the threshold of my office. Some were green, artificially aged probably by being drenched in urine and buried for a few months. Even from afar his dozen dials made the ʿAbd al-Aʾimma forgeries look like masterpieces, and the Hebrew astrolabe was among the worst in the lot. He had brought the dozen, he claimed, from an itinerant Persian peddler who had a suitcase full of them. We talked for a while about astrolabes, and he was most astonished to discover that the great majority of genuine ones are already recorded in a computerized census made by Derek de Solla Price and his associates at Yale University.

Not surprisingly, the peddler could not be traced, so the dealer returned to inquire what price he ought to charge for his pieces. 'If they were genuine, the prices would run upwards of a few thousand dollars each,' I told him, 'but as for the prices of modern decorative metalwork from the Middle East, that is something fine arts dealers are supposed to know.' Actually, I had been hoping to buy the least wretched example from his box, because a reasonable fake can be very instructive, and I thought his best one might puzzle my students for a few minutes before they caught on. 'Well,' he said, 'I was thinking of $500 apiece.' I nearly fell off my perch and quickly declined his offer.

It would have been instructive to have photographs of a few of them, simply because they were so bad. An opportunity to record one of the contemporary fake Hebrew astrolabes finally did come along, however, and the results are shown in Figure 17.5. The owner had

Figure 17.6 Here 15 (!) zodiacal signs comprise the ecliptic band on a fake astrolabe. It is spuriously dated under the rectangular grid on its reverse side (inset) as 1021 AH=1612 AD.

been led to believe that he was acquiring not an astrolabe but some unspecified different kind of navigational device. Clearly, his instrument shows all the superficial characteristics of an astrolabe, but its inaccuracies would render any kind of measurements useless.

Eventually, Harvard was able in an unexpected way to acquire an impressive fake for teaching purposes. From time to time owners of astrolabes bring them around for inspection, and usually they go away disappointed after finding that their wonderful old instrument wouldn't even work for an astrologer, much less for an astronomer. But occasionally fine instruments, previously unrecorded, turn up in this way, as when Maxwell Rimler contacted me in 1974. His astrolabe proved to have been made by Muhammad Muqim, the uncle of the maker of the Lahore instrument described in Chapter 16. He was quite pleased to have his instrument pronounced genuine and dated to the seventeenth century.

Several years passed before I heard from Rimler again. He had the chance to buy another astrolabe, and he sent along Xerox copies of it. This time he was lucky again, not in buying a second good astrolabe but in investigating before he plunged, since his prospect was incorrect in every aspect. He wrote back thanking me for saving him from a serious blunder, and he mentioned that the astrolable was with Mr X, who did not realize that he had a forgery on his hands.

Although Mr X is one of the most famous and reputable rare book dealers in New York, he had never handled an astrolabe before. He was rather indignant when I demolished all credibility for his instrument without even deciphering the Arabic inscriptions. By now readers of this chapter should be able to point out a whole handful of faults in the astrolabe, which is shown in Figure 17.6. For starters, the symmetrical rete without any marked stars is an obvious giveaway. The 15 (count them!) zodiacal signs on the off-centered ecliptic circle add a touch of hilarious absurdity. And readers of Arabic will probably double up in laughter at the collection of nonsense in the outer band. The rest of us can be content to notice the complete lack of what is surely the *sine qua non* of a respectable astrolabe: a calibrated angular scale on its perimeter.

On the reverse side, calibrations of the shadow square are completely lacking. What passes for a calibrated angular scale is just a series of radial scratches, too far from the rim to fool anyone knowledgeable about real astrolabes.

Needless to say, such a fraudulent instrument was completely out of place in one of the finest rare book showrooms in America, but I could offer a happy solution: donate the instrument to a university where it could be used, not as an elegant exhibition piece, but as an unusual teaching tool. Thus the instrument has filled an instructive niche both on these pages and at Harvard. Indeed, when you can spot a fake astrolabe, you are well on your way to understanding what this ingenious astronomical device is all about.

Notes and references

For further details, see Owen Gingerich, David King, and George Saliba, 'The ᶜAbd al-A'imma Astrolabe Forgeries,' *Journal for the History of Astronomy*, **3** (1972), 188–98. See also Sharon Gibbs, Janice Henderson, and Derek de Solla Price, *A Computerized Checklist of Astrolabes* (Yale University, 1973).

18 Newton, Halley, and the comet

In 1696 Britain faced a coinage crisis. For years the irregular edges of hand-struck coins had been systematically clipped by thieves, who made a precarious living by selling the tiny silver chips. Finally, the government introduced a new style of machine-pressed coin, which on larger denominations included a legend around the edge: *DECUS ET TUTAMEN*, 'a decoration and a protection.' The letters and mill marks prevented the edges of the coins from being trimmed.

Naturally, it was very difficult to get enough new change into circulation at the beginning of the coinage reform. Yet one reason why the new system finally succeeded was that the newly appointed Warden of the Mint took the work very seriously. He did not accept the post merely as a well-deserved sinecure.

The new officer was none other than the former Cambridge University professor, Isaac Newton. His great *Principia* and the invention of calculus behind him, the eminent mathematician-physicist had apparently tired of science, university politics, and life in a provincial college town. He was ready for a change of pace and threw himself into the work at the mint, then located in the Tower of London. There his day began at 4 a.m. with a mighty din when horse-powered presses began stamping out 50 coins per minute.

Yet it was during those years at the mint that two interesting Newtonian scientific achievements occurred. The first concerned a challenge problem issued by the Swiss mathematician Johann Bernoulli in June, 1696. Later called the brachistochrone problem, the challenge was to find the curve by which a bead, sliding down a frictionless wire, would get from point A to point B in the shortest time (see Figure 18.3). Its solution requires what is now termed the calculus of variations, and Bernoulli and his friend Leibnitz supposed that only their form of the newly-invented calculus was powerful enough to find the solution. Although Bernoulli addressed the test problem 'to all the mathematicians of Europe,' the real quarry was apparently Newton himself. Perhaps Bernoulli and Leibnitz doubted that the strange man who had abandoned science to run the mint was really such a genius after all.

Newton received his copy of the challenge from the Royal Society on 29 January. That day, recounted his niece, 'Sir I. N. was in the midst of the hurry of the great recoinage [and] did not come home until four from the Tower very much tired, but did not sleep till he had solved it.' The next morning her uncle sent the solution to the society, and it was printed anonymously. When Bernoulli saw the solution he exclaimed, '*Ex ungue leonem!*' – 'By the claw, the lion is revealed.' There was no question as to the author's identity.

In the same year that Newton became Warden of the Mint and in which Bernoulli issued the challenge, another Newtonian triumph was in the making. At the Royal Society meeting of 3 June 1696, member Edmond Halley produced the results of a calculation on the

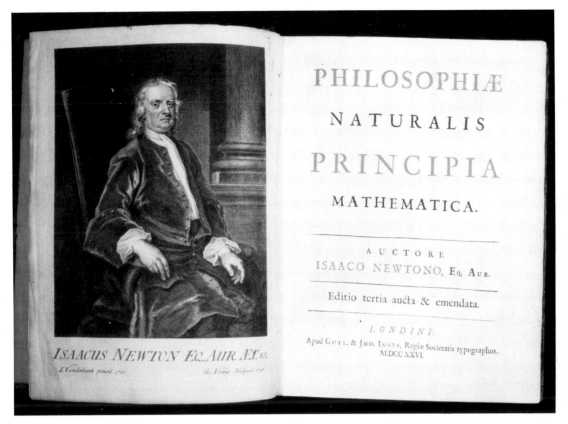

Figure 18.1 Sir Isaac Newton graces the frontispiece of the third edition of his Principia – *a masterpiece that might never have been written had it not been for Edmond Halley.*

Figure 18.2 Halley as depicted in the frontispiece of his Astronomical Tables.

comets of 1607 and 1682, 'which are in all respects alike, as to the place of the Nodes and Perihelia, their Inclinations to the plain of the Ecliptick and their distances from the Sun, whence . . . it was highly probable, not to say demonstrative, that these were but one and the same Comet, having a period of about 75 years, and that it moves in an Elliptick Orb about the Sun.'

Newton and Halley were stamped out of widely different molds. The former was an absent-minded professor, a near recluse, with (in Wordsworth's lines) 'a mind forever voyaging through strange seas of thought, alone.' The other was a congenial *bon vivant* and an astronomer who, in the opinion of straitlaced John Flamsteed, the Astronomer Royal, 'swore like a sea captain.' What course of events brought these gifted geniuses together?

Newton was born at Woolsthorpe Manor near Grantham on 25 December 1642. At age 18 he entered Trinity College in Cambridge, where he was what we might today call a work-scholarship student. Then came the plague year, when the university shut down, and during that enforced 18-month vacation Newton planted all the seeds of his future fame: mechanics, optics, the calculus, and gravitation. 'I was in the prime of my age for invention,' he later said, 'and minded Mathematics and Philosophy more than any time since.'

When the university reopened, Newton returned to Cambridge and began a swift climb up the academic ladder. First he was elected a fellow at Trinity College; then, astonishingly, his teacher Isaac Bar-

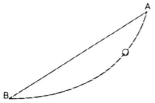

Figure 18.3 To test the newly discovered calculus, Johann Bernoulli posed this challenge problem to Isaac Newton: For a bead being acted upon by gravity, find the path by which it descends along a wire most swiftly from point A to point B. Along the cycloidal path, rapid initial acceleration is followed by a fast sideways motion. The bead reaches B more quickly than if it followed the straight-line path, where the acceleration is slower.

row resigned his mathematics professorship in favor of his former student!

Meanwhile, Newton's investigations into optics led him to invent the reflecting telescope. Shown to members of the Royal Society, the telescope created a sensation and brought about his immediate election. In return, Newton offered a paper on optics, which he described as 'the oddest if not the most considerable detection which hath hitherto been made in the operations of Nature.'

Newton's paper, on light and colors, is one of the most remarkable in the entire history of optics, but it did not escape criticism by Robert Hooke and others. If there was one thing Newton could not stand, it was criticism. He felt cheated and victimized and later claimed that 'Science is such a litigious lady that a man had as good be engaged in lawsuits as to have to do with her.' Newton tried to resign from the Royal Society, but others paid his dues for him. Nevertheless, he no longer played an active role in it.

In 1679 Hooke became the secretary of the society, and he soon attempted to draw Newton back into its circle by asking for his views on some questions of motion. Now this was also a topic that had already attracted Newton's attention. He replied at once, but all too hastily, for he blundered in making a sketch to show how a falling body might spiral to the center of the earth. Although Newton had invited objections, his pleasure 'evaporated like dew in August' (in the words of Newton's recent biographer, Richard S. Westfall) when Hooke publicly pointed out the error. Once more Newton withdrew to his private studies, especially of theology and alchemy.

In 1680 and 1681 the long hours with the books of Trinity College were broken by the appearance of two bright comets. Only Flamsteed believed they were one and the same, first seen approaching the sun and then seen being repelled by it. Over three years were to pass before Newton agreed with Flamsteed, and only then after some unexpected stimulation from Halley.

As a teenager Halley briefly assisted Flamsteed, who in 1675 had just been appointed Astronomer Royal. Then at age 20 Halley sailed to the island of St Helena, where he made a catalog of southern stars. He also created a new constellation, Robur Carolinum (Charles' Oak), named after the tree in which the monarch hid after the battle of Worcester. Naturally flattered, King Charles II ordered Oxford to waive the residency requirements and give young Halley his MA forthwith.

Promptly elected a member of the Royal Society, Halley was straightaway caught up in a controversy that secretary Hooke was having with the Danzig astronomer Johannes Hevelius concerning the efficacy of naked-eye observations. Although relatively inexperienced, Halley was sent to the Baltic to compare his own telescopic observations with those made simultaneously by the naked-eye master. The result was something of a standoff – though Halley was invariably a scintillating source of ideas, he seldom shone as an observer.

In 1682 Halley returned to England, married, and set up his household and an observatory in Islington, a suburb of London. Before he

Figure 18.4 At left is a silver
half crown circulated during
Newton's lifetime, revealing
edges trimmed by thieves. To
prevent this practice,
machine-stamped half
crowns (right) with milled
edges were minted during
Newton's term as Warden of
the Mint.

was really ready, another bright comet appeared. Halley's rather messy observations are preserved in his notebook, but when he came to study the comet in detail some 14 years later, he had to beg for better observations. To get those from Flamsteed, he sought Newton's intervention because 'he will not deny . . . you, though I know he will me.'

By living in the London area, Halley managed frequent contacts with members of the Royal Society like Hooke and the astronomer-turned-architect Christopher Wren. On one occasion the three of them discussed what mathematical form would describe the force drawing the planets to the sun. The trio agreed that it was likely to go as the reciprocal of the distance squared, the so-called 'inverse square' force, but none of them could prove it. Hooke was sure he could derive the relationship but claimed that the others would appreciate it more if they tried to solve the problem themselves. Wren thereupon offered a prize – a book worth 40 shillings – to whomever could solve the problem first. But it was not until eight months later, in August, 1684, that Halley finally went to Cambridge and put the question to Isaac Newton.

In Newton's later recollection, 'the Dr [Halley] asked him what he thought the curve would be that would be described by the planets supposing the force of attraction towards the sun to be reciprocal to the square of their distance from it. Sir Isaac replied immediately that it would be an ellipsis, the Doctor struck with joy and amazement asked him how he knew it, why saith he I have calculated it, whereupon Dr Halley asked him for his calculation without any further delay.' But Newton, having been twice bitten, was thrice shy and claimed not to be able to find the proof among his papers. Only after he had a chance to reconsider and expand his proof did he show his results to Halley.

Stimulated by this outside interest, Newton began to analyze the planetary motions in one draft after another. Near the end of 1684 the drafts change significantly: Newton finally realized that the planetary orbits would be only approximate ellipses, for the mutual attractions of the planets would perturb their paths. Comets, too, would be attracted, and surely the comets of 1680 and 1681 were but a single blazing star in a parabolic path with a sharp hairpin turn near the sun. At age 41 Newton had finally discovered *universal* gravitation.

Now driven at almost demonic intensity, Newton completed his *Principia* in about 18 months. Nevertheless, publication required cajoling, and eventually financing, from Halley. Thus the young astronomer from Islington played midwife to the greatest single book in the history of physical science.

Included within a dazzling array of applications of the laws of motion and gravitation was the parabolic or elliptical orbit of the comet of 1680, one that we would now call a sun-grazer. So, though the general outline of a geometrical approach was known, a great deal of skill and perseverance would be needed to work out the details for the better-observed comets of historical times. Halley, who had his hands full of meteorological, navigational, astronomical, and other scientific investigations, eventually decided to take up the task.

In September, 1695, he wrote to Newton saying that he was more and more convinced that the comet of 1682 had appreared three times since 1531. Since the intermediate appearance, in 1607, gave a period of 76 years for the first interval and only 75 for the second (from 1607 to 1682), Halley soon asked Newton to consider 'how far a Comet's motion may be disturbed by the Centers of Saturn and Jupiter, particularly in its ascent from the Sun, and what difference they may cause in the time of Revolution of a Comet.' Then, late in October, Halley wrote: 'I have almost finished with the Comet of 1682 and next you shall know whether that of 1607 were not the same, which I have more and more reason to suspect.' As we have seen, Halley became convinced enough of his results to announce them to the Royal Society the following spring.

But just as Halley was making brilliant progress with the comet computations, Newton became Warden and eventually Master of the Mint. In turn, he recommended Halley for an appointment at the Chester Mint, a supposedly lucrative post that mostly just gave Halley a headache. Halley soon became involved in other affairs, including two voyages to the South Atlantic as captain of the *Paramore Pink* to investigate variations of magnetic compass bearings, and a pair of secret diplomatic missions to the court in Vienna. These efforts, in turn, gave Halley highly placed patrons to assure his election as Savilian Professor of Astronomy at Oxford, a position once held by his elder friend Wren. And it was at Oxford in 1705 that Halley finally got around to writing up his comet researches. In a large but thin pamphlet first published in Oxford, Halley analyzed about two dozen comets, including those of 1531, 1607, and 1682, and announced his prediction that another would appear in 1758.

Halley was lucky to discover that the orbital elements of the three comets were similar. The comet that would later bear his name is the *only* really bright one among the 100 or so periodic comets now known. It represented Halley's only chance for finding multiple returns of a comet from the old naked-eye observations.

His discussion of the three appearances and his prediction comes off the best in his posthumously printed *Astronomical Tables*: 'You see therefore an agreement of all the elements in these three, which

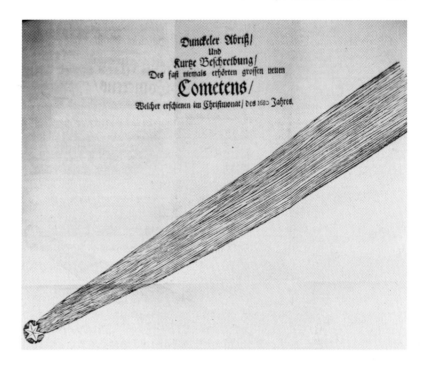

Figure 18.5 The Comet of 1680 as drawn on a broadside.

would be next to a miracle if they were three different comets. . . . Wherefore if according to what we have already said it should return again about the year 1758, candid posterity will not refuse to acknowledge that this was first discovered by an Englishman.'

Indeed, the comet was recovered on Christmas night, 1758, by Johann Palitsch, a Dresden amateur astronomer, and a few months later, in a session of the French Academy of Sciences, the astronomer Nicholas Louis de Lacaille referred to it as 'Halley's Comet.' Candid posterity has kept that name ever since.

Notes and references

This chapter is based on the author's public lecture given at the Charlottesville, Virginia, meeting of the American Astronomical Society. A particularly useful and authoritative source is Richard S. Westfall's *Never at Rest: A Biography of Isaac Newton* (Cambridge, 1980). The Newton–Halley letters are found in the multivolume *The Correspondence of Isaac Newton* (Cambridge, England, 1959–77). See also Colin Ronan, *Edmond Halley* (New York, 1969). I thank Craig Waff for the information that Lacaille was the first to use the name Halley's comet, in May, 1759.

Predicting the details of a solar eclipse requires not only a fairly good idea of the motions of the sun and moon, but also an accurate distance to the moon and accurate geographical coordinates. Rough determinations of eclipse occurrences became possible after the work of Claudius Ptolemy (around AD 150), and diagrams of the eclipsed sun are found in medieval manuscripts and in the first printed astronomical books.

However, the much more laborious calculation of an actual eclipse path across the surface of the earth did not become common until the eighteenth century. In fact, the maps shown in Figures 19.1 and 19.2 are the earliest such charts that I know about. Such sheets, printed on only one side, are called 'broadsides.' They first became common in the sixteenth century, when they were hawked by itinerant 'newsboys' much as today's newspapers are sold on city streets. Whereas in the sixteenth century the broadsides often contained the marvels and curiosities of nature (ranging from two-headed calves to frightening comets), by the early 1700s they contained rather more sophisticated information, sometimes of considerable scientific interest.

The two eclipse broadsides shown were prepared by Edmond Halley in 1715. The first alerted the English public to the detailed circumstances expected for the forthcoming total eclipse on 22 April. Thanks to Halley's promotional efforts, the eclipse was well observed, and it became possible afterward to pinpoint the actual path with considerable accuracy. The second broadside, issued 5 months later, showed where the path of totality had actually gone and also included a prediction of the eclipse of 1724. It is interesting that the actual path of the 1715 eclipse differed about 20 miles from Halley's prediction. The observations of the 1715 eclipse remain of value even today, and they have recently been used to demonstrate that the solar radius has apparently contracted by a third of an arc second since that time.

I am inclined to believe that Halley invented the idea of graphing the eclipse path on the earth, rather than showing the appearance of the sun in the sky from a particular place. One reason is that none of the competing English broadsides for the 1715 eclipse came in this particular form whereas the technique was used by others in 1724.

In 1724 both Halley and William Whiston (who had for some years been Newton's successor as Lucasian professor at Cambridge) published broadsides showing the eclipse path of 11 May. Their maps differed in the exact placement of the path, with Halley's track roughly 25 miles farther north than Whiston's. At least one observer noted his displeasure with Halley's prediction: this was the Philadelphia scholar and bookman James Logan, who was in London on a business trip at the time. Logan went out to Windsor Castle to view the eclipse, but as he noted on the corner of his copy of the Whiston broadside:

Figure 19.1 Edmond Halley's original prediction of the 1715 eclipse path, possibly the first printed eclipse track.

Clouds interposing about 3 or 4 in ye Afternoon of the 11th of May 1724 vast numbers who crowded out of London Westwd were disappointed and amongst others my Self who . . . rode out of town that day to Windsor to view the same on the Terrace of that Castle where we expected without fail to find it total, but tho the Sun was beclouded we were certain it was not total there, as Dr Halley then the Kings Astronomer had by his Map given ye world to expect, and therefore this Map of W Whiston's is by much the truest.

Figure 19.2 Note the differences in the 1715 eclipse path as actually observed, compared with Halley's prediction shown in Figure 19.1. The two broadsides were originally bought and dated by Narcissus Luttrell, the famous eighteenth-century collector of ephemera.

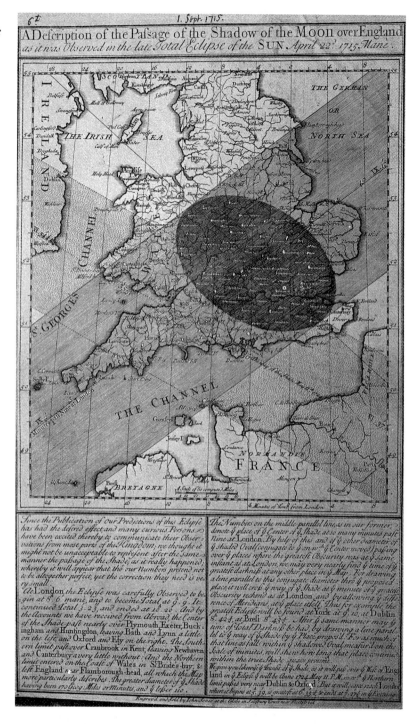

Later in the same week Logan visited the Royal Society, where he saw the venerable and shaky Newton in the presidential chair, and where, he remarked, 'Dr Halley, I observed, seem'd very well pleased, that by the sky being overcast, none of the Astronomers there had any account to give of it, only I having recvd a letter from my brother at Bristol mentioning the total darkness there, it was read

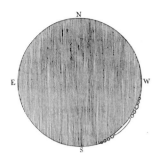

Figure 19.3 Baily's beads as observed by Samuel Williams from Islesboro, Maine, at the 1780 eclipse. From the Memoirs *of the American Academy of Arts and Sciences.*

publicly.' This was particularly ironic in light of the closing words on Halley's broadside: 'We wish our Astronomical Friends a Clear Sky.' Logan noted in the corner of the Halley broadside, somewhat cynically, that the *Philosophical Transactions* took no notice of the eclipse 'doubtless' out of courtesy to Halley. Despite Logan's caustic judgment, recent calculations made for me by Charles Kluepfel show that the actual edge of the eclipse path lay about midway between Halley's and Whiston's published positions.

Although cloudy weather diminished the impact of the 1724 event, eclipse observations in the eighteenth century generally proved of considerable value because they furnished a method for establishing the longitude differences between distant locations, and hence provided the foundation for better navigational charts. Finding longitude at sea was a major navigational problem at that time, and the English parliament had offered a large prize for a practical solution. The major reward of £20 000 went to John Harrison for the invention of the marine chronometer. However, an additional prize of £3000 was awarded for the lunar tables calculated by Tobias Mayer, because the fourth Astronomer Royal, Nevil Maskelyne, believed that lunar sightings from shipboard offered a legitimate solution to the longitude problem. In reality, the chronometer provided the practical solution, but Mayer's *Tables of the Motions of the Sun and Moon*, as published in 1770, did make eclipse prediction more precise.

Given the importance of eclipse measurements for longitude determinations, plus the intrinsic rarity and interest of total solar eclipses, it is not surprising to learn that intellectual leaders among the American colonists made a serious effort to observe such events as passed through their territory. The Philadelphia astronomer David Rittenhouse had hoped to go to Williamsburg, Virginia, for the eclipse of 24 June 1778, but wartime conditions prevented his journey. Shortly afterward Thomas Jefferson wrote to him that the observers had been clouded out.

The prospect of observing an eclipse from Maine in October, 1780, aroused great interest at Harvard College, in the newly founded American Academy of Arts and Sciences, and in the Massachusetts legislature. The situation was, however, complicated by the Revolutionary War then in progess and by the fact that the British had captured Maine. Nevertheless, a scientific truce was arranged long enough for a Harvard expedition to sail to Penobscot Bay where, on the day of the eclipse, the party was dismayed to find themselves just outside the edge of totality! The expedition leader, Prof Samuel Williams, claimed that false maps and an error in latitude accounted for the botch – a rather strange story considering that it is generally pretty easy to establish latitude but far harder to find longitude.

Precisely what went wrong has become the center of renewed attention because of the recent bicentennial of that event. Last September a small group of students and early-instrument buffs transported the original sextants, clocks, and telescopes from the Harvard Historical Scientific Instruments Collection to Williams' original observing site on Islesboro in Penobscot Bay. There they reenacted the original observations though, of course, without a solar

Figure 19.4 Williams' original records are no longer extant, but these observations of the clock corrections and of the partial eclipse by an unidentified observer in Massachusetts are still preserved in the academy archives.

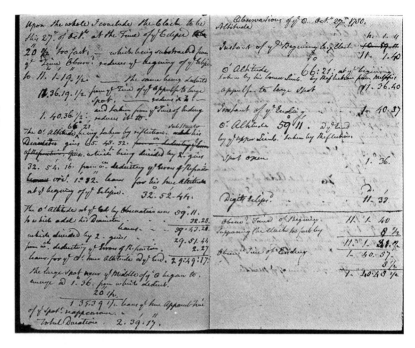

eclipse. They suffered some of the same difficulties that Williams must have encountered in getting the instruments set on stable foundations in a limited time, and in coping with the vagaries of the weather during their attempts to establish their latitude and longitude.

One of the participants in the reenactment, Robert Rothschild, a Harvard alumnus and resident of Islesboro, Maine, became particularly intrigued by Williams' failure to get his expedition within the path of totality. Williams was eventually driven out of Harvard because he tried to falsify some documents relating to a debt. With respect to the eclipse, was Williams also a scoundrel engaged in a cover-up to conceal his own incompetence? In the absence of any better explanation, Rothschild conjectured as much in an article in the January–February, 1981, issue of *Harvard* magazine.

Meanwhile, I had also become interested in several of the eighteenth-century eclipses and had asked Alan Fiala of the US Naval Observatory to recompute the 1780 eclipse path. His calculations showed that the 75-mile wide eclipse path indeed missed Islesboro, by a good 20 miles.

Next, I made a survey of the maps that might have been available to Williams in 1780. The best maps of the area – and they are very good indeed – were a set of harbor and coastline charts made in 1776 by Joseph Des Barres, but in light of what happened in 1780, I have to conclude that Williams probably never saw these. By far the most common map with enough detail to serve Williams' purposes was the so-called Mitchell's map of British North America. I first examined a copy of this map in the Library of the American Philosophical Society in Philadelphia.

When I looked at the coordinates for Penobscot Bay and for

Figure 19.5 The instruments and participants at the bicentennial reenactment of the 1780 solar eclipse. From left: Eben Gay, his wife, Gordon Schiff, Barbara Hughes, and Patri Pugliese.

Islesboro (then called Long Island) I was taken by surprise, for the map placed them near to the correct longitude, but about 20 miles too far north, just contrary to the expected nature of the errors. My surprise was all the greater because Rothschild had also used Mitchell's map, even reproducing a section of it in his article, and he had not noticed such a problem. A careful comparison showed that the map Rothschild used, from the New York Public Library, was printed from the very same plate, but with a different state of the engraving, so that Penobscot Bay was actually shown at differing places in the two printings. Back at Harvard, I checked its two copies of Mitchell's map, and found that both of them agreed in placing Penobscot Bay too far north.

Supposing that Williams had used this form of the map. I plotted the Naval Observatory's latitude and longitude coordinates of the eclipse path (see Figure 19.6). On Mitchell's map, the path fell directly across the northern part of Penobscot Bay, just barely missing Islesboro. Now it seemed that the missing key lay in the computations Williams would have made. If his calculations had placed the eclipse about 20 miles farther to the southwest, and if he had used the faulty Mitchell map, then Islesboro would have been an obvious destination for the expedition.

In 1780 the proper basis for eclipse calculations would have been the 1760 *Tabulae Lunarum* of Tobias Mayer, which were in use in America at that time. For example, the copy at the American Philosophical Society contains contemporary notes about the eclipse of 1778, and Houghton Library at Harvard holds a manuscript copy made by Joseph Willard, probably before he became president of the college in 1781.

Rothschild, realizing the importance of recomputing from Mayer's tables if the mystery were ever to be properly resolved, persuaded Kluepfel to carry out the task. I supplied the latter with a copy of the tables, and with astonishing speed he came up with the results: according to Mayer's tables, the eclipse path would fall 20 miles southwest of the modern calculations. As the reader can immediately

Figure 19.6 On this reproduction of Mitchell's map, courtesy Harvard Map Room, are plotted the central line and the edges of the path of totality for the October, 1780, solar eclipse as derived from modern elements (right) and from those in Tobias Mayer's Tabulae Lunarum *(left).*

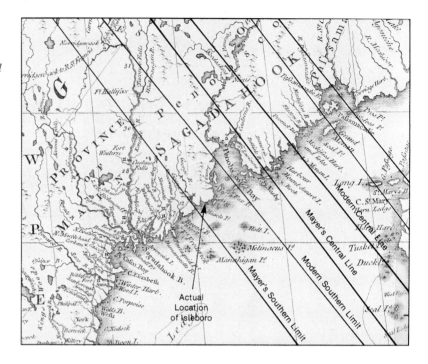

see on the reproduction of the Michell map, this puts Islesboro well within the eclipse path.

In reporting on the eclipse in the first volume of the *Memoirs* of the fledgling American Academy of Arts and Sciences, Williams claimed that the maps had misled him to a position too far south. Given the great likelihood that Williams chose his destination from Mitchell's map, then his claim appears vindicated, although he perhaps exaggerated the size of the error. Once on Islesboro, a week before the eclipse, he must have appreciated the latitude error as soon as he took a noon sight of the sun, but his hands were tied. After arriving in Penobscot Bay, Williams wrote: 'The vessel was directed to come to anchor in a cove on the east side of Long Island. After several attempts to find a better situation for observations, we fixed on this place as the most convenient we had reason to expect.' The British commander had given him strict instructions to stay put, and in the minimum time remaining he would scarely have had a chance to renegotiate the position of his observing site.

Furthermore, the calculations from Mayer's tables indicated to Williams that although he would be off the central line, he would by no means be outside the path of totality. As things worked out, Williams had no way of calculating the longitude until the eclipse itself. 'I could have wished to have had some observations of the eclipses of Jupiter's satellites, and of the occultations of the fixed stars by the moon. But no observations of this kind could be made.' Besides, the version of Mitchell's map that Williams was undoubtedly using also placed Islesboro about 10 miles too far east, which made the situation look more favorable than it actually was. We can imagine Williams' chagrin when he realized that his expedition had narrowly missed totality despite his carefully laid plans!

There remains still one mystery. Williams' published report describes the circumstances seen with the 90-power Short reflecting telescope:

After viewing the sun's limb about a minute, I found almost the whole of it thus broken or separated in drops, a small part only in the middle remaining connected. This appearance remained about a minute, when one of my assistants, who was looking at the sun with his naked eye, observed that the light was increasing. [Within about 20 seconds] it was evident that the broken parts of the sun's limb began to increase and unite.

This is the phenomenon that later came to be called the Baily's Beads. Yet his observing site lay about 20 miles southwest of the edge of totality as given by modern calculations, seemingly too far for the beads to be observed at all. Furthermore, all of the Islesboro observers reported beginning and ending times about 40 seconds later than given by these modern calculations. Perhaps there is still something to be learned about where the eclipse of 27 October 1780 actually fell.

Notes and references

Most of the information about Logan is found in Edwin Wolf's *The Library of James Logan of Philadelphia* (Philadelphia, 1974). The two broadsides annotated by Logan are today at the Pennsylvania Historical Society. For a modern use of the 1715 observations, see 'Observations of a Probable Change in the Solar Radius Between 1715 and 1979,' by David W. Dunham *et al.*, *Science*, **210** (1980) 1243–5.

I acknowledge with thanks the numerous discussions of the 1780 eclipse expedition held with Gordon Schiff, who wrote his senior honors thesis at Harvard on Williams, and with Robert Rothschild; the calculations made especially for this article by Alan Fiala of the US Naval Observatory and by Charles Kluepfel; and the help afforded by William Forbush, the American Philosophical Society Library, the Harvard Map Room, and Harvard's Houghton Library.

An error in our map, pointed out by Robert Rothschild in 'Where Did the 1780 Eclipse Go,' *Sky and Telescope,* **63** (1982), 558–60, has been corrected in this reprinting.

20 *The 1784 autobiography of William Herschel*

By the summer of 1780, the German immigrant William Herschel already had a considerable local reputation as a versatile musician in the English resort city of Bath. Fourteen years before, he had come as organist for the fashionable Octagon Chapel, but his musical talents included more than playing the organ. As his sister Caroline wrote in her diary,

He composed glees, catches, &c., for such voices as he could secure . . .; sometimes he gave a concerto on the oboe, or a sonata on the harpsichord. . . . He took great delight in a choir of singers who performed the cathedral service at the Octagon Chapel, for whom he composed many excellent anthems, chants, and psalm tunes.

Herschel's talents were not only musical; by 1780 he had also become an avid natural philosopher, amateur astronomer, and telescope maker. Elected to the Bath Philosophical Society, he regularly contributed papers on a wide variety of topics. His answer to a musical mathematical challenge question had been printed in the *Ladies Diary*.

But Herschel's merely local reputation in the summer of 1780 contrasts markedly with the international acclaim he achieved unexpectedly the following year. By the end of 1781 he had been elected a fellow of the Royal Society, had been awarded one of its gold medals, and had begun a correspondence with men of science throughout Europe. In addition, he was well on his way toward receiving a royal stipend from King George III. What had transformed Herschel into a world celebrity was his discovery of a new planet, called by the English 'Georgium Sidus' or 'the Georgian,' and by the rest of the world Uranus.

Needless to say, people were filled with curiosity about this hitherto little-known amateur astronomer and his wonderful discovery, which he had accomplished with a telescope of his own making. Late in 1784 Charles Hutton, a mathematics professor at the Royal Military Academy and editor of the *Ladies Diary*, notified Herschel that another magazine had commissioned a portrait and proposed to publish the picture together with an account of Herschel himself. It was its custom, Hutton reported, 'not to satyrise or abuse Gentlemen, but to do them honour.' Hutton warned Herschel that publication could not be prevented and that he should therefore furnish them with accurate particulars of his life rather than letting the magazine fill its columns with such anecdotes as it might happen to find.

Fortunately for the readers of *The European Magazine*, and for us today, Herschel chose to respond. Although his original letter to Hutton and the magazine has long since vanished, William and

Figure 20.1 An engraving of William Herschel from The European Magazine.

Caroline made a fair copy to keep for their own records. A few years ago this precious manuscript was acquired by the Houghton Library at Harvard University, and a full transcription has been published in the *Harvard Library Bulletin*.

Here is an extended excerpt of this fascinating account, written by the 46-year-old Herschel shortly after he had abandoned his musical career and moved to Datchet (in August, 1782) to be nearer the royal household at Windsor. Entitled 'In a letter to Dr Hutton,' it reads in part as follows:

I was born in the City of Hannover November 15, 1738. My Father was a musician & I was the second of four Sons whom he all brought up to the same business from their very infancy. Having also two daughters, so numerous a family would not permit my Father, with his scanty circumstances, to bestow much on the education of his children & the utmost he could do was that besides the common learning of a school such as reading, writing & arithmetic, he provided me with a private instructor for the French language. The person who undertook this care, finding me a very ready memory so as to give him no trouble in that line, and being himself a man of Science, chiefly devoted his time to encourage the taste he found in his pupil for the study of philosophy, especially Logic, Ethics & Metaphisics, which were his own favourite pursuits. To this fortunate circumstance it was undoubtedly owing that altho' I loved Music to an excess & made a considerable progress in it, I yet determined with a sort of enthusiasm to devote every moment I could spare from business to the pursuit of knowledge which I regarded as the sovereign Good, & in which I resolved to place all my future views of happiness in life. Thus several years were spent in study & in the practise of my profession, to which I was introduced at a very early period, till the troubles in the Electorate of Hannover, during the last war, made my situation there very uncomfortable.

At this time George II, a Hanoverian, was on the English throne; Herschel's father was a musician in the Hanoverian Guards. In the 1740s Europe was plunged into the War of Austrian Succession, owing in the first instance to the ambition of the new Prussian king, Frederick the Great. After a brief respite in the early 1750s, the Seven Years' War began, in which Hanover (and England) sided with Prussia against France and Austria. Fear of a French invasion became so great in England that the Hanoverian regiment (including the band) was briefly sent to England in 1755. The following year young William was actually under fire in the battle of Hastenbeck near Hanover.

The known encouragement given to Music in England determined me to try my fortune abroad & accordingly about the year 1759 I came to settle in this country where I had before passed some months in company with my Father & eldest brother being then in his Majesty's Service. The difficulty of succeeding in London induced me to visit some places in the country and after some years in Newcastle, Leeds &c: I was chosen Organist at Hallifax in Yorkshire. During all this time, tho' it afforded not much leisure for study I had not forgot my former plan but had given all my leisure hours to the study of languages. After I had improved myself sufficiently in English I soon acquired the Italian, which I looked upon as necessary to my business; I proceeded next to Latin & having also made a considerable progress in that language I made an attempt of the Greek, but soon after dropt the pursuit of that as

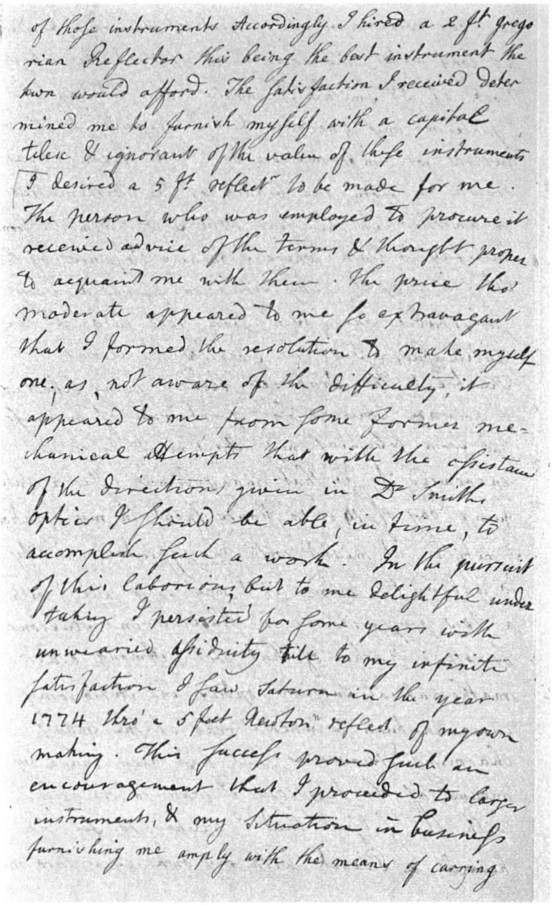

*Figure 20.2 A portion of
Herschel's manuscript
autobiography.*

*Figure 20.2 A portion of
Herschel's manuscript
autobiography.*

leading me too far from my other favourite studies by taking up too much of
my leisure.

The theory of music being connected with mathematics had induced me
very early to read . . . all what had been wrote upon the subject of Harmony;
& when, not long after my arrival in England the valuable book of Dr Smith's
Harmonics came into my hands I perceived my ignorance & had recourse to
other authors for information by which means I was drawn on from one
branch of the mathematics to another.

In the year 1766 I was removed from Hallifax to Bath where I became
Organist of the Octagon Chapel. My situation proved a very profitable one, as
I soon fell into all the public business of the Concerts, the Rooms, the Theatre
& the Oratories, besides many Schollars & private Concerts. The great run of
business far from lessening my attachment to Study encreased it, so that
many times after a fatiguing day of 14 or 16 hours spent in my vocation I
retired at night with the greatest avidity to unbend the mind (if it may be so
called) with a few propositions in Maclaurin's Fluxions or other books of that
sort.

Among other mathematical subjects, optics and astronomy came in turn, &
when I read of the many charming discoveries that had been made by means
of the telescope I was so delighted with the subject that I wished to see the
heavens & planets with my own eyes thro' one of those instruments. Accordingly, I hired a 2-ft Gregorian reflector, this being the best instrument the
town would afford. The satisfaction I received determined me to furnish

myself with a capital telescope & ignorant of the value of these instruments, I desired a 5-ft reflector to be made for me. The person who was employed to procure it received advice of the terms & thought proper to acquaint me with them. The price, tho' moderate, appeared to me so extravagant that I formed the resolution to make myself one; as, not aware of the difficulty, it appeared to me from some former mechanical attempts that, with the assistance of the directions given in Dr Smith's optics, I should be able, in time, to accomplish such a work.

In the pursuit of this laborious, but to me delightful, undertaking I persisted for some years with unwearied assiduity till to my infinite satisfaction I saw Saturn in the year 1774 thro' a 5-foot Newton reflector of my own making.

In fact, Herschel's rather terse diary from these years showed that he first bought some lenses and tin tubes in the spring of 1773, then for several months rented the 2-foot (focal length) reflector. On 22 September he bought tools and a metal casting for making his own mirror. Such was his success that he soon made reflectors of 7-foot, 10-foot, and finally 20-foot focal lengths. Indeed, he made over 200 mirrors of the 7-foot variety, until 'at length I obtained one that would bear any power I could apply to it.'

All this while I continued my astronomical observations & nothing seemed now wanting to compleat my felicity than suficient time to enjoy my telescopes, to which I was so much attached that I used frequently to run from the Harpsichord at the Theatre to look at the stars during the time of an act & return to the next Music. To this perseverence at length was owing the discovery of the Georgium Sidus, which happened on the 13 of March 1781.

It has generally been supposed that it was a lucky accident that brought this star to my view. This is an evident mistake. In the regular manner [that] I examined every star of the heavens, not only of that magnitude but many far inferior, it was that night *its turn* to be discovered. I had gradually perused the great Volume of the Author of Nature & was now come to the page which contained a seventh planet. Had business prevented me that evening, I must have found it the next, and the goodness of my telescope was such that I perceived its visible planetary disk as soon as I looked at it; And by the application of my micrometer determined its motion in a few hours.

Herschel's amateur astronomical program was an ambitious one. To professionals of the day, the sky was primarily a two-dimensional backdrop against which the planets and comets ran their courses. Herschel, however, hoped to discover the three-dimensional structure of the Milky Way. He planned to use double stars in an ingenious scheme to establish the actual scale of the sidereal system, so he carefully inspected star after star looking for new doubles. Because he was using a high magnification (so high that most of his contemporaries were incredulous), he immediately recognized the unusual disklike appearance of the new planet.

By this time George III, a well-known patron of the arts and sciences, was on the throne, and thus Herschel sagely named his discovery after him.

In the Spring of the next year his Majesty wished to see my telescope &, ever ready to encourage arts & sciences, was pleased with his Royal Goodness to take me from my former employment that I might devote myself intirely to

Astronomy. In Jan[uary] 1781 I began to make a 30-ft Newtonian reflector & soon after twice cast a speculum of 36 inch diameter but neither of them succeeded at that time. The interruption also of my removal from Bath [to Datchet] put a temporary stop to the work. . . . I am however still in hopes soon to be able to resume & perhaps to go beyond my former attempt of a larger instrument.

Herschel's letter to Hutton is his major autobiographical account and, along with Caroline's diary, is the primary source documenting his transition from professional musician to professional astronomer. It ends just as he was on the threshold of a series of brilliant astronomical investigations made with his 20-foot reflector. The larger telescope alluded to in his closing paragraph – a 40-foot instrument with a 48-inch speculum mirror – was eventually completed in 1789.

By then Herschel had turned his back on a musical career; his decision to abandon music for astronomy was a clear choice between competence and genius. In time he not only discovered satellites of Uranus and Saturn, the direction of the sun's motion in space, infrared radiation, hundreds of double stars (and their true binary nature), and thousands of nebulae, but also actually succeeded (at least qualitatively) in delineating the form of the Milky Way galaxy.

Today Bath maintains its festive air, an arcaded Georgian city with its Roman baths and a two-century-old pump room, and the Herschel Society has rehabilitated the old family residence on New King Street. The Houghton Library's Herschel autobiography brings alive the events taking place there just over 200 years ago and gives us a graphic glimpse into the remarkable career change of the greatest observational astronomer of the eighteenth century.

Notes and references

This chapter is primarily a condensation of my article in the *Harvard Library Bulletin*, **32** (1984), 73–82. Here Herschel's text has been slightly abridged, somewhat modernized with respect to punctuation and capitalization, and quoted with the permission of Houghton Library. Among the other Herschel manuscripts at Harvard are the fair copies of the contributions that Herschel prepared for the Bath Philosophical Society, and 'Account of a Comet' from 22 March 1781, Herschel's earliest description of the discovery of Uranus.

Caroline Herschel's quotation is from Mrs John Herschel, *Memoir and Correspondence of Caroline Herschel* (New York, 1876), p. 36. An interesting discussion of Herschel's career as a musician by Sterling E. Murray, together with the scores for three of his symphonies, is found in *The Symphony 1720–1840*. Series E. Vol. III (Barry S. Brook, editor, New York, 1983).

21 *The Great Comet that never came*

In 1857 the populace of Paris and much of France was anticipating, with increasing dread, the imminent end of the world. Newspapers contained forecasts of the impending doom, scheduled for 13 June, when a comet was expected to collide with the earth.

In retrospect, this farcical episode was set in motion, quite unwittingly, by the eighteenth-century English astronomer Richard Dunthorne and the French cometographer A.-G. Pingré, though some elements of the story go back to the Middle Ages. One of the grandest comets ever recorded had appeared in the middle of 1264. Both European and Chinese historians regarded it with wonder and astonishment. A truly brilliant comet, its tail stretched over 100° from the eastern horizon past the meridian.

Almost 292 years later, in 1556, another spectacular comet, rivaling Jupiter in brightness, crossed the skies. Again, both Chinese and European astronomers documented its progress. Although these pre-telescopic observations were not of the precision attained in later centuries, it was still possible for Edmond Halley to include a rough orbit for the 1556 comet among the 24 listed in his *A Synopsis of the Astronomy of Comets*.

In 1751 Dunthorne was persuaded to try to determine the orbit for the comet of 1264; to his surprise, the results were similar to those Halley had given for the comet of 1556. He immediately concluded that the comets were one and the same, and he surmised that a reappearance might be expected in 1848. Around 1760 Pingré collected all the accounts he could locate about the comet of 1264, finding many unknown to Dunthorne. His conclusions matched and greatly strenghthened those of his English predecessor, and he identified these two comets with another remarkable object that had appeared 289 years earlier, in 975. This led Pingré to predict the comet's return in about 1848.

The next episodes of this intriguing story took place as the time of the expected return approached. The English astronomer J. R. Hind investigated anew the question of identity of the comets, convincing himself that those of 1556 and 1264 were one and the same. He noted, however, that the comet's path had been perturbed by Saturn and Neptune and that its return would be delayed. When it had failed to return by 1849, a fresh determination was made by the Belgian astronomer B. Bomme with rather contradictory results. Writing in *The Comets* (1852), Hind looked forward to a confirmation ('or otherwise') of the calculations sometime between 1856 and 1860.

Then, a German astrologer, who as yet is unidentified, not only predicted the comet's return in 1857 but insinuated that it would collide with the earth on 13 June. It is difficult to know just how much mass hysteria was produced, or why it seems to have been concentrated in Paris. Three months before the anticipated event, the *Illustrated London News* reported from Paris that:

Figure 21.1 What went wrong when the Great Comet of 1857 failed to appear? In 1856 the Viennese astronomer Karl von Littrow discovered a copy of the rare pamphlet shown here, a contemporary account of the comet of 1556 written by Joachim Heller. With these additional observations, M. Hoek proved conclusively that the comets of 1556 and 1264 were not *the same, though perhaps of the same comet family.*

At the last reception at the Tuileries, the coming comet was one of the subjects of conversation. Her majesty the Empress seeing M. Leverrier, the well-known astronomer, among the guests, determined to make fun of the unsuspecting savant, and, feigning great alarm at the impending destruction of our globe, which this 'extravagant and erring' luminary is, according to a German stargazer, to accomplish on the 13th of June, consulted him on the subject. M. Leverrier, to the great amusement of the guests, entered into a long refutation of this notion; and his embarrassment in endeavouring to avoid accusing her Majesty, who ill-naturedly would not be persuaded, and his scientific enthusiasm made the evening pass much more merrily than is usually the case with Imperial soirées – generally solemn stiff affairs.

In America, *Harper's Weekly* took note of the 'recent prophecy that this earth of ours is to be annihilated by a comet.' It used the occasion for a long review of the 'already ascertained facts' concerning comets in general and concluded with the authority of the French celestial mechanican Jacques Babinet, who said: 'the earth, in coming into a collision with a comet, would be no more affected in its stability than

Figure 21.2 'La Comète, tableau de la fin du monde, 13 juin 1857,' drawing by C. Mettais in L'Illustration *for 21 March 1857. The grapes and cask of wine in the foreground refer to the particularly good vintage associated with the comet of 1811.*

would a railway train coming into contact with a fly.'

The newspaper went on to recount the history of this particular prediction:

Meantime, a Belgian almanac-maker took the matter in hand, and predicted that this comet would strike the earth on the 13th of June. From this gentleman's almanac-prediction have come all the rumors, the alarms, and excitments of the present year. A Paris correspondent writes: "For a fortnight we have not been able to step out without hearing the cry, 'Here is the end of the world! a full description of the comet of June 13, only one sou!' " Women have miscarried; crops have been neglected; wills have been made; comet-proof suits of clothing have been invented; a cometary life insurance company (premiums payable in advance) has been created; and our 'Man About Town' has fancied himself walking the streets with a veritable blazing star – all because an almanac-maker of Liege thought proper to insert, under the week commencing June 13, 'About – this – time – expect – a – Comet.'

Figure 21.3 'Mister, I am leaving your service. . . . I am returning to my home. . . . I don't want to be found here at the end of the world.' No. 4 of the comet series, published 13 March 1857, shows Daumier's sharp-witted perceptions of Parisian types.

Figure 21.4 'The German astronomer loosing a famous canard.' Here Daumier puns on the French word 'canard,' which means both a duck and a hoax. No. 6 of the comet series, 17 March 1857.

In Paris the 'Universal Journal' *L'illustration* produced a marvelous full-page tableau of the end of the world in its issue of 21 March. The picture, shown in Figure 21.2, includes an allegorical allusion to the popular superstition that comets made for good wine. Even the philosopher G. W. F. Hegel wrote, 'I once made Mr Bode sigh by saying that we know from experience that comets are followed by good vintage years, as was the case in 1811 and 1819, and that this double experience was just as good, if not better, than that regarding the return of comets.' Hegel was underscoring his belief in the flimsiness

Figure 21.5 'Believing it to appear,' No. 10 in the comet series, published in Le Charivari, *23 March 1857.*

Figure 21.6 In a final touch, when a real naked-eye comet appeared in 1858, Daumier poked fun at the astronomer J. Babinet and the others who looked for the comet that never came. Published 22 September 1858.

of empirical observations, and he no doubt would have taken the events of 1857 as further confirmation of his scepticism.

L'illustration, in fact, made little mention of the anticipated calamity, only pausing to remark in its 13 June issue that:

The alarmists are furious, no one believes any more in the end of the world tomorrow, and the spectacle is indefinitely postponed. Where is the comet that would play the leading role in this extraordinary performance? Pursued by the telescopes of learning, the wandering star has escaped from them into the solitude of the heavens; and when ignorance in its turn begs M. Hume to invoke the phenomenon, this celebrated medium had the modesty to recognize that his phantasmagoria can do nothing. We Parisians, however, are not reduced to waiting for a catastrophe for our diversion: our summer holidays can pass without the cooperation of any comet.

Without doubt the most significant legacy of the Great Non-Comet of 1857 was a series of cartoons drawn by the caricaturist Honoré Daumier for the witty tabloid newspaper *Le Charivari*. Daumier is today best known for his incisive portrayal of mid-nineteenth-century French society in his thousands of satirical lithographs. Between 4 March and 25 March, Daumier contributed to *Le Charivari* a numbered series of ten full-page cartoons entitled 'La Comète de 1857.' He had already alluded to the comet in three other cartoons in February, in his ongoing series 'Croquis Parisiens' and 'Actualités.'

In April and May Daumier lampooned the French Academician Babinet concerning a prediction of the sun's eventual demise, and some of these were combined with references to the still-expected comet. Babinet took the brunt of this satire not so much because he was a famous celestial mechanician and physicist, but because he was well known to the public through his popular science writing in Parisian newspapers. By 15 June 1857, there was clearly no comet to poke fun at, and Daumier's comet references came to an end – almost.

In 1858 a really bright comet *did* appear, but it had nothing to do with those of 1264 and 1556. Daumier did not necessarily publish his cartoons in the order in which he drew them, so we cannot help but wonder if he had another lying in reserve. Soon after Donati's comet became a naked-eye splendor, Daumier published what has become one of his most famous cartoons, 'Monsieur Babinet anticipating the visit of the comet with his portable telescope.' In this 22 September lithograph, Babinet peers off to the left with his refractor, totally oblivious to the brilliant spectacle being pointed out by an excited Parisian woman who is tapping him on his shoulder. There can be little doubt that Daumier is taking his last sly laugh at all those who were looking so hard for the great comet that never came.

Notes and references

I am particularly grateful to Philippe Veron and Gustav Tammann for sharing their enthusiasm for this story and the results of their sleuthing; their account of this comet is found in 'Astronomical broadsheets and their Scientific Significance,' *Endeavour*, New Series, **3** (1979), 163–70. Thanks also to Brian Marsden, Barbara L. Welther, and Donald Yeomans.

22 *Unlocking the chemical secrets of the cosmos*

In 1844, two years after completing his famous multivolume *Course of Positive Philosophy*, the French philospher Auguste Comte published a further treatise on the philosophy of astronomy. At the very outset he proclaimed his views on the limitation of this science: 'The stars are only accessible to us by a distant visual exploration. This inevitable restriction therefore not only prevents us from speculating about life on all these great bodies, but also forbids the superior inorganic speculations relative to their chemical or even their physical natures.' He thus echoed his earlier claim that 'Men will never encompass in their conceptions the whole of the stars.'

Little did Comte suspect that, within another generation, unanticipated new scientific discoveries would prove his declarations false. These surprising developments took place with astonishing swiftness in the German university town of Heidelberg, beginning late in 1859. Some of the initial excitement is still conveyed by a letter written on 15 November that year by the chemist Robert Bunsen to his English colleague H. E. Roscoe:

At present Kirchhoff and I are engaged in an investigation that doesn't let us sleep. Kirchhoff has made a wonderful, entirely unexpected discovery in finding the cause of the dark lines in the solar spectrum, and he can increase them artificially in the sun's spectrum or produce them in a continuous spectrum and in exactly the same position as the corresponding Fraunhofer lines. Thus a means has been found to determine the composition of the sun and fixed stars with the same accuracy as we determine strontium chloride, etc., with our chemical reagents. Substances on the earth can be determined by this method just as easily as on the sun, so that, for example, I have been able to detect lithium in twenty grams of sea water.

Bunsen had met Kirchhoff in Breslau in 1851, where the young physicist had the equivalent of an assistant professorship. In the following year Bunsen moved to Heidelberg, where he became director of the Chemical Institute, and two years later he arranged for Kirchhoff to take up the physics professorship there. Just as Faraday has been called Davy's greatest discovery, it is sometimes said that Bunsen's was Gustav Kirchhoff.

During the 1850s Bunsen developed his well-known burner. Because of its colorless flame, it quickly became a standard piece of equipment for identifying elements by the colors that various chemicals imparted to the flame. Bunsen at first used filters for distinguishing the characteristic tints, but Kirchhoff suggested using a prism to produce a spectrum. In essence, they invented the spectroscope by combining, for the first time in a single piece of equipment, a collimator, prism, and eyepiece. It was with the Bunsen burner and the new spectroscope that optical chemical analysis really began. With these tools the Heidelberg scientists found that each element gave its own

Figure 22.1 This portrait was taken in 1862 and shows, from left to right, G. Kirchhoff, R. W. Bunsen (seated), and H. E. Roscoe. Reproduced from the latter's The Life and Experiences of Sir Henry Enfield Roscoe *(London, 1906).*

spectral pattern of bright colored lines, each pattern as unique as an individual fingerprint.

By a brilliant series of experiments undertaken in the fall of 1859, Kirchhoff extended the analysis from the laboratory to the cosmos itself. Neither Bunsen nor Kirchhoff ever published any personal account concerning the origin of these discoveries, and only scattered anecdotes remain. However, David Robinson, one of my undergraduate students a few years ago, pointed out a particularly suggestive note that was published in a 1902 issue of *Nature*.

Heidelberg lies nestled along the Neckar River on the fringe of the Black Forest, and from the surrounding heights it is possible to look across the Rhine plain toward Mannheim, a bustling port city about 10 miles to the west. One evening as Kirchhoff and Bunsen were looking out of their laboratory window, they noticed a fire raging in Mannheim, and with the spectroscope they could detect the presence of barium and strontium in the flames. Some time later, when they were on one of their frequent strolls along the 'Philosopher's Walk' through the wooded hills north of the Neckar, Bunsen mused that if they could analyze the fire at Mannheim, why could they not do the same for the sun? 'But,' he added, 'people would think we were mad to dream of such a thing.'

Quite possibly this incident was the origin of Kirchhoff's researches into the nature of the dark Fraunhofer lines in the solar spectrum. Much earlier in the century, when Fraunhofer had mapped these dark lines, he had noticed the coincidence of a conspicuous dark line in the yellow region of the solar spectrum (which he designated with the letter D) and the bright yellow line that almost always turned up in the spectrum of a flame. The laboratory origin of the yellow flame was elusive, however; it was not then known that the yellow color arose solely from the sodium in common salt. The situation was confused because salt was a ubiquitous contaminant. However, Bunsen, being an excellent chemist, had been able to prepare much purer reagents; hence both he and Kirchhoff realized that the yellow line was a specific characteristic of sodium.

Kirchhoff decided to test as directly as possible whether the bright sodium line obtained in the laboratory was really coincident with the dark solar line, and so he placed a colored flame in a weak beam of sunlight. He was then able to fill in and thus cancel the dark line, therefore proving the coincidence of wavelength. But when he increased the intensity of the solar spectrum, to his great astonishment he found that the dark D line became darker when the colored flame was interposed. In a note to himself, Kirchhoff concluded that this was either nonsense or 'a very great thing.'

By the next morning, Kirchhoff had concluded that it might indeed be important, and so he replaced the beam of sunlight with a lamp giving a continuous spectrum. As the excitement mounted, Bunsen is said to have exclaimed, 'If the dark line really appears, I think I will go mad!' It is quite possible that the dark line did not at first appear, because the temperature of the Bunsen burner flame was too high compared to that of the Drummond limelight providing the continuous spectrum. But Kirchhoff persisted and soon found that with a cooler alcohol burner he could artificially produce the dark sodium line.

Kirchhoff promptly prepared a 600-word article that was read before the Berlin Academy on 27 October 1859. His announcement included the fact that the flame had to be cooler than the source producing the continuous spectrum, and in his very next paper he announced his famous law of radiation, that the ratio of the coefficients of emission and of absorption is a function of wavelength and temperature, and is the same function for all bodies. A third paper, in January, 1860, developed the theory more rigorously, but meanwhile he had also begun a meticulous mapping of the solar spectrum together with the laboratory spectra of a variety of elements. By the following year Kirchhoff had detected sodium, calcium, magnesium, iron, chromium, nickel, barium, copper, and zinc in the sun.

Kirchhoff's explanation of the Fraunhofer lines had barely been announced when he and Bunsen added to their triumph the spectroscopic discovery of a new terrestrial element, named cesium after its brilliant blue spectral lines. The following year, in 1862, they

Figure 22.2 The Drummond limelight, as depicted in H. Schellen's Spectrum Analysis *(London, 1872), produced its brilliance by the incandescence of a baked-lime cylinder heated by an oxyhydrogen jet flame. This is the famous theatrical illumination that gave rise to the expression, 'being in the limelight.' On the screen undispersed light would fall at d¹ and the spectrum extends from r (red) to v (violet).*

announced yet another new element, named rubidium for its red lines.

In retrospect, it is fascinating to notice that the same critical experiment that Kirchhoff performed in Heidelberg in the fall of 1859 had already been performed in Paris a decade earlier, by the physicist Léon Foucault. Like Kirchhoff, Foucault had wanted to check the coincidence of the solar D line with the bright yellow line that appeared in the spectrum of a carbon arc, and he found quite unexpectedly 'that the arc, placed in the path of a beam of solar light, considerably strengthens the solar D line when the two spectra are exactly superimposed.'

Nevertheless, Foucault failed to exploit this curious finding. Apparently he had no clue that the bright D line came from sodium contamination of his carbon arcs. Furthermore, unlike Kirchhoff, Foucault had no explanation for what he saw, no suggestion that the terrestrial absorption line and the solar absorption line were due to the presence of the very same chemical element.

A few of the English physicists became aware of Foucault's finding, but they, too, in the absence of the colorless Bunsen burner flame and the pure salt-free reagents, did not recognize the profound significance of the observation. It was very much the fortunate combination in Heidelberg of a skilled chemist and a physicist seeking generalizations that in 1859 gave birth to astrophysics – or, perhaps more properly, astrochemistry.

Kirchhoff's major memoir on his spectrum researches, including his detailed chart of part of the solar spectrum, appeared in 1861, and Roscoe promptly published an English translation. In a section entitled 'On the reversal of the spectra of colored flames,' Kirchhoff remarked that 'when an incandescent gas is penetrated by rays of sufficient intensity from a continuous spectrum, the spectrum of the incandescent gas must be reversed.' This seems to be about as close as he ever came to stating any of the three 'laws of spectrum analysis' that now commonly bear his name in elementary astronomy text-

Figure 22.3 The 'cigar box' spectroscope, the first one made by Bunsen following Kirchhoff's suggestion. This illustration is reproduced from Roscoe's Spectrum Analysis *(London, 1885).*

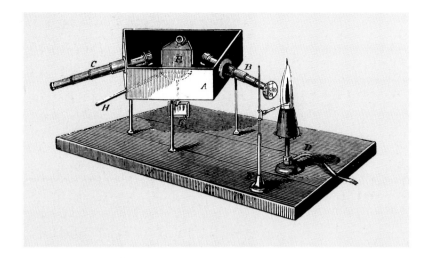

books. Kirchhoff envisioned the sun as a glowing solid body surrounded by an outer atmosphere of glowing gas that produced the Fraunhofer lines, a zone that quickly became known as the 'reversing layer.'

At the end of the decade a series of easily accessible total solar eclipses provided an opportunity for astronomers to test the validity of Kirchhoff's model, the idea being to watch the spectrum in the moments before totality when only the outermost solar layer could be seen – a layer that would presumably show an emission spectrum, a reverse copy of the normal dark Fraunhofer lines.

The eclipse of 1868, which crossed India and the Malay peninsula, was chiefly memorable for having inspired P. Janssen and N. Lockyer with a method for observing solar prominences spectroscopically without an eclipse; however, the line reversal was not observed. Again in 1869 the phenomenon went unobserved as the moon's shadow swept across America, although Charles A. Young and William Harkness independently found the characteristic bright green line in the solar corona. Although the few minutes of totality in 1869 sufficed for Young to measure the green coronal line, the seeming brevity of the event caused him to remark, 'I cannot describe the sensation of surprise and mortification, of personal imbecility and wasted opportunity, that overwhelmed me when the sunlight flashed out.'

In the following year, Young was more experienced and better prepared, and this time, at Jerez de la Frontera in Spain, he caught the fleeting reversal phenomenon, the so-called 'flash spectrum.' 'The dark lines of the spectrum and the spectrum itself gradually faded away,' he tells us, 'until all at once, as suddenly as a bursting rocket shoots out its stars, the whole field of view was filled with bright lines more numerous than one could count. The phenomenon was so sudden, so unexpected, and so wonderfully beautiful as to force an involuntary exclamation. Gently and yet very rapidly they faded away, until within about two seconds, as nearly as I can estimate, they had vanished.' With this brilliant *experimentum crucis*, Young had verified Kirchhoff's interpretation of the Fraunhofer lines and the physical constitution of the sun's atmosphere.

Today, on a picturesque old edifice on the main street in Heidelberg, there is a plaque reading (in translation): 'In this building KIRCHHOFF together with Bunsen founded Spectral Analysis, turned it to the sun and stars, and with it unlocked the chemistry of the universe.' With Kirchhoff's work, what had seemed forever beyond our grasp had become an astounding reality.

Notes and references

This chapter is based in part on a senior honors thesis prepared at Harvard under my supervision by David Robinson, 'The Development of Spectroscopy and the Phenomenon of Multiple Nondiscovery,' 1976. Other references include: A. Comte, *Traité philosophique d'astronomie populaire* (Paris, 1844), p. 109; Henry E. Roscoe, *Spectrum Analysis* (London, 1885), and his *The Life and Experiences of Sir Henry Enfield Roscoe* (London, 1906); and Agnes Clerke, *A Popular History of Astronomy During the Nineteenth Century* (London, 1902). Harlow Shapley and Helen Howarth, *A Source Book in Astronomy* (New York, 1929), pp. 279–82, reprint from Kirchhoff's principal memoir the description of the original experiment.

23 The discovery of the satellites of Mars

The planet Mars makes its appointed rounds from one close approach to the next in about two years and two months. At those favorable times of opposition the planet is a blazing ruby in the midnight sky, shining nearly ten times as brilliantly as its average light. But the oppositions themselves are not equal; owing to the eccentricity of its orbit, an opposition occurring on 27 August will bring the planet twice as close as one that occurs in February. Particularly favorable approaches occur every 15 or 17 years, and one of these took place about a century ago, in September of 1877. In fact, it was the closest approach by Mars since 1845, and better than the favorable oppositions in 1860 and 1862.

Between the favorable opposition of 1862 and that of 1877, important progress had taken place in astronomical instrumentation. In 1862, the largest refractors were the twin 15-inch Mahler and Merz telescopes, one at Harvard College Observatory and the other at Pulkovo in Russia. Fifteen years later the recently installed Clark 26-inch refractor at the Naval Observatory in Washington, DC, reigned as queen of the lenses.

This combination, a particularly close approach of Mars and the world's largest refractor, were essential ingredients in the successful discovery of the Martian companions. But to say that this discovery was merely a routine piece of sharp-eyed observing under favorable conditions would underestimate the fundamental understanding brought to the problem by astronomer Asaph Hall.

Before filling in the circumstances of the detection of the Martian satellites in August of 1877, let me digress to the uncanny prehistory of the Martian moons. According to Jonathan Swift's *Gulliver's Travels*, written almost exactly 150 years earlier, the scientists of Laputa had discovered two satellites revolving about Mars; 'whereof the innermost is distant from the center of the primary planet exactly three of his diameters, and the outermost five, the former revolves in the space of ten hours, and the latter in twenty-one and a half; so that the squares of their periodical times, are very near in the same proportion with the cubes of their distance from the center of Mars; which evidently shews them to be governed by the same law of gravitation, that influences the other heavenly bodies.'

The most unsettling part of the prediction is the short, ten-hour period for the inner satellite – considerably shorter than the 42 hours for Io, the fastest of the ten satellites known in Swift's day, and roughly approximating the 8-hour period that Phobos actually has. Table 23.1 compares the discovery data with Swift's predictions. I hasten to add that it gives me certain satisfaction to say that Swift was not as clairvoyant as the first glance suggests. Table 23.2 gives a basis for guessing the distances of the satellites in terms of planetary diameters. The choice of three and five planetary diameters for the

Table 23.1.

Mars' satellites	Actual		Swift's predictions	
	$a/d_{\male}$	P (hrs)	$a/d_{\male}$	P (hrs)
Phobos	1.4	7.6	3	10
Deimos	3.5	30.3	5	21½

Table 23.2.

Jupiter	$a/d_{\jupiter}$	P (hrs)	Saturn	$a/d_{\saturn}$	P (hrs)
Io	3.0	42	Tethys (1684)	2.5	45
Europa	4.8	85	Dione (1684)	3.2	66
Ganymede	7.8	172	Rhea (1672)	4.5	108
Callisto	13.6	400	Titan (1655)	10.5	383
			Japetus (1671)	30.6	1900

distances of two satellites very nearly matches the corresponding distances for Jupiter's Io and Europa.

More puzzling, however, is Swift's prediction of ten hours for the period of the first satellite. Even if Jupiter had been chosen as the model for satellite spacing, the periods would not follow by direct analogy. If Mars were as dense as the earth, then the first satellite at three planetary diameters should revolve in roughly one day; if the density were more like that of a Jovian planet, the period should be closer to two days. A possible solution to this problem was pointed out to me by N. T. Roseveare of Chelsea College in London. Roseveare has noted a relevant passage in Newton's *Principia* that states 'The smaller the planets are, they are, other things being equal, of so much greater density.' Now the diameter of Jupiter is about 22 times that of Mars, and if we take what now seems an absurdly high density for Mars, 22 times that of Jupiter, then the inner satellite should have a period of ten hours. Kepler's harmonic law was well known by 1726, and that Swift used it correctly should occasion little surprise, but all in all it appears likely that Swift had some professional help.

That Mars should have a pair of satellites was widely believed in Swift's day. After all, the earth had one satellite, Jupiter four, and Saturn five, so by a rhythmic progression, Mars ought to have two. Voltaire expressed this line of reasoning in his *Micromégas* (1752), where he states, '[the voyagers] would see the two moons which belonged to this planet, and which have escaped the searches of our astronomers. I know quite well that P. Castel will write, and even rather pleasantly, against the existence of these two moons; but I am in agreement with those who reason by analogy. The best philosophers know how difficult it would be for Mars, which is next from the sun, to have less than two moons.' (Here Voltaire was making a little joke at the expense of the voluminous anti-Newtonian writer, Father Louis-Bertrand Castel.) Even earlier Fontenelle had mentioned the possible Martian satellites in his *Conversation of the Plurality of Worlds*. There the pupil argues, 'Because Nature hath given so many Moons to *Saturn* and to *Jupiter*, it is a kind of proof that *Mars* cannot be in want of Moons.'

The inspiration for two Martian satellites may well have derived from Johannes Kepler, who repeatedly argued from archetypal principles based on harmony or analogy. In a letter to Galileo, Kepler wrote, 'I am so far from disbelieving in the discovery of the four circumjovial planets, that I long for a telescope, to anticipate you, if possible, in discovering two round Mars (as the proportion seems to require), six or eight round Saturn, and perhaps one each round Mercury and Venus.'

A curious trap awaited Kepler because of his belief in the possibility of a pair of Martian satellites. Shortly after Galileo had published his *Sidereus Nuncius*, he made yet another discovery that was announced to Kepler in an anagram. Kepler subsequently published this (as well as another Galilean anagram temporarily concealing his discovery of the phases of Venus) in the *Dioptrice*, his book on the theory of the telescope. Kepler, who once published an anonymous work in which three different anagrams of his own name appeared on the title page, promptly rose to the bait. He transposed the letters to read

Salue umbisteneum geminatur Martia proles.
Hail, twin companionship, children of Mars.

Kepler's ingenious but erroneous deciphering hinged on the word *umbistineum*, apparently the Latinization of a German word *umbeistehn*. In fact Galileo's anagram had nothing to do with the discovery of two satellites of Mars, but with the peculiarities of Saturn that we now understand to be its ring system. As it happened Kepler's *Dioptrice* was reprinted in England twice in the seventeenth century, so his proposed Martian satellites were probably better known there than many of his other ideas.

But let us now return to our centenary. Asaph Hall belonged to an old and once properous New England family. His early schooling was rather irregular, but eventually, hoping to become an architect, he enrolled in Central College in McGrawville, New York. There, according to the *New York Tribune*, students could meet part of their expenses by manual labor, something that appealed to the 25-year-old Asaph, who had already for six years been a journeyman carpenter. In McGrawville he found a motley crowd of idealists and adventurers who cared little for a classical education. There, however, he met Chloe Angeline Stickney, a frail but determined suffragist, who taught mathematics while completing her own senior year. Hall was among her pupils, and she soon became his fiancée. After their marriage they went to the University of Michigan, where Hall began a study of astronomy. Firmly resolved to become an astronomer, he proceeded to Cambridge, where, in spite of Director George Bond's admonition that he would starve, he took a low-paid job at the Harvard College Observatory. A few years later he became an assistant at the US Naval Observatory, where one of his early memor-

Figure 23.1 Asaph Hall,
Edward S. Holden and
Simon Newcomb.

able experiences was playing host to an unexpected nighttime visit from President Abraham Lincoln.

In 1875 Hall was placed in charge of the still new 26-inch Clark equatorial. His first discovery with this telescope, in December, 1876, was a white spot on the planet Saturn, which he measured through more than 60 cycles, thus finding the first reliable period of Saturn's rotations since Herschel's determination in 1794.

At the time of the particularly favorable approach of Mars in August, 1877, Hall undertook a systematic search for possible satellites. In the first section of his discovery monograph he described some of the circumstances, mentioning the thorough search made in 1862 by Professor D'Arrest at the Copenhagen Observatory. Hall goes on to say 'In his statement in the *Astronomische Nachrichten*, D'Arrest assumes a distance of Mars from the earth equal to 0.52, and with an assumed value of the mass of the planet in a given number of days. He shows that a satellite at an elongation of 70' would have a period greater than the period of Mars around the sun, or greater than 687 days, and hence infers that it is useless to search beyond the distance of 70'. The fact that D'Arrest, who was a skillful astronomer, had searched in vain was discouraging; but remembering the power and excellence of our glass, there seemed to be a little hope left.'

A decade later, Professor E. C. Pickering sent the following query:

'Is the rule thus assumed by D'Arrest of general application and correct? If so, it would serve as a guide in looking for additional satellites of other planets. Would a satellite placed at a greater distance from its planet continue under any circumstances to accompany the planet, or would it become an independent member of the system? . . . Presuming that you have had occasion to make yourself familiar with this subject, I take the liberty of asking whether you can conveniently give me the information above requested.'

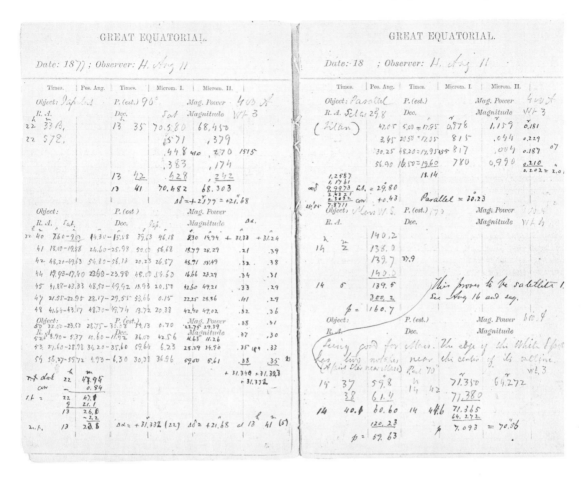

Figure 23.2 The first observation of Deimos in the record book of the 26-inch refractor for 1877.

A few days later, on 14 February 1888, Hall responded, disclosing the theoretical background for his search:

In the Spring of 1877, when I began to think of searching for a satellite of Mars, a little rough calculation convinced me that this planet could have no moon even at half the distance D'Arrest assumes as a limit. [That is, 70']. . . . The disturbing force of the sun on a satellite is the difference of its action on the planet and on the satellite. . . . Computing the forces by the expression μ/r^2, μ being the mass of the sun, or of the planet, I found the disturbing force of the sun more than twice as great as the attraction of the planet on the satellite. Hence, we would at once reduce the elongation of 30', and this being a *limit* the probable elongation would be much less. A little trial, and the analogy of other planetary systems, led me to search very near the planet.

Later Hall wrote: 'The chance of finding a satellite appeared to be very slight, so that I might have abandoned the search had it not been for the encouragement of my wife.' Angeline Stickney Hall was an enthusiast, and Angelo, the third of the four Hall sons, claimed that she 'insisted upon her husband's discovering the satellites of Mars.' Hence there is indeed some justice that a larger crater on Phobos has now been named 'Stickney.'

Hall first glimpsed the object that was eventually named Deimos at 2:30 a.m. on the night of 11–12 August. His observing book for that date, in none-too-legible writing, says 'Seeing good for Mars. The edge of the white spot has two notches near the center of its outline,' and parenthetically 'A faint star near Mars.' Hall subsequently has added the note above: 'This proves to be satellite 1.'

The following nights were cloudy and on August 15 the seeing conditions were very poor. On 16 August Hall's first observation was a rough measurement of 'Star near Mars.' On the following night the first page of the record book closes with the remark that 'The Mars Star observed tonight is a *fixed star* and not the object observed last night.' It is plain from this remark that Hall was already convinced that his object was a new satellite. Later that night he recorded for the first time two satellites, each ambiguously labeled 'Mars Star.'

The situation clearly changed by the following night, for Hall was joined in the dome by D. P. Todd, Simon Newcomb, and William Harkness, all of whom made measurements, and in the course of the observing the expression 'Mars Star' becomes 'Mars-Satellite.' The first remark is in Hall's hand: 'Images very poor at $9^h 40^m$, but saw the satellite immediately.' This is followed by four lines signed by D. P. Todd: "Seeing extremely bad: still I saw the companion without any difficulty. 'Halo' around the planet very bright, and the satellite was visible in this halo."

An interesting sidelight to the Mars satellites discovery was provided 11 years later in the same letter to E. C. Pickering that I have quoted previously. Hall wrote:

In the case of the Mars satellites there was a practical difficulty of which I could not speak in an official Report. It was to get rid of my assistant. It was natural that I should wish to be alone; and by the greatest good luck Dr Henry Draper invited him to Dobb's Ferry at the very nick of time. He could not have gone much farther than Baltimore when I had the first satellite nearly in hand.

The assistant Hall so much desired to get out of the way was none other than Edward S. Holden, a young protégé of Simon Newcomb's, and the man who later became the first director of Lick Observatory. Newcomb himself was in charge of the Nautical Almanac Office and, for all practical purposes, the scientific director of the Naval Observatory. Many years later, in a letter to Seth Chandler, Hall described Holden's interest in possible Martian satellites:

There are several points about the discovery of the satellites of Mars that have not been noticed. Thus Newcomb and Holden had the 26-inch Telescope, for the first two years and they tried to make discoveries. . . . After two years Newcomb got tired of the night work and offered the instrument to me. He had made good determinations of the masses of Uranus and Neptune. Procyon had been examined very carefully for the disturbing companion. Of course one of the first things I did was to find out what my predecessors had been doing. I found in a drawer in the Equatorial room a lot of photographs of the planet Mars in 1875. From the handwriting of dates and notes probably Holden directed the photographer, but whoever did the pointing of the telescope had the satellites under his eyes. All that was needed was the right way of looking, and that was to get rid of the dazzling light of the planet. The satellites might have been found at Harvard in 1862 very easily.

Hall's apprehensions that Holden would try to get into the act were promptly confirmed. On August 28 Holden wrote from Dobb's Ferry that he and Dr Henry Draper had detected a third satellite of Mars on 26 and 27 August, and a month later Holden claimed a fourth discovery, but both turned out to be as spurious as the canals observed elsewhere at the same opposition.

After Hall's death in 1907, his biographers lauded him as a great observer. Yet, as this episode attests, many of the greatest observations are spurred on by theory. Asaph Hall's discovery of the two satellites of Mars cannot be written off simply as the good luck of a keen observer who had the world's largest refractor at his command, for the record is firm both that his search was deliberate and guided by gravitational theory, and that others not so guided had failed in similar endeavors. It was, I think, appropriate to immortalize Angeline Hall née Stickney on Phobos, but I am happy that Asaph Hall himself is even more prominently commemorated there.

Notes and references

When I first came to the Harvard College Observatory as a summer assistant, Harlow Shapley assigned me the task of investigating Swift's astonishing prediction of the periods and distances of the two satellites of Mars, this in response to an inquiry from the noted parapsychologist J. B. Rhine. At the time, in 1949, I was unable to come up with any satisfactory explanation. This chapter is a belated solution to the Shapley–Rhine puzzler! It was written for the celebrations of the US Naval Observatory that commemorated the centennial of Hall's discovery of the two satellites of Mars. I drew on my earlier article published in Journal for the *History of Astronomy* 1 (1970), 109–15 and my entry on Asaph Hall in *Dictionary of Scientific Biography*, Vol. 6 (New York, 1972). The Pickering–Hall correspondence is in the Harvard University Archives, and some of the other material came to me from the Hall family and is used with their permission.

24 *The first photograph of a nebula*

September 30, 1880, marked one of the great turning points in the development of astronomical techniques. On that night the faint glow of a nebula was for the first time recorded on a photographic emulsion. Faint and fuzzy as the image was, it nonetheless provided the initial step in transfering the study of these diffuse glowing patches from visual observers to astronomical photographers.

This feat was accomplished by a wealthy New Yorker who was neither a professional astronomer nor a professional photographer, but a medical doctor. Yet Henry Draper's private observatory in Hastings-on-Hudson, and the physical laboratory at his Madison Avenue home in New York City, could well have been the envy of many a professional astronomer. The observatory boasted an 11-inch Clark refractor, with a special correcting lens for photographic work, as well as a 28-inch silver-on-glass reflector of his own construction. He had published a monograph on his $15\frac{1}{2}$-inch reflector, which had already become a standard reference for telescope making.

Dr Draper's background in photography could also have been the envy of professional astronomers, had they been much interested in this new-fangled technique. His father, John William Draper, took the first daguerreotype of the moon in the winter of 1839–40, using the method publicly announced by Louis Daguerre less than a year before. He was, by the way, also one of the first to daguerreotype the human face, a portrait of his sister Dorothy Catherine Draper. A few years later the elder Draper was apparently the first to photograph a spectrum, using a diffraction grating. As early as 1850 his 13-year-old son Henry was busily helping him photograph slides through a microscope.

At the age of 17, the precocious young man entered medical school, and by 1857 he had completed a thesis illustrated with daguerreotype microphotographs. But since he was only 20, and still too young to receive his medical degree, he toured abroad for a year. In Ireland, Draper visited Lord Rosse's observatory with its giant reflector and there hit upon the idea of combining photography and astronomy.

That very same year, George Phillips Bond, the second director of Harvard College Observatory, actually used wet collodion plates and the Harvard 15-inch refractor to photograph Alcor and Mizar. He wrote, 'There is nothing, then, so extravagant in predicting the future application of photography on a most magnificent scale. ... What more admirable method can be imagined for the study of orbits of the fixed stars and for resolving the problem of their annual parallax?' Despite Bond's enthusiasm, most astronomers were cool to the new possibilities. As Agnes Clerke wrote in 1885, concerning the elder Draper and his pioneering experiments of lunar photography, 'but slight encouragement was derived from them, either to himself or

Figure 24.1 Henry Draper (1837–82).

others' and she noted that Alcor and Mizar were 'a comparatively easy naked-eye object.'

The transit of Venus in 1874 offered the opportunity of the new art of photography to make its mark, and national astronomical committees in England, France, and Germany, as well as America, enlisted knowledgeable photographers. By that time Henry Draper had become professor of physiology and then dean of the medical faculty at New York University, though he had also made scores of telescope mirrors and had taken hundreds of photographs of the moon and of the solar spectrum, it is not surprising, therefore, that Draper became director of the photographic department of the US commission to observe the transit. Although the commission asked Congress to strike a special gold medal for Draper, the scientific results were generally disappointing.

Without photography, the advance of modern astronomy would have been impossible. Today even a short exposure records more than the eye can hope to see, but in Draper's lifetime this was never the case. Photography's principal advantage then was permanency – an advantage counterbalanced by the comparative nuisance of the technique and unfamiliarity with it. As John Lankford has shown, following the transit of Venus flop, professional astronomers once again relegated photography to the hands of skilled amateurs.

It is against this background that Henry Draper's success in recording the Orion nebula must be appreciated. The wet photographic plates of his day were very slow, and long exposures had been impossible because they dried too quickly. But as the result of a visit to William Huggins in England in the spring of 1879, Draper discovered the somewhat increased sensitivity of the new dry gelatino-bromide plates. Furthermore, these made possible for the first time the long exposures required for faint astronomical objects. Draper was able to make a 51-minute exposure, long enough to capture some of the faint nebulosity.

Draper promptly sent off a brief description of his success to England, France, and the *American Journal of Science*. He remarked only that his accomplishment had been achieved with the 11-inch Clark triplet, using his own mounting and clock drive, that his exposure was 50 minutes, and that he intended 'at an early date to publish a detailed description of the negatives.'

But how could he adequately describe his really very fuzzy negative? Using the old adage that a picture is worth a thousand words, Draper decided to distribute actual enlargements of his plate rather than publish a technical description. The prints were mounted on a card approximately 6 by 8 inches, with a printed caption reading 'FIRST PHOTOGRAPH OF THE NEBULA IN ORION, TAKEN BY PROFESSOR HENRY DRAPER, MD.' Unfortunately, Draper appears never to have recorded the brand of emulsion, though earlier he had specifically praised those made by Wratten and Wainwright of London as being the most sensitive.

Perhaps another reason why Draper never published further details of his pioneering achievement was that he soon surpassed it with a

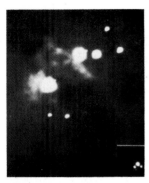

Figure 24.2 Henry Draper's 51-minute exposure taken 30 September 1880, is reproduced from a copy supplied by G. de Vaucouleurs from an original print at Paris Observatory. Draper's inset is a 5-minute exposure, showing clearly the four Trapezium stars. The one at bottom left seems curiously faint. It is, in fact, the Algol-type eclipsing variable BM Orionis, which has a range of 0.7 magnitude in blue light and remains at mid-minimum for about 6 hours. Is it possible that Dr Draper photographed an eclipse?

much longer exposure of 104 minutes. After his initial success, he wrote to E. S. Holden saying 'The exposure of the Orion nebula required was fifty minutes; what do you think of that as a test of my driving clock?' By March, 1882, he had obtained an exposure of 137 minutes which, of course, showed far richer detail. This remarkable picture was reproduced by photolithography as an addendum to Holden's 1882 *Monograph of the Central Parts of the Nebula of Orion.* Again, the 11-inch Clark and gelatino-bromide plates were used. 'The night was clear but cold and windy. The mean temperature was 27° Fahr; the wind NNW and in gusts, the strongest pressure being 5 pounds per square foot about nine o'clock; the whole travel of the wind during the exposure was 35 miles. The variation in the force of the wind is one reason why the stars show some ellipticity under this magnifying power; the gusts of course displaced the telescope somewhat, though the mounting is firm and the clock-work strong.'

Except for the addendum, Holden's hefty monograph dealt with visual observations of the Orion nebula; it vainly tried to settle the then-vexing question as to whether the nebula had exhibited evolutionary changes during two centuries of visual observations. The answer – that the nebula does *not* show conspicuous changes on such a short time scale – would be given by photography, and perhaps Holden himself sensed that he stood on the threshold of a new era. He wrote,

Although it is still too soon to give a final discussion to the photographic results attained by Dr Draper, I cannot refrain from pointing out some of the conclusions which may be drawn from this marvelously perfect representation of the nebula.

If we compare [the engraving of G. P. Bond's visual observations with Draper's photograph], we shall be able best to appreciate the important advance which has been made. Bond's engraving is the most accurate drawing that has been made, even as a map, and as a picture it is decidedly the best representation of a single celestial object which we have by the old methods. The work of observing alone extended over years and consumed many precious hours. I have before said how much labor was spent upon the mechanical execution of the steel plate; scores of revises were criticised and read.

Dr Draper's negative was made in 137 minutes, and for nearly every purpose is incomparably better than the other. The color and tint of the nebula, which is wonderfully preserved in Bond's engraving, is lost in the photograph; and yet, if the latter is held up between the eye and a window, the pictorial effect is most striking.

The amount of preparation for the two works is not to be estimated by years or hours, but it may be left out of account in a comparison. It required the best efforts of each observer to attain the results.

Two months after taking his 137-minute exposure Draper wrote to Holden, 'I think we are by no means at the end of what can be done. If I can stand 6 hours exposure in midwinter, another step forward will result.' With this in mind Draper undertook the plans for a new form of mounting to permit a continuous exposure of six hours. Unfortunately, Draper died before another winter came.

His widow, Anna Palmer Draper, established the Henry Draper

Figure 24.3 Dr Draper's 137-minute exposure was taken 14 March 1882, about 1½ years after the one above.

Memorial at Harvard to further the photographic researches he had pioneered. The 11-inch Clark went to the observatory as well, where it took thousands of plates before being sent on long-term loan to Canton, China, in 1947.

Meanwhile in England, A. A. Common began his own series of nebula photographs, and early in 1883 he secured a still more splendid photograph of the Orion nebula with his 36-inch reflector. This plate demonstrated a powerful new advantage for astronomical photography, because it registered stars too faint for the eye. No longer could this technology be left solely in the hands of gifted amateurs.

Copies of this beautiful photograph were pasted in as the frontis-piece of Clerke's *History of Astronomy during the Nineteenth Century* (beginning with the second edition in 1887), and her comments echoed Holden's: 'Photography may thereby be said to have definitely assumed the office of historiographer to the nebulae; since this one impression embodies a mass of facts hardly to be compassed by months of labour with the pencil, and affords a record . . . of the stupendous object it delineates, which must prove invaluable to the students of its future condition.' The visual observer with his pencil, laboriously drawing the nebula, had become obsolete.

As Joseph Ashbrook wrote 'The introduction of a powerful new observing technique often ends a chapter in astronomical history.

Figure 24.4 From visual studies with Harvard College Observatory's 15-inch refractor between 1859 and 1863, G. P. Bond prepared this careful rendering of the Orion nebula.

Figure 24.5 A. A. Common's extraordinary photograph of the Orion nebula, taken in 1883.

This happened when Henry Draper took the first successful photograph of the great Orion nebula on 30 September 1880.' In less than three years, with the continued work of Draper and Common, astronomical photography had truly come of age.

Notes and references

Barbara Welther first showed me an original enlargement of the Draper photograph. These original mounted enlargements are now exceedingly rare, and I would be interested to learn how many observatory archives still contain copies. G. de Vaucouleurs has kindly lent for reproduction the copy of the Draper photograph appearing in his book *Astronomical Photography* (New York, 1961). See also Dorrit Hoffleit, *Some Firsts in Astronomical Photography* (Cambridge, 1950), and John Lankford, 'The Impact of Photography on Astronomy,' Chap. 2 in *The General History of Astronomy*, edited by Owen Gingerich (Cambridge, 1984), pp. 16–39. Joseph Ashbrook's 'The Visual Orion Nebula' appeared in *Sky and Telescope*, **50** (1975), 299–301. For biographical information on Henry Draper, see the entry by Charles A. Whitney in *Dictionary of Scientific Biography*, Vol. 4. (New York, 1970).

25 The Great Comet and the 'Carte'

A dramatic and important event in the history of astronomical photography took place just over a century ago. A brilliant sun-grazing comet made an unexpected appearance in the first weeks of September, 1882. It was independently found by several southern hemisphere observers but remained unknown in the north until its daytime rediscovery by A. A. Common only a few hours before its perihelion passage on 17 September. (This near-debacle was catalytic in creating a world clearinghouse for comet information at Kiel Observatory.)

On the morning following perihelion passage, the comet displayed, in the words of David Gill, Her Majesty's Astronomer at the Cape of Good Hope, 'an astonishing brilliancy' as it rose behind the mountains. 'The Sun rose a few minutes afterwards, but to my intense surprise the comet seemed in no way dimmed in brightness. . . . It was only necessary to shade the eye from direct sunlight with the hand at arm's-length to see the comet with its brilliant white nucleus and dense white, sharply-bordered tail. . . .'

For the next few days the Great September Comet remained bright enough to be observed during the daytime. Even in twilight it was difficult to estimate the tail's length because of the sky brightness. As the comet moved into the darker morning sky, however, its tail proved to be spectacular indeed. By 4 October several photographers in South Africa had managed to obtain images of the long tail with ordinary photographic apparatus, and finally Gill himself resolved to try to record the phenomenon.

He called upon E. H. Allis, a photographer from a neighboring village who had some experience with the new-fangled dry plates, and together they fastened a portrait camera to the declination counterpoise of the 6-inch equatorial at the Cape Observatory. The lens, belonging to Allis, had a diameter of $2\frac{1}{2}$ inches and a focal length of 11 inches. With this apparatus they were ready to work when the comet appeared in the predawn sky on the morning of 19 October. Their 30-minute exposure captured the brilliant coma and a 15° tail. Similar results were obtained on the following two mornings, but then moonlight prevented further attempts.

Had this been the end of the story, Comet 1882 II would have gone down in history as a splendid and even memorable member of a group of sun-grazing comets – but little more, not even as the first to have its picture taken. However, the comet had begun to develop an 'anomalous' sun-pointing tail. Today this phenomenon is well understood in terms of dusty cometary debris combined with an unusual geometry. In the last century, however, the development of a sun-pointing tail was mysterious and unexpected. Perhaps Gill decided to make a fresh attempt at comet photography, using longer exposures, so as to record this curious feature. On 7 November, as soon as the moon had waned sufficiently, he tried a 100-minute

Figure 25.1 (a) David Gill's 100-minute exposure of the Great September Comet on the morning of 7 November 1882, seven weeks after perihelion. The tail extends about 18°, and slightly trailed stars are readily visible on the original reproductions. (b) To aid in the identification of the field in Hydra, a portion of the Tirion Sky Atlas 2000.0 is shown at the same scale.

(a)

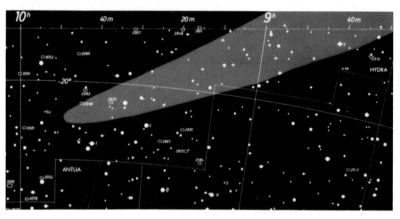

(b)

exposure guided (not too successfully) on the comet's nucleus. A few nights later he attempted one of 80 minutes (on 13 November) and the next night one of 120 minutes, both guided on the stars.

The two longest exposures showed not only the short sun-pointing tail and the normal tail of nearly 18°, but also a striking number of stars, to nearly tenth magnitude. As it happened, this was exactly the limiting magnitude of the *Bonner Durchmusterung* star charts and catalog, a great visual observing project that had been completed earlier in the century for the northern skies to declination −24°. Gill immediately saw the possibility of extending the BD photographically to the far-southern sky, and he persuaded the Royal Society to initiate support for this project.

Gill sent copies of his comet pictures to Europe, where others also promptly saw the potential of the new dry plates for celestial documentation. Common, who had found the Great September Comet visually, swiftly turned to photography and within a matter of a few months had recorded more stars in a photograph of the Orion nebula than he could see visually with his 3-foot reflector. In France, Rear Admiral Ernest Mouchez, director of Paris Observatory, received a set of comet prints from Gill. On his staff were two brothers, Paul and Prosper Henry, who had been very successful at discovering minor planets and who were proposing to make a detailed celestial map of the zodiacal regions. They, too, realized that photography could come to their aid.

The gelatino-bromide dry plates, which had become commercially

Figure 25.2 The brothers Henry with the 13-inch astrographic telescope they designed, and which was adopted for the Carte du Ciel *celestial mapping program.*

available only in the late 1870s, were blue sensitive, whereas astronomical refractors were corrected for the yellow–green portion of the spectrum. To overcome this problem, the brothers Henry set out to construct a photographic lens corrected for blue–violet wavelengths, and at last they succeeded in figuring a 13-inch lens that for its day was a veritable triumph of optical workmanship. Mouchez gave them every encouragement, according to Oxford's H. H. Turner in *The Great Star Map*, 'but their workshop was after all a mere shed. I have often heard Dr Common speak with amusement of his visit to the workshop which had turned out to the admiration of the world the first successful photographic refractor – the modest building and humble appliances were so surprising.'

Turner went on to say that it was the work of the lens produced by the Henrys that led to the inception of the 'great star map,' the so-called *Carte du Ciel*. 'Correspondence between Sir David Gill – under whose direction the comet photographs had been taken – and Admiral Mouchez, who had encouraged the work of the Henrys, led ultimately to the assembling of a great international conference at Paris in 1887.' Fifty-six delegates from nearly 20 countries sat down on 16 April to lay the plans for an immense photographic sky chart requiring over 22 000 plates, with the work to be divided into zones and parceled out among cooperating observatories. During a crowded week and a half the delegates agreed to adopt the Henrys' telescope as a standard and to produce not only a photographic map to magnitude 15.0 but also an *Astrographic Catalogue* of stars through the eleventh magnitude.

Thus began an international collaboration that in a sense was the forerunner of today's International Astronomical Union, and in a more subtle, inverse way one of the contributors to the rise of American astronomy at the turn of the century. What happened was the initiation of a vast program of photographic mapping and cataloguing that tapped the strength of observatories throughout the world – except in America – for decades to come.

Precisely how much credit (or blame!) do Gill and the Great September Comet of 1882 deserve in this matter? Gill's biographer, George Forbes, writes with some exaggeration that:

[Gill's] visit, in 1887, to Europe was most eventful for astronomy. It would have been so had nothing come of it except the inspection and delivery of the great 7-inch heliometer for the Cape Observatory, which was destined, in the hands of probably the finest observer in the world, to furnish results of unparalleled accuracy in problems beyond the capacity of most observers. [Gill used the heliometer for measurements of stellar parallax and of minor planets to determine a solar parallax of 8".80; this entered into Simon Newcomb's weighted average, which also came out to be 8".80 and which was used in all national ephemerides until 1968.]

It would have been equally eventful for astronomy if nothing had come of it except the Astrographic Congress, with the initiation of the International *Carte du Ciel* and Catalogue. David Gill was elected its *President d'honneur*, acclaimed by ballot, and he proved himself in the sequel to be the greatest organizer of astronomical joint undertakings known to his generation.

Unfortunately, this glowing accolade seems at least partly

Figure 25.3 Portions of the original provisional agenda for the 1887 Astrophotographic Congress, sent by Mouchez to E. C. Pickering, director of Harvard College Observatory. He responded in detail to the questions but did not attend the conference in Paris.

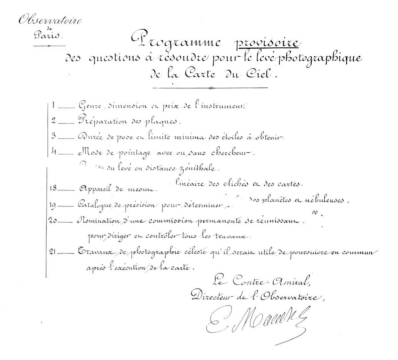

unfounded. The *proces-verbaux* of the congress was published by the French Academy of Sciences, and the detailed record shows that Mouchez was designated by acclamation as '*President d'honneur,*' with Gill playing a rather low-keyed role. None of the flowery opening speeches mentioned Gill, nor does he occupy a special place in the group photograph. And a 107-page booklet issued by Mouchez in advance of the conference, in which he outlined the history of astronomical photography, barely mentioned Gill except to say apropos of the comet pictures, that 'this success by M Gill shows that we have finally arrived at the moment when we can construct the photographic *Carte du Ciel* as had been foreseen 25 years ago by M de la Rue.'

Even Gill himself, speaking at a Friday evening meeting of the Royal Institution several weeks after the conference, handed the credit to the French admiral: 'Lastly, I would add that the good will which pervaded the meetings, the general success of the Congress as a whole, were in no small degree due to the genial influence of the single-hearted, earnest-minded man who convened it – Admiral Mouchez.' Yet, despite even his own reluctance to claim a key role in proposing the conference, we can find in the French *Bulletin Astronomique*, then just in its third volume, Gill's correspondence with Mouchez suggesting precisely such an international congress to be held in April, 1887.

Probably the best balanced account is provided in Agnes Clerke's *A Popular History of Astronomy during the Nineteenth Century*. Her exposition was issued in four editions: 1885, 1887, 1893, and 1902. The first two quote Gill's description of the comet on the morning following the perihelion passage but say nothing about its photography. In

Figure 25.4 David Gill in his study at the Cape Observatory in October, 1888. His wife Isobel sits at the left, and across the desk is the astronomical historian Agnes Clerke. The man standing is probably Gill's secretary-assistant, H. Sawerthal, who had discovered a bright comet the previous February. Gill wrote to E. M. Clerke, 'Your sister sits opposite me in the study with a pile of books on either hand, which is gradually growing till she seems to be coming through a gate with rather badly built pillars on either side. At night she is to be found in the equatoreal – weather permitting, engaged in flirting with the spectra of variable stars. But, alas, the weather has not been very favorable – and Mr. Sawerthal and she play duets in the evening, or my wife reads aloud.'

1888 Clerke paid a visit of several weeks to the Gills at the Cape, and in the third edition of her history she added a full-page collotype of four of Gill's photographs of the comet, together with a brief history of the *Carte du Ciel* project. Clerke mentioned, in turn, the photography in 1882 at the Cape of Good Hope and the 'Southern Durchmusterung,' next the Henrys' telescope, and then she states, 'These results suggested the grand undertaking of a general photographic survey of the heavens, and Dr Gill's proposal, 4 June 1886, of an International Congress for the purpose of setting it on foot, was received with acclamation, and promptly acted upon.' (Actually, the generally accurate Clerke had picked the wrong letter; she should have said 1 March 1886.)

The connection between the Great September Comet and the *Carte du Ciel* seems firmly established. As David Evans has written in Vol. 4A of *General History of Astronomy (GHA)*, 'Gill's greatest influence on the whole world of contemporary astronomy began in 1882 and, for better or worse, was to continue for eighty years. . . . The subsequent history of the [Astrographic Congress], which dragged on until the 1960s, makes one wonder whether it should be regarded as Gill's greatest triumph or his grandest failure.'

While astronomical photography made impressive strides, the *Carte du Ciel* and its associated *Astrographic Catalogue* languished. By tying up the resources of two dozen observatories around the world, it fossilized their efforts into a program that was overtaken by improved methods. Meanwhile, the Americans, who never joined the project, remained free to concentrate not only on larger and more efficient forms of instrumentation, but also on the more exciting problems of astrophysics and stellar evolution. 'Lack of American participation notwithstanding,' writes John Lankford in the same

GHA volume, 'the Astrographic Congress and the *Carte du Ciel* ushered in a new epoch in the history of astronomical photography.' The on-going efforts of this project educated astronomers to the value of the new technology, which made possible the vast sidereal explorations of the twentieth century.

Notes and references

Joseph Ashbrook touched on this topic in two articles in *Sky and Telescope*, 'The Great September Comet of 1882,' **22** (1961), 331–2, and 'The Brothers Henry,' **17** (1958), 394, 399. For Gill's viewpoint, see his *A History and Description of the Royal Observatory, Cape of Good Hope* (London, 1913) and George Forbes, *David Gill, Man and Astronomer* (London, 1916). Gill included actual photographic prints of his six plates of the 1882 comet in *Annals of the Royal Observatory, Cape of Good Hope*, Vol. 2, Part 1, 1886, and Agnes Clerke published four of them in her *A Popular History of Astronomy during the Nineteenth Century* (third edition, London, 1893).

An independent English view of the Astrophotographic Congress is found in H. H. Turner's *The Great Star Map* (New York, 1912), and a French prospectus in Rear Admiral E. Mouchez' *La Photographie Astronomique et la Carte du Ciel* (Paris, 1887). Detailed minutes of the congress proceedings were published by the Institut de France, Académie des Sciences, *Congrès Astrophotographique International* (Paris, 1887); a translated abridgment and a few additional comments are given by Albert G. Winterhalter, *The International Astrophotographic Congress*, Appendix I of *Washington Observations for 1885* (Washington, 1889).

26 *James Lick and the founding of Lick Observatory*

To such astronomical philanthropists as Yerkes, Hooker, Carnegie, and Rockefeller, American science owes a great debt. And yet I cannot imagine celebrating their anniversaries in quite the same way that we can commemorate James Lick, who applied his San Francisco real-estate fortune to many scientific and educational enterprises. For with Lick, the building of a great telescope seems to have been his own idea, and although it was fostered and encouraged by others, he set the specifications and determined the site in his own idiosyncratic way. That Lick Observatory remains one of the great astronomical institutions of the world owes much not only to Lick's fortune, but to the inspiration that he instilled in the non-astronomers who actually built the observatory. Lick Observatory was the first mountaintop observatory, and played a key role in bringing American observational astronomy into worldwide ascendancy.

The story of James Lick's life is remarkable by any standard. He was born in Pennsylvania in 1796, when George Washington was serving his last year as President. In 1819 the young Lick began his years of wandering, temporarily settling in Baltimore as an apprentice piano maker. Migrating to South America, he established himself as a master piano maker, first in Buenos Aires, then in Valparaiso, and finally in Lima.

After nearly a decade in Peru, Lick became increasingly eager to try his luck in California. James K. Polk had been elected President on an expansionist platform, promising to extend the Union from the Atlantic to the Pacific, and Lick felt sure that California would become part of the United States. On 7 January 1848 Lick sailed through the Golden Gate. He had brought along his workbench and tools, but never again would he hang out his shingle as a piano builder, for the 51-year-old Lick had quickly decided to gamble his future on San Francisco real estate. Within 20 years, when a local newspaper made a survey of the wealthiest men in town, Lick ranked first with $750 000.

How Lick became interested in building the world's largest telescope is a bit of a mystery. In 1860 a young Portuguese-American, George Madeira, was lecturing on astronomy and geology throughout California. Lick attended one of his lectures in San Jose and afterward invited the young man to visit his ranch for several days. Together they observed the heavens through Madeira's telescope and at one point the young man is said to have exclaimed, 'Why, if I had your wealth, Mr Lick, I would construct the largest telescope possible.' Probably these words planted the seeds for Lick's greatest monument.

On the other hand, Joseph Henry, then secretary of the Smithsonian Institution, liked to think that *he* had persuaded Lick to build a

Figure 26.1 James Lick, who amassed a fortune in San Francisco real estate, gave it away almost entirely to educational and cultural institutions, of which one of the best known is the Lick Observatory, atop Mount Hamilton in California.

great observatory when he met the Californian in the Lick House in 1871. Two years later, when Prof. George Davidson, President of the California Academy of Sciences, called upon Lick to thank him for the valuable lot he had just deeded to the Academy, Lick told Davidson of his intention to bequeath a large sum of money for a telescope 'larger and more powerful than any existing.' Naturally Davidson was most interested in such a significant scientific venture, and the Academy president offered his encouragement and advice. Davidson described the ensuing encounters as follows:

James Lick originally intended to erect the Observatory at Fourth and Market Streets. His ideas of what he wanted and what he should do were of the very vaguest character. It required months of careful approaches and the proper presentation of facts to change his views of location. He next had a notion of locating it on the mountains overlooking his millsite near Santa Clara, and thought it would be a Mecca, but only in the sense of a show.

Gradually I guided his judgment to place it on a great elevation in the Sierra Nevadas. . . . At the same time, by my presentation of facts and figures of the cost and maintenance of other observatories, he named the sum of $1 200 000 in one of his wills, as the sum to be set aside for founding the James Lick Observatory, and for its support.

In making him acquainted with the size and performance of the telescopes of the larger observatories, I naturally mentioned the great reflector of Lord Rosse. That seemed to fire his ambition and at the next interview he insisted on a refractor of six feet in diameter. It required long and patient explanations to get him down to forty inches, which was the diameter we finally adopted.

In addition to his own interventions on behalf of the observatory, Davidson urged Joseph Henry to encourage the strong-minded millionaire. In the correspondence, which is preserved in the Smithsonian Archives, the Smithsonian secretary wrote:

While there are thousands of enterprising men in our country who have talents for accumulating wealth there are but a few like yourself who have the wisdom and enlightened sympathy to apply it as you have done. There is in most men an instinct of immortality which induces the desire to live favorably in the memory of their fellow men after they have departed this life, and surely no one could choose a more befitting means of erecting a monument to himself more enduring or more worthy of admiration than that which you have chosen.

Henry went on to say:

It should be recollected, however, that, besides a suitable building, funds are required to sustain, properly, an establishment like that of the Academy. A curator will be necessary and the means for publishing the proceedings. Furthermore, an establishment of the kind ought to have the means of consecrating to science any one who may be found in the country possessed of the peculiar character of mind in a marked degree for original investigation.

By the following year Henry was even more specific; he not only urged Lick to appoint the best available astrophysicist, but even singled out the English astronomer Norman Lockyer, the self-trained spectroscopist and founder of the journal *Nature*. Lockyer was a dynamic individual who is personally credited with far more scientific contributions than Edward Holden, the man who in fact became the first director of Lick Observatory. There is no disputing the tremendous scientific contribution made indirectly by Holden in bringing together a truly remarkable staff in the first decade of the Lick Observatory's operation. It is probably idle to speculate whether Lockyer could have done as well in staffing the mountain. Both Holden and Lockyer seemed always immersed in controversy, and one might imagine that the newspapers would have found as much to criticize in Lockyer as they were to find for the much-harassed Holden.

Meanwhile, back in San Francisco, the aging Lick was quite irked to discover that he could not will his money directly to the state for the construction of an observatory. He was obliged to place his fortune in the hands of a trust commissioned to carry out his wishes. The Deed of Trust specified the construction, on the borders of Lake Tahoe, of

a powerful telescope, superior to and more powerful than any telescope ever yet made, with all the machinery appertaining thereto and appropriately connected therewith, or that is necessary and convenient to the most powerful telescope now in use, or suited to one more powerful than any yet constructed, and also a suitable observatory connected therewith. Provided, however, if the site above designated shall not, after investigation, be deemed suitable by said Trustees, or a majority of them shall select a site on which to erect such telescope, but the same must be located within the State of California.

Toward the end of 1874 Lick became more and more impatient about the slow progress the trust had made in the few months of its existence. In principle Lick had signed away his entire fortune, and when he resolved to revoke his trust deed and write a new one, it was judged legally impossible. Lick simply hired an astute lawyer to break the trust, expecting years of litigation. Without going into court directly, the attorney took the matter to the public press and managed to form a 'public opinion' so that the first five trustees willingly resigned.

As chairman of his second board of trustees, Lick named Captain Richard F. Floyd, a dashing southerner who had acquired distinction in San Francisco business and social circles. For two or three years he had commanded steamers on the West Coast; hence the title 'Captain.' His wife was the heiress to a fortune, so he retired at an early age to his estate on Clear Lake about 100 miles north of San Francisco. In front of his pretentious dwelling was moored a 72-foot steamer as well as a yacht and several smaller boats. When Captain Floyd planned this home, he might well have anticipated a life of luxurious ease in idyllic surroundings. But the course of his life was abruptly altered when he was introduced to James Lick in 1875. Upon their first meeting Lick had taken an immediate liking to Floyd, and with uncanny acumen he exclaimed that this was the very man he wanted to build his observatory. In retrospect one must wonder if such a successful observatory would have been established without Captain Floyd's dedicated leadership. F. J. Neubauer, in his articles

Figure 26.2 Captain Richard F. Floyd, who abandoned retirement plans to build the observatory.

on the history of Lick Observatory in the old *Popular Astronomy*, wrote that 'From all that can be gathered from the records, it is not an exaggeration to state that the Lick Observatory, as it finally came into being, was the brain child of Captain Floyd.'

The second trust brought about several important changes. First, Lick reserved the right during his lifetime to hire or fire the trustees. Second, the contemplated observatory was deeded to the Regents of the University of California rather than the state. Finally, Lick was persuaded to relocate the Observatory from Lake Tahoe to Mount Hamilton, a move that pleased him since the mountain was located in his favorite Santa Clara County. D. D. Murphy, mayor of San Jose and a member of the second trust, influenced the County Supervisors to build the necessary road to the top of the mountain. He also took the necessary steps to obtain the land, which was done through an act of Congress.

Although happy over the selection of Mount Hamilton as the site for his observatory, Lick was still distressed at the slow pace of his trustees. On a Sunday morning in August, 1876, Lick astonished his friend Charles Plum by announcing his intention to discharge the trustees except for Captain Floyd. Lick ordered Plum to return at 3 p.m. with a new slate, but as Plum departed, he realized that such a mission could not be legally transacted on a Sunday. The next morning, however, the new board of trustees was appointed. When the changes in trustees became known, rumors flew that the old man was insane. Lick feared that his will might be attacked by his Pennsylvania heirs and his trust broken. He ordered his trustees to assemble a jury of the most eminent doctors to examine him and to issue a written report on his mental condition. All nine agreed that he was in sound mind and 'perfectly conscious that one who leaves so much money, if found vulnerable, would be plutarchic carrion for litigious vultures.' But for Lick himself the end was very near.

Early in the morning of 1 October 1876 James Lick died quietly in his room at Lick House. For three days the body of James Lick lay in state in Pioneer Hall while thousands of San Franciscans passed the elegantly draped casket. On the afternoon of 3 October the funeral procession, attended by a full quota of dignitaries, moved slowly to Masonic Hall for the elaborate services. The casket was then placed in a vault at the Masonic cemetery; ten years later it was moved to its final resting place under the great telescope of Mount Hamilton.

A certain amount of litigation hung over Lick's estate, and several years elapsed before all the claims were cleared and the properties could be sold. This proved quite fortunate, because in 1876 the country was in the throes of a depression, and the sale of the real estate at that time would not have supported all of Lick's charities.

The problems facing the third board of trustees included the fact that at the time of Lick's death they had not been formally elected, and Lick's properties were threatened by the necessity of raising $40 000 for taxes. The secretary to the trustees, Henry E. Matthews, took it upon himself to collect the sum. The secretaryship was a highly responsible position that required initiative and an endless amount of tact. Captain Floyd had chosen Matthews to fill this

Figure 26.3 Lick Observatory shortly after its completion.

Figure 26.4 Thomas E. Fraser, superintendent of construction of the observatory.

important position, and he performed with unusual efficiency and harmony. Matthews was a professional photographer for a time before entering the services of the trust, and many of the early photographs of Lick Observatory were taken by him.

The third member of the trio that brought about Lick Observatory was Thomas E. Fraser, the superintendent of construction. A native of Nova Scotia, he had become the trusted manager of the Lick properties in Santa Clara County a few years before Lick's death. It was Fraser who introduced Captain Floyd to Lick. He was a self-educated construction engineer, and the highest paid employee of the trust. When the observatory was all but completed he resigned, dying just a few years later at the early age of 41. His friend, Captain Floyd, had died almost exactly a year earlier, at the age of only 47 years.

Mary Shane, in her unpublished history of the Lick Observatory, has written:

in eight years from the beginning of active work, the finest observatory of its time was built, equipped and in operation. This was primarily the achievement of three remarkable men; Floyd, Matthews, and Fraser. None of these men had astronomical training, but they were completely dedicated to making the Lick Observatory the finest institution of its kind. They achieved this through ingenuity, preseverence and such arduous work that the endeavor essentially cost Floyd and Fraser their lives.

Although the Mount Hamilton site had been selected and secured, progress in obtaining the 'largest possible telescope' was considerably slower. Lick himself had inquired about the cost of such an instrument from the obvious expert, Alvan Clark, who had built the 26-inch refractor for the US Naval Observatory. Clark's estimate of $200 000 outraged Lick, who believed the Cambridge telescope maker was taking advantage of him. Only after James Lick's death could the trustees proceed with a contract for the 36-inch lenses to be

Figure 26.5 Preparations for the interment of James Lick, which took place on 11 January 1887 in a vault beneath the 36-inch refracting telescope.

figured by Alvan Clark and Sons. The glass disks were ordered from Feil and Company of Paris. The flint glass arrived safely at the Clark's factory in 1882, but the crown glass cracked during packing. The Feil brothers had immense difficulty producing another disk, succeeding only after 19 failures, and at a cost that drove the firm into bankruptcy. Finally in October, 1885 the crown glass was delivered to the Clarks, and the finished objective arrived safely on Mount Hamilton on 29 December 1886. Almost a year later the last carload of tubing arrived from Cleveland, accompanied by Ambrose Swasey who personally supervised the mounting of the great telescope.

Although the taming of the mountain and the building of the physical plant had been carried out by truly dedicated non-astronomers, the astronomers, too, were pioneers facing rugged and extreme conditions. The early history of the Lick Observatory is to some extent the history of 25 years in the life of Edward Singleton Holden. Holden was a man of enormous energies. For example, in his first five weeks as director of the observatory he wrote no fewer than 504 letters in his own hand. Holden was first recommended for the job when he was still a junior staff member at the US Naval Observatory. There he had become an assistant on the new 26-inch refractor, then the largest in the world. Holden worked under Simon Newcomb, who was at that time America's leading astronomer, and a year later Newcomb recommended Holden to the first Lick trust as a possible director for the planned observatory. Since Holden was not yet 28 years old and had had only a year of practical experience in observatory work, this tells a great deal about Newcomb's high esteem for Holden.

In 1876 the Navy sent Holden to England to examine scientific instruments, and in London Holden first met Captain Floyd. From

Figure 26.6 Tourists visiting the observatory.

this time on there must have been some tacit agreement that Holden would eventually take over the observatory directorship. As the observatory headed for completion, the Regents of the University of California appointed Holden President of the University, and he served rather uneventfully for a year and a half. Holden became Director of Lick Observatory on 1 June 1888, and perhaps the best testimonial to his leadership as director is the list of senior staff that he had assembled by this time: S. W. Burnham, J. M. Schaeberle, J. E. Keeler, and E. E. Barnard. The fact that ultimately Holden was unable to hold his staff together at Lick Observatory tells more about the lonely conditions on an isolated mountain, especially difficult for families, rather than any failure in Holden's personality.

Some of these difficulties are graphically recounted by Holden in his retiring Presidential Address of the Astronomical Society of the Pacific, an institution that Holden was largely instrumental in founding:

Every necessity of life at Mt Hamilton must be provided by individuals, except water. That is furnished by the Observatory. To distribute this, we have a system of four reservoirs, with several miles of pipes. ... All the motive power used in revolving the great dome, or in raising its floor, depends on the water-supply, and the slightest accident to the wind-mill, to a reservoir, or to the pipes stops the work of the great telescope. After every snow-storm a whole day's work, and sometimes more, is necessary to get the revolving parts of the dome into satisfactory working order. . . . The reservoir capacity is not sufficient to store enough water to carry us through the dry season. Hence, every year it has been necessary to use for domestic purposes some of the rain-water collected during the winter and stored for use as power. All this water has passed many times through the hydraulic rams, and is therefore covered with a heavy film of oil, and is really unfit for use, and produces more or less illness when it is used. But it must be used. There is no

Figure 26.7 Early projects of the 36-inch refractor, largest in the world at the time of its completion, included a photographic atlas of the moon, and a study of the apparent radial velocities of nearby stars, leading to the conclusion that the solar system is moving toward Vega at 12 miles per second.

other. There is absolutely no present remedy. It will be necessary to provide a greater storage or a greater supply. Either of these things can readily be done; but either will require an expense which there is no present way of meeting.

Holden was a prolific writer and first-class publicist, always willing to supply readable articles to news magazines. He wrote on topics as diverse as 'The Picture Writings of Central America,' 'Our Country's Flag,' and biographies of William Bond and William Herschel. Perhaps he wrote too much, because for some reason the popular press took a dislike to him, and his lengthy rejoinders seldom helped his local reputation. Simon Newcomb, in his autobiography, addressed the question of Holden's administration. He wrote:

To me its most singular feature was the constantly growing unpopularity of the director. I call it singular because, if we confine ourselves to the record, it would be difficult to assign any obvious reason for it. . . . Nothing happens spontaneously, and the singular phenomenon of one who had done all this becoming a much hated man must have an adequate cause. I have several

Figure 26.8 The lens components of the 36-inch, cast in France and finished in New England, arrived at the observatory carefully packed and escorted, on 27 December 1886.

Arrival of the 36″ O.g. at Lick Observatory, 12.30 P.M. Dec. 27. 1886. — Beautiful Day. —

times, from pure curiosity, inquired about the matter of well-informed men. On one occasion an instance of maladroitness was cited in reply.

'True,' said I, 'it was not exactly the thing to do, but after all, that is an exceedingly small matter.'

'Yes,' was the answer, 'that was a small thing, but put a thousand small things like that together, and you have a big thing.'

Finally, at the end of 1897, Holden felt obliged to resign. However, I do not wish to end on this negative note, but rather with a brief survey of the scientific results achieved in that first decade. This can be done in part by looking at the early publications of the Lick Observatory, and in part by reviewing the interests of each staff member.

S. W. Burnham, the oldest man on the original staff, had led a remarkable double life. During the day he had served as a court stenographer in the US Circuit Court in Chicago, but during the night he had acquired an international reputation as a double-star astronomer. His Chicago observatory had been a backyard affair, but it boasted a particularly fine 6-inch Clarke lens. From 1866 until he came to the Lick Observatory he also had access to the $18\frac{1}{2}$-inch

Figure 26.9 Front page of the San Jose Daily Mercury *of 28 June 1888, reporting the dedication of the Lick Observatory.*

Dearborn refractor, with which he made some of his famous observations. It was Burnham who was invited in 1879 to bring his 6-inch telescope to Mount Hamilton in order to conduct a site test. Burnham used the opportunity to detect some new double-star systems, and he had concluded that, 'So far as one may judge from the time during which these observations were made, there can be no doubt that Mt Hamilton offers advantages superior to those found at any point where a permanent observatory has been established.' At Lick, Burnham pursued his double-star observations, which fill Volume 2 of the Lick Observatory *Publications*.

Burnham was very popular with the public, and his resignation in

Figure 26.10 E. S. Holden, first director of the Lick Observatory.

1892 brought considerable unjust blame on Holden. It was simply a matter that Burnham could not see any sense in maintaining two establishments, one on the mountain and the other in San Jose where his children attended school, and he could return to the Circuit Court in Chicago at a much higher salary than he received at Lick Observatory.

F. J. Neubauer reports that John Schaeberle was a 'born bachelor, that kind who attends the YMCA gymnasium regularly, taking care of his health in a common sense way.' He had graduated from the University of Michigan, had made several successful telescope mirrors, and had discovered several comets. Holden placed him in charge of the meridian circle at Lick Observatory, but he is probably most famous for the 40-foot camera lens of his own design used to obtain large-scale photographs of solar eclipses. Since Holden already had a large permanently-sited observatory, he risked the criticism of other professional astronomers by investing so much in solar eclipse expeditions to distant parts of the world. However, the eclipse team achieved an impressive record of results. Schaeberle and Burnham first used the long camera in December 1889 at Cayenne in South America. It was used again in Chile in 1893 and in Japan in 1896.

At 31, Edward Emerson Barnard was the youngest member of the original staff at Lick Observatory. He was an expert professional photographer in his home of Nashville, Tennessee, but at the same time he had become a dedicated amateur astronomer. In 1877 Barnard met the great Simon Newcomb, with whom he discussed his ideas about photography as applied to astronomical observations. Newcomb encouraged the young man, and by 1881 Barnard announced his first discovery of a comet. At Lick Observatory he specialized in stellar photography, and Volume 11 of the Lick Observatory *Publications* contains over 100 plates of star fields and comets taken by him. His most acclaimed observation at Lick was the discovery on 9 September 1892 of the fifth and innermost satellite of Jupiter. Back in Chicago, Burnham proudly displayed a telegram from his young friend, and he extravagantly proclaimed to the press that 'the discovery of this satellite is the greatest astronomical achievement of the century.' Barnard stayed with Lick Observatory until 1895 when George Ellery Hale persuaded him to accept a professorship at the new Yerkes Observatory.

It is no exaggeration to say that James Keeler was the best trained scientist on the original staff. After receiving a BA from Johns Hopkins, he became an assistant at Allegheny Observatory. He went abroad for a year, studying physics in Heidelberg and Berlin, returning to Allegheny Observatory until his appointment as assistant to the Lick Trust two years before Lick Observatory formally opened. With the 36-inch telescope and a grating spectroscope, he measured the wavelengths of the bright lines in nebular spectra. His accuracy was sufficient to show that – like stars – gaseous nebulae have measurable motions toward or away from the earth. In 1891 Keeler left Lick to become Director of the Allegheny Observatory in Pittsburgh. During that time he designed a spectrograph with which he confirmed that the rings of Saturn are meteoritic in nature, and he became cofounder

Figure 26.11 The Crossley 36-inch reflecting telescope in its original mount. Studies of spiral nebulae made with the Crossley paved the way for the development of much larger reflectors.

with George Ellery Hale of the *Astrophysical Journal*. In 1898 he was recalled to Lick Observatory to succeed Edward Holden as director.

Holden, throughout his administration, was remarkably successful in securing special funds and gifts for various projects. In 1895 he had convinced an English amateur, Edward Crossley, to donate his three-foot mirror, together with its mounting and dome, to Lick Observatory. Subsequently he persuaded the Wells Fargo Express Company and the Southern Pacific Railroad to donate the freight charges to bring the telescope from Britain to Mount Hamilton, so that Lick Observatory would have the largest reflector in America.

The mounting for the Crossley telescope was very clumsy and designed for a much higher latitude in England. It was not until Keeler took an active interest in the Crossley that successful photographs were made with it. His pictures revealed how greatly spiral nebulae outnumbered all the other hazy objects detectable in the sky, and his success with a reflecting telescope paved the way for the successful design and application of the larger reflectors that were soon to dwarf the 36-inch Crossley. In 1900 Keeler was quite unexpectedly cut down with a heart attack, bringing his directorship and his astronomical career to a premature end at age 43. His marvelous photographs of the nebulae were posthumously published in Volume 8 of the Lick Observatory *Publications*.

To end this account of Lick Observatory in the nineteenth century, let me quote a poem penned on an early page of the Lick Observatory logbook by the superintendent of construction, Thomas Fraser:

Here pride and arrogance will sink to mites
Glassed with a billion suns and stellar lights;

Sordid villainy will all its courage loose;
And wealth learn how its treasures it may use.
How use them for a noble lasting end
As here is done by California's friend.
No words are meet to praise the illustrious Lick
Through ages long his high panegyric,
The Sun and all yon glittering Stars will write
With penciled rays of there unfading light
And comments coursing through the Milky Way.
Sound his high praise till Earths remotest day.

Notes and references

This chapter is based on the James Lick Centennial Lecture given at the University of California, Santa Cruz, 11 October 1976.

I wish to acknowledge the invaluable assistance given by the late Mary Lea Shane, Curator of the Lick Observatory Archives, in the collection of the material and for the numerous illustrations used in the original lecture. Dorothy Schaumberg of the Mary Lea Shane Archives of the Lick Observatory has graciously provided the subset of photographs used for this chapter. I deeply appreciate Donald Osterbrock's unflagging helpfulness on matters of Lick Observatory history.

27 Atget's eclipse watchers

'Le père Atget' the young artists in Montparnasse called him. Shabbily dressed, trudging through the streets of Paris, with his heavy antiquated camera, he was a picturesque character obsessed with the vision to document on film a great city that was rapidly disappearing. He never exhibited in salons nor published in magazines, and when he offered his pictures to shopkeepers, artists, or museums, he sold them as historical records. Between 1898 and his death in 1927, at age 70, Eugène Atget took 10 000 photographs documenting Paris and its environs.

Today, Atget's prints are treasured as the work of a perceptive and dedicated artist, and several books of his photographs are currently available. Frequently included in such collections is one of his more unusual street scenes, showing a score of Parisians intently watching a solar eclipse (and one young girl impishly watching the photographer). Butchers and merchants, shoppers and schoolchildren are clustered together at the open center of the Place de la Bastille, taking advantage of a noon break and a spring day 'almost like summer' to observe a great rarity: the passage of the moon's shadow almost directly across a major metropolitan area.

Atget's eclipse watchers greatly intrigued me when I first saw a reproduction of the photograph, and eventually I was able to obtain one of the copies from the original glass plate, reprinted in 1956 by Berenice Abbott, a prominent photographer who recognized Atget's remarkable legacy and who saved many of his negatives from oblivion. Many of Atget's photographs are undated, but in this case the picture can be placed in time precisely: around noon on 17 April 1912, when the central path of an annular eclipse passed a few miles to the west of Paris.

Recently, I began to wonder just how rare it was for a major eclipse to graze Paris so closely; the answer involves the geometrical circumstances of solar eclipses.

On the average, the length of the moon's shadow at new moon is 373 530 km and the distance to the nearest point of the earth's surface 378 030 km. This means that when the moon passes in front of the sun, the shadow will usually miss the earth by some 4500 km and the eclipse will be annular, with a ring of sunlight still visible around the opaque moon.

Of course, total eclipses do occur, because the distance of the new moon varies between 350 400 km from the earth's surface and 399 900 km, on account of the eccentricity of the moon's orbit. And this leads us to the question concerning how often an annular or total eclipse can be expected at a given point on earth.

Since the distance to the moon varies, the width of the eclipse path differs from one eclipse to another. This width will change even during a single eclipse, because different parts of the earth lie at different distances from the moon and also because of geometrical

Figure 27.1 A crowd of Parisians viewing the deep partial phases of the 17 April 1912, solar eclipse, as captured by Eugène Atget. This reproduction is from print No. 4 in a numbered edition of 100 made by Berenice Abbott in 1956.

effects as the shadow falls at an oblique angle onto the surface. These conditions make the calculation rather involved, but some readers may wish to try to recompute the answer given by Russell, Dugan, and Stewart in their classical textbook *Astronomy*: they state that, on the average, an eclipse will occur at some arbitrary station 'only once in about 360 years.'

These authors also indicate that the maximum width of the moon's shadow on the earth is 167 miles, which can occur when the subsolar point is directly on the line between the centers of the earth and moon, the moon is at perigee, and the earth is at aphelion. They do not give the full details of their subsequent reasoning, but we can reconstruct it as follows. When such an eclipse sweeps to the edge of the earth, its width there will be only 130 miles, because the edge of our globe is farther from the moon than the sublunar point. On the average (45° in longitude from the central point), the width of the eclipse path will be: $(167-130) \times \sin^2(45°) + 130 = 148$ miles. (The geometrical factor is squared because two dimensions, latitude and longitude, are involved.)

However, the moon will not always be at perigee, nor the earth at aphelion; at some combinations of distances the width of the path of totality will go to zero (and for other distances the eclipse will become annular). Again, the average will go roughly as $\sin^2 45°$ (squared because two effects are involved), so that the mean width should be around 70 miles; Russell, Dugan, and Stewart give 60–70 miles, and presumably their slightly smaller number results from a better approximation than sin² 45°.

Given this result, we now imagine the sunward hemisphere of the earth covered with 120 east–west eclipse paths, each 66 miles wide and stacked one above the other (since 120×66 miles=7920 miles, the diameter of the earth). We need another 120 eclipses to cover the other hemisphere, and since total solar eclipses occur every 1½ years on the average, it would take 360 years to blanket the earth with 240 optimally arranged eclipses. This is the number that Russell, Dugan, and Stewart reported.

However, the earth is not stationary during an eclipse, and since its rotation is in the same direction as the eclipse path, the optimally stacked eclipses will not cover an entire hemisphere after all. This effect is greatest at the equator, with a reduction by $\frac{4}{5}$. Using a geometrically weighted average, the mean time between eclipses at one location should be increased by about 14 percent, to 410 years. Note that although the rotational reduction is greatest at the equator, total eclipses are nevertheless more frequent there because this part of the earth is, on the average, closer to the moon.

In my original 'Astronomical Scrapbook' in *Sky and Telescope*, I only hinted at the Russell–Dugan–Stewart calculation, but the puzzle challenged veteran eclipse computers Jean Meeus in Belgium and Charles Kluepfel in New York to see if the actual frequency of eclipses came close to matching the theoretical approximations. One of the first things they found was that at the present time total eclipses are considerably more frequent in the northern hemisphere than in the southern. More total eclipses occur when the earth is at its greatest distance from the sun (aphelion), because the moon's shadow is longer then. During the present millennium aphelion occurs in July, when the northern hemisphere has the larger share of sunshine. Meeus' calculations show that at latitude 40° north, one can expect a total eclipse every 333 years, whereas at latitude 40° south the interval is 427 years. The same line of reasoning shows that annular eclipses are now more common in the southern hemisphere.

Overall, Meeus found that the mean frequency for a total eclipse at any point on the earth's surface is once in 375 years, and the more common annular eclipses have a mean frequency of 224 years. Working independently, Kluepfel came to the same conclusions.

Without retracing those calculations, there is another way to check the validity of their answer, and this occurred to me as I was contemplating Atget's photograph of the 1912 eclipse. Suppose we take the ten largest population centers and ask how many eclipses cross them at any time in the twentieth century. This is equivalent to asking how many eclipses will fall at a single point over the course of 1000 years. The 375-year statistical interval indicates that we would most probably find three eclipses taking place, provided we consider just the center of each city. But since they are modestly larger than a point, we can expect a somewhat larger number.

The *Canon of Solar Eclipses* by J. Meeus, C. Grosjean, and W. Vanderleen provides an approximate way to answer this question.

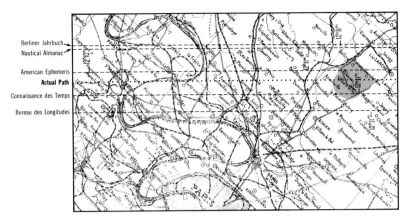

Figure 27.2 Note the two conspicuous prominences at top as well as the inner corona, recorded in this picture of the 1912 eclipse by J. Quenisset at Juvisy, France.

Figure 27.3 Seven decades ago, forecasting the geographical track of a solar eclipse was a much less precise art than it is today. The dashed lines are the central lines for the 17 April 1912, eclipse predicted by the sources identified at left. The dotted line shows the actual path, crossing the city of Saint Germain-en-Laye some 20 kilometers west of Paris. The hatched area indicates where observers in a balloon saw the annular phase.

Unfortunately, a practical difficulty is that their maps are not quite detailed enough for us to be sure whether an eclipse is a direct hit or a near miss. Here is an approximate tally for the largest population centers listed in the *Times Atlas* (London).

New York	1925
Tokyo	
London	
Shanghai	1948, annular (close)
	1987, annular
Mexico City	1908, annular; 1991
Paris	1912, annular (close); 1999
Buenos Aires	1918, annular
Osaka	
Sao Paulo	
Peking	

Three of the eclipses are total, and five are annular, reflecting the higher frequency of eclipses when the moon is too distant to eclipse the sun totally. It is interesting to note that Calcutta, number 12 on the population list, was near the 1980 path of totality and even closer to the paths of the annular eclipse in 1933 and the total eclipse in 1995. More precise calculations made by Kluepfel show that the 1999 eclipse will be a near miss in Paris, and even the 1925 New York eclipse is a problematic statistic: the eclipse was total in northern Manhattan, but not at the zero point of the *Times Atlas* position. The fluctuations in the statistics of small samples have worked against us, and while the results are consistent with the theoretical predictions, they don't confirm the calculation to any high degree.

Incidentally, the 1912 event was one of those unusual hybrids where the eclipse was total over only a small part of its path and annular throughout the rest. At the ends of the path, the distance to the moon was too great (owing to the curvature of the earth) for its shadow to touch the earth's surface. Along the path in Spain there was a brief moment of totality, but in Paris the eclipse was only annular – though complete enough for Parisians, peering through their smoked glasses, to observe prominences and the innermost corona, as shown in Figure 27.2.

One evening a few years ago, while in Paris on a research trip, I strolled along the Left Bank, across the Pont Sully, to the Place de la Bastille, the site of Atget's memorable photograph. The few remaining ruins from the infamous fortress of the days of the French Revolution are now found underground in the Metro station; in the Place itself is a mighty column commemorating the populist uprising of July, 1830. It was in front of the iron-and-marble balustrade of the monument that Atget recorded his eclipse watchers.

The same buildings, very little changed except for their advertising signs, still frame Rue St Antoine toward the west. It would be splendid, I thought, to come here on 11 August 1999, and take a new version of Atget's scene. But who wants to be under a partial eclipse when the path of totality is only 20 miles away?

Notes and references

A nicely printed and comparatively inexpensive collection of Atget's photographs is found in the Aperture series (Elm St, Millerton, NY, 12546, 1980); a more lavish edition is *A Vision of Paris: the Photographs of Eugène Atget, the Words of Marcel Proust* (New York 1980). Numerous contemporary accounts of the 1912 eclipse are found in *L'Astronomie, Bulletin de la Société Astronomique de France*, **26** (1912).

After the original article was printed, Charles Kluepfel and Jean Meeus joined me in a letter to the editor of *Sky and Telescope* in which we described the frequency of eclipses in more detail, and that letter is now incorporated into the foregoing text. Note also Meeus', 'The Frequency of Total Annular Solar Eclipses for a Given Place,' *Journal of the British Astronomical Society* **92** (1982), 124–6; he neglected to mention that it was published in response to this article.

28 *Faintness means farness*

In 1543 Nicholas Copernicus published the first modern blueprint of the planetary system. He took a radical step, declaring that the earth, which everyone had assumed was solidly fixed in the middle of the universe, was actually in swift motion around the sun. In his revolutionary heliocentric cosmology the sun rested at the center, seated, Copernicus said, as if on a throne governing the planets that wheeled around it. 'For who in this most beautiful temple would place the lamp anywhere except where it can illuminate everything at the same time?'

But placing the earth in motion had a potential problem: as it swiftly bypassed Mars, it made Mars appear briefly to go backwards in the sky. Well and good: that matched the observations of Mars' so-called retrograde motion. Jupiter moved backwards, too, a little less, and Saturn also, but still less. But why did not the starry background also have an annual shift back and forth as the earth pursued its yearly course around the sun? Copernicus' solution was to say that the stars were simply too far away: 'So vast, without any question, is the divine handiwork of the almighty Creator!'

Nearly three centuries were to elapse before geometrical triangulations to a few nearby stars proved how right Copernicus really was. The tiny annual oscillation of these stars against the sky was roughly the angle of a small coin seen three miles away. It is no wonder that it took until 1838 for this subtle effect to be detected. Converted into distances, the closest star lies at about 25 million million million miles or 40 million million million kilometers. These results allow us to envision a scale model of our part of the sidereal firmament. If we choose a half-inch ball-bearing to represent the sun here in the Royal Institution's lecture theatre, then in our model the nearest stars would be in Dublin or Amsterdam.

Although our knowledge of the whole distance scale of the universe rests on these so-called trigonometrical parallaxes, the geometrical method can plumb only a small fraction of the universe before the angles become too small to be measured satisfactorily. Therefore, in order for astronomers to fathom stars still more remote, they are obliged to turn to a different physical principle: the diminution of light with distance.

The way the intensity of light drops off with increasing distance from the source is readily understood with a pyramidal diagram such as Figure 28.1. Suppose that we first measure the intensity of the source at unit distance; at twice this distance the light is dispersed through four times the surface area, and hence the intensity diminishes to a quarter of its original value; at three times the distance, the intensity drops to a ninth of its original value; and so on. We can set up the following table, where the original intensity is arbitrarily called L:

Figure 28.1 The geometrical diminution of light with distance. At three times the unit distance, each panel receives ⅑ as much light.

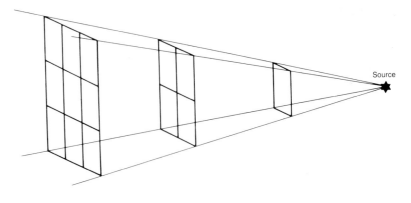

Distance	Intensity
1	*L*
2	*L/4*
3	*L/9*
4	*L/16*

From this we see that the measured intensity decreases as the inverse square of the distance. We can conveniently epitomize this principle with the aphorism *faintness means farness.*

Actually, this physical principle had already been used to get a rough idea of the distances of stars long before trigonometrical parallaxes were measured. In the latter part of the seventeenth century James Gregory, Isaac Newton, and Christiaan Huygens all worked out the approximate distances of the nearby star Sirius by the 'faintness means farness' principle. Huygens, for example, placed a small aperture in the window of a darkened room, and adjusted the lenses in the opening until the tiny beam of sunlight appeared to match the brightness of the star Sirius as he remembered it. He then calculated that he was viewing a round speck with 1/27664 of the diameter of the sun. This fraction squared is his estimate of the ratio of light coming through his aperture to that of the full sun. Since the relative distances of the objects go according to the square root of his quantity, Sirius was 27 664 times as far away as the sun.

'And what an incredible distance that is,' exclaimed Huygens. 'For if 25 years are required for a bullet out of a cannon, with its utmost swiftness, to travel from the sun to us, then for Sirius we shall find that such a bullet would spend almost 700 000 years in its journey between us and the fixed stars. And yet when in a clear night we look upon them, we cannot think them above some few miles over our heads. What I have here enquired into concerns the nearest of them. And what a prodigious number must there be placed in the vast spaces of heaven! For if with our bare eyes we can observe over a thousand, and with a telescope can discover ten or twenty times as many, what bounds of number can we set to those that are out of reach even of these assistances, especially if we considered the infinite power of God! Really, when I have been reflecting thus with

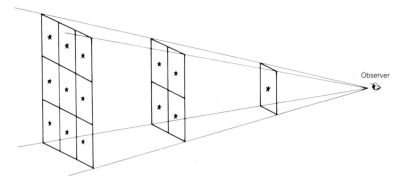

Figure 28.2 If stars are distributed uniformly in space, the observer would receive the same amount of light from each succeeding shell because the increased number of stars in each shell exactly compensates for the decreased light from each individual star.

myself, methoughts all our arithmetic was nothing, and we are versed but in the rudiments of numbers, in comparison to this great sum.'

The perceived vastness of that starry frame soon led astronomers to a curious conundrum. Let us turn back to our light pyramid once more, but this time we will place an observer at the apex and suppose that the pyramid has stars scattered uniformly throughout – in other words, this represents a narrow sample of the universe itself (Figure 28.2). At the first unit distance we have placed a single star. Twice as far away the space opens up and we can place four stars. The apparent brightness of each of these stars is only one-quarter as much as the first star, but there are four of them, so the second zone contributes as much light as the first. Similarly the third zone, with nine stars each one-ninth as bright as the first star, in sum contributes as much light once again. It is easy to see that if the pyramid is infinitely extended, the sky would be ablaze with light. It would not be infinitely bright, because the nearby stars would block out some of the more distant starlight, but since every line of sight would eventually intersect with a star, the entire heavens would shine with the brilliance of the solar disk, and the earth would swiftly vaporize.

The argument was first set out in a form something like this by the Swiss astronomer Loys de Cheseaux, who mentioned it in his treatise on the comet of 1744. De Cheseaux proposed a solution, that space was not perfectly transparent so that the light of more distant stars was increasingly absorbed. Unfortunately his solution is wrong. The absorbed energy must go somewhere, and in fact the absorbing material would heat up and glow just as brightly as the stellar surfaces. Nevertheless, it was an interesting suggestion, which will reappear in another context presently.

Another way out of the difficulty is to suppose that the framework of stars has a natural limit or edge. Such a solution is found – to a degree – in the work of William Herschel on the structure of the Milky Way. Herschel, who eventually became one of the greatest observational astronomers who ever lived, started out life as a musician. He first came to England as an oboist in the Hanoverian Guards, and later settled here, becoming organist and music director at the Octagon Chapel in the fashionable Georgian city of Bath. His excess energies were increasingly given over to his hobby of astronomy and telescope making, and after he had the wonderful good fortune to discover the planet Uranus, he was able to devote himself entirely to astronomy. Among his ingenious accomplishments was a brilliant pioneering attempt to delineate the structure of the Milky Way.

Herschel knew that the great milky band of light girdling the sky was actually formed by the confluence of numerous stars too faint to be resolved individually by the naked eye, but readily visible in his huge telescopes. He suspected that these stars were arranged in a giant disk, with the sun as one star near the center. When the line of sight ran edgewise through the thin disk, the numerous stars crowded together to give the impression of the Milky Way. Herschel further assumed that within the Milky Way disk the stars were uniformly distributed right up to the edge of the disk, so that if he simply counted all the stars in some part of the sky, he could easily reckon the relative extent of the Milky Way in that direction. His result is shown in Figure 28.3, a cross-section of the Milky Way's ragged disk with the sun near the center.

Had Herschel concerned himself with the riddle of the dark night sky, he could have argued that, because the Milky Way had an edge, the stars did not pile up along the line of sight without limit. Such an argument would have been fair only until Herschel began to believe that beyond the Milky Way lay nebulous island universes, vast aggregations of stars, for then the infinite number of such nebulae would fill the night sky with light. As far as we know, Herschel was simply unaware of the problem.

Ingenious and novel as Herschel's observational approach to the problem of the sidereal structure was, it rested on the bold but dubious assumption that the distribution of stars in space throughout the Milky Way was uniform. Herschel had already noticed that in some directions, where there were relatively few stars (and hence by his assumption the Milky Way's edge was relatively close), the stars were quite faint. According to the principle of 'faintness means farness,' such stars (and therefore the edge) were far away, in contradiction to his assumption of a uniform stellar distribution.

One reason that Herschel and his immediate successors did not use the 'faintness means farness' principle in this problem was the great difficulty of measuring quantitatively stellar brightnesses, especially for the fainter stars. Today the great variety of electronic photodetectors gives considerable precision over a wide range of brightnesses, but not until the final decades of the last century did a suitable physical method emerge for measuring apparent stellar magnitudes. Then photography came to the rescue. The first person to exploit the new abundance of photographic data for the problem of sidereal structure was a Dutch astronomer, Jacobus Cornelius Kapteyn. To understand his approach, we must briefly mention the magnitude system that astronomers have traditionally used for specifying the brightnesses of stars.

Since antiquity, the stars have been classified so that the brightest stars are of the first magnitude, and the faintest that can been seen with the naked eye are of the sixth. With a typical pair of binoculars you can penetrate to the tenth magnitude, and the largest telescopes

Figure 28.3 William Herschel's cross-section of the Milky Way derived from his star counts.

Figure 28.4 A uniform distribution of stars, with exponentially spaced shells corresponding to successive magnitude intervals.

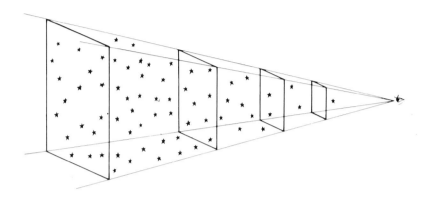

can, with sophisticated electronic detectors, record stars as faint as the twenty-fourth magnitude. About a century ago, when astronomers placed the ancient magnitude system onto a physical scale, a difference of five magnitudes was defined as a factor of 100 in brightness, a fact that will be relevant to us shortly.

If all stars had the same intrinsic brightness, then clearly the fainter stars would be the more distant ones, as expressed by our aphorism, faintness means farness. However, a typical second magnitude star is not simply twice as far away as a typical first magnitude star. If we think about the inverse square pyramid once again (Figure 28.1), we notice that a star at distance 2 is a quarter as bright as a star at distance 1, an easily discernible difference in intensity. But if we imagine a star at distance 101 compared with a star at distance 100, the ratio in intensity is only $(100)^2/(101)^2 = 10\,000/10\,201$, an amount undetectable by eye. To have a magnitude system in which each step appears to have a uniform decrease in brightness requires that the *ratio* of intensity be constant, and therefore equal steps in magnitude do not correspond to equal steps of distance. Instead, the arrangement goes multiplicatively, and looks something like the divisions of the pyramid shown in Figure 28.4.

To gain some insight into Kapteyn's investigations of the sidereal structure, we will first hypothesize that the stars are distributed uniformly and inquire how many more stars are visible with each additional magnitude step. Imagine a light pyramid such as the one in Figure 28.2, but 10 units long. In the first interval there is one star, while in the entire 10-unit pyramid there are $10 \times 10 \times 10$ or a thousand stars. The stars 10 times farther away will be 1/100 as bright as the star in the first zone, and this ratio of 1/100 corresponds to five magnitudes. Now if there are four times more stars for each magnitude step, then we would have

first magnitude step	4
next magnitude step	16
next magnitude step	64
next magnitude step	256
fifth magnitude step	1024

This simple numerical demonstration shows that if we have about four times more stars for each additional magnitude step, then we come out with the expected number for a uniform distribution in our pyramid.

Now let us compare our hypothesis to the actual sky. There is one brilliant star of magnitude −1, Sirius. If we go one magnitude step fainter, to magnitude 0, we find Canopus, Rigel, and Arcturus, and together with Sirius we have the expected four. (Stars of magnitudes −1 and 0 can be considered super first magnitude stars, since they are several times brighter than ordinary +1 magnitude stars. This situation arose from the attempt to match the majority of stars with the ancient magnitudes of Ptolemy, which had grouped stars of a considerable brightness range all in the first category.) Magnitude +1 adds a dozen more, bringing our total to 16 − astonishing agreement with our table, especially considering that we must except some statistical scatter. But the situation is not so regular when we carry on to fainter magnitudes:

Mag	Stars	Ratio
−1.0	1	
0.0	4	4.0
1.0	16	4.0
2.0	47	2.9
3.0	162	3.4
4.0	509	3.1
5.0	1616	3.2
6.0	4936	3.0
7.0	14278	2.9
8.0	40233	2.8

The numbers for the still fainter stars, which are known from Kapteyn's massive star counts made around the turn of the century, bear out this deficit. What are we to make of this? One subtle thing that may have gone wrong is our assumption that all the stars have the same intrinsic brightness, but in fact, the ratio of 4 holds even when the stars have a distribution of intrinsic brightnesses, provided the distribution itself holds uniformly throughout space. Kapteyn spent a huge effort showing that this was true enough. A more obvious conclusion from the star counts might be that our initial hypothesis was wrong, and that the stars are not uniformly distributed in space. This was essentially what Kapteyn concluded, and

Figure 28.5 The 'Kapteyn Universe' of 1922.

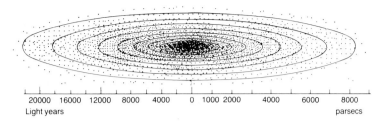

20000 16000 12000 8000 4000   0 1000 2000   4000   6000   8000
Light years   parsecs

he published a schematic diagram of the Milky Way (Figure 28.5) showing the sun near the center and the stars dropping off to a diffuse boundary about 10 000 light years away.

At the same time that Kapteyn was working at his astronomical laboratory in Groningen, Holland, in the late 1910s, a brash young American named Harlow Shapley had begun to exploit the largest telescope in the world, the 100-inch reflector on Mount Wilson in California. Quite unexpectedly, his seemingly unrelated field of research – variable stars in globular clusters – led to a new picture of the Milky Way, one quite different from Herschel's or Kapteyn's. And once again the principle 'faintness means farness' was intimately involved.

The globular clusters, Shapley's chosen topic, are immense congeries of tens of thousands of stars in relatively compact spherical swarms (Figure 28.6). From the tropics a few globular clusters can be seen with the naked eye, and from the north temperate latitudes several can be found with a pair of 7×50 binoculars. Altogether the Milky Way includes about a hundred globular clusters. Among the most luminous stars in these globulars are a few that pulsate in brightness, the Cepheid variable stars, named after a prototype in the constellation Cepheus. From a dozen relatively nearby Cepheids, Shapley was able to deduce their intrinsic brightnesses, and then with the 'faintness means farness' principle and the reasonable assumption that these variables are the same everywhere, he was able to calculate the farness of the very faint Cepheids in several representative globular clusters.

The results of Shapley's measurements were staggering. The typical globular cluster seemed to be about 50 000 light years away, much farther than the bounds of Kapteyn's Milky Way. This might not have been too worrisome, except that Shapley noticed that the globulars had a peculiar distribution over the celestial sphere. Virtually all were in one half of the sky, and fully one-third were concentrated in the 5 percent of the sky near the constellation Sagittarius in the summer southern Milky Way. Shapley made the daring conjecture that the globulars were centered on a distant and invisible nucleus of our Milky Way, which lay about 50 000 light years away in the direction of Sagittarius. Kapteyn's universe, according to Shapley, was merely a local star cloud within a much vaster system.

The older astronomers steeped in the Kapteyn tradition of star counting and stellar statistics were not about to accept the new hypothesis without a fight. They challenged Shapley's calibration of the Cepheid luminosities and his bold extrapolation to an unseen

Figure 28.6 Globular Star cluster M13 in Hercules.

Figure 28.6 Globular Star cluster M13 in Hercules.

nucleus. But gradually more evidence accumulated, and when Continental astronomers showed that the stars in our part of the Milky Way were rotating about a distant center in the direction of Sagittarius, the debate seemed settled. Still, there remained a glaring discrepancy between Kapteyn's comparatively small universe and the giant Milky Way envisioned by Shapley.

Once more an answer came from a seemingly unrelated field of research, this time carried out by Robert Trumpler, an astronomer studying open clusters at the Lick Observatory in California. Open clusters, as their name suggests, are much more loosely bound than the rich globulars; typically they have a few hundred stars rather than tens of thousands. The Pleiades is a fine example of a relatively nearby open cluster.

Despite the variety of open clusters, Trumpler managed to sort them into a number of simple categories. Then, within a particular category, he could establish their relative distances by not one but two different principles. First, he could use 'faintness means farness,' so that the fainter clusters were naturally at greater distances. But, secondly, he could use another principle, 'smallness means farness,' because the more distant clusters subtended smaller angles in the sky. It was necessary to use nearly a hundred clusters in order to average out the roughness of his categories, and when he looked at the results, he found the two methods disagreed. How could Trumpler make sense of this? One way out of the difficulty, and it soon became clear that this was the right answer, was to assume that the space in the plane of the Milky Way was not perfectly transparent, but that it was sparsely scattered with fine dust that absorbed the distant starlight. The angular sizes of the clusters were not affected, but the fogginess of space made the more distant clusters appear fainter and hence even farther away.

Trumpler's discovery of interstellar absorption immediately provided the key to understanding the discrepancy between Kapteyn's and Shapley's results. The Dutch astronomer had unwittingly been looking at a foggy universe, and the stars at great distances appeared to thin out not because there were fewer of them, but because the dust in space was simply obscuring the view. Of course, there really is an edge to the Milky Way, but it is much further than Kapteyn had supposed. However, it was not quite so large as Shapley had argued, because it now became apparent that the dust had caused him to overestimate the distances to the globular clusters. Thus, the fogged-out center of the Milky Way lies at 25 000 light years rather than the 50 000 originally determined by Shapley.

If the entire sidereal universe consisted only of the Milky Way, de Cheseaux's riddle of the dark night sky would be solved because the distribution of stars simply did not extend to infinity. However, the story of celestial discoveries at the Mount Wilson Observatory did not end with Shapley's explorations. During the 1920s Edwin Hubble, another of the California astronomers, observed several Cepheid variables in the great Andromeda Nebula, a huge spiral-shaped pattern of faint stars and nebulosity. When he applied the 'faintness means farness' principle, the resulting distance was even more breathtaking than those found by Shapley. The Andromeda Nebula was nearly a million light years away. Hubble had found an island universe rivaling the Milky Way itself.

The Andromeda Galaxy appeared as the largest, and presumably one of the closest, of thousands of spiral nebulae that had by then been photographed (Figure 28.7). In the Andromeda Galaxy the Cepheid variables were already close to the limit of visibility, so that it was almost hopeless to find such stars in the still fainter spirals. Nevertheless, the 'faintness means farness' and the 'smallness means farness' principles were the only tools available, and both depended on finding objects that were likely to be intrinsically similar. Hubble and his colleagues therefore turned to clusters of galaxies, which at least had the merit of showing samples of nebulae and not just isolated individuals that might be tiny dwarfs or vast giants. They could hope that the fifth brightest galaxy in one cluster would be similar to the fifth brightest in another, even if the first or second brightest cluster members happened to be extraordinary and non-typical specimens. Such comparisons soon showed galaxies 100 million light years away, and the universe seemed to stretch forth without limit.

Armed with such hypotheses, Hubble soon made another spectacular discovery. When he examined the spectra of the distant galaxies, he found that certain characteristic features had been shifted toward the red end of the rainbow compared to laboratory measurements. This meant that the galaxies were racing away from us, and he found that the farther they were, the faster they were moving, with typical recession speeds measured in the tens of thousands of kil-

Figure 28.7 The great spiral galaxy M31 in Andromeda with satellite galaxies M32 (above) and NGC 205 (lower left).

ometers per second. The universe was getting larger! Not only that, but it was possible to calculate when this colossal expansion all started – a value now given as approximately 15 billion years ago.

An infinite universe of galaxies poses once more the riddle of the dark night sky. What prevents even more starlight, in the form of galaxies, from crowding the heavens with light? Today the riddle can be solved, by considering not only the vast space of the universe, but time as well. The role of time in the solution can be better appreciated if we examine one further component of the universe, the quasars, which were discovered and interpreted in the 1960s.

First called 'quasi-stellar radio objects,' they were just that: faint objects that looked like stars and were powerful sources of radio noise (and therefore first noticed by the radio astronomers). They proved to have spectra unlike any stars or galaxies ever seen. Eventually Maarten Schmidt of the California Institute of Technology showed that these were spectra exhibiting enormous red shifts, and he concluded that the objects lay at immense distances, almost beyond the realm of the galaxies. In the past two decades thousands of these quasars have been found. Nevertheless, they are extremely rare compared with stars. For example, there are at least 10 million fifteenth-magnitude stars, and only about 40 fifteenth-magnitude quasars. It we look at the counted numbers of quasars, we find a surprising distribution:

Mag	Quasars	Ratio
15	41	
16	412	10
17	4120	10
18	29174	7

Now if the quasars were distributed uniformly throughout space, we would expect to find four times as many for each succeeding magnitude step, just as in the case of stars. Here, however, we find at each step an enormous overabundance of quasars. This result cannot come from the dustiness of space, for fogginess would diminish rather than augment the numbers at each step. Space cannot be super-transparent. The only conclusion remaining is that there really are more quasars at the more distant environments – it is as if we are living in a peculiar region of the universe where the quasars are curiously depleted. Such a view is difficult to accept until we realize that not only space but also time is involved. As we look out into the realm of the quasars we are looking back around ten billion years in time, to an epoch closer to the beginning of the universe than to our own age. The quasars have long since expended their brilliant fireworks, and in our region of space and recent time they remain only as glowing embers from a distant past.

Our journey has carried us from the solar system, as first envisioned by Copernicus, into the starry realms of our galaxy, and then to the galaxies and to the distant quasars. But our journey through space has carried us backward in time, toward the beginning of time itself, perhaps 15 billion years ago. Here, then, is the real solution of de Cheseaux's riddle – not a paradox at all, but an observational consequence of the fact that the universe had a beginning and therefore an edge. If we consider the distribution of stars in our part of the galaxy, and ask how far we must go to find enough stars so that every line of sight intersects a star – so that the sky would then be ablaze with light – the answer is 300 million million light years. This is so far beyond the beginning that we can easily say that the universe isn't big enough, or that it isn't old enough, to fill the sky with light. And in most of space stars are spread far more sparsely than in our Milky Way, so that the distance becomes even larger.

The galaxies fill up space much more efficiently than stars. For example, if I use a pound coin to represent the Milky Way, then the coin symbolizing the Andromeda Galaxy in my scale model would not lie in a distant city, but less than 2 feet away. Because the stars within the galaxies are spread out so sparsely, the galaxies are very dark and contribute very little light to the night sky. When we look at a typical photograph of a spiral galaxy, we really get a misleading impression. It is a time exposure, in which the photons of light have been built up to an unnaturally high level. In reality the realm of the galaxies is a very dark and cold region.

In contrast we live in a bright, and astonishingly complex environment, and we can indeed be grateful to the Creator that we can enjoy such a rich habitat. The paradox is not de Cheseaux's, but another very different one. The high technology that has helped us discover and understand galaxies and quasars is also powerful enough and awesome enough to change and irrevocably alter, in a matter of minutes, this magnificent planetary home. Let us pray that we are ourselves bright enough to preserve this precarious environment not only for ourselves but for generations to come, who may find still deeper surprises in our aphorism *faintness means farness*.

Notes and references

This chapter was presented as a Friday Evening Discourse at the Royal Institution in London, 21 February 1986.

   I am greatly indebted to Michael A. Hoskin for a series of illuminating discussions on these topics, and to Antony Hewish for inspiring me to undertake this lecture. The Copernicus quotations are from Book I, Chapter 10, of his *De Revolutionibus*, and the Huygens quotation from near the end of his *The Celestial Worlds Discover'd* (London, 1698). De Cheseaux's riddle, commonly known as Olbers's Paradox, is discussed in Edward Harrison's *Darkness at Night: A Riddle of the Universe* (Cambridge, Mass., 1987). Willem DeSitter's *Kosmos* (Cambridge, Mass., 1932) contains a chapter on Kapteyn and his system of the universe.

29 *The mysterious nebulae, 1610–1924*

On the night of 28 August 1758, while observing the comet of that year in the constellation Taurus, a 28-year-old apprentice astronomer atop the Tour de Cluny in Paris glimpsed what appeared to be a second, equally bright comet about 5° farther south. His pulse must have quickened, for he had not yet found a comet of his own, and such a discovery would have brought recognition and perhaps a promotion. It was already after midnight, but no doubt he waited until dawn, anxiously hoping for the telltale movement that would designate a comet. We can imagine the high anticipation with which the young Charles Messier waited for the next night, and the crushing disappointment when his first comet turned out to be a mere nebula.

A mere nebula indeed! Little could he know that his nebula would come to be far more fascinating than all of the comets he ever discovered. 'A whitish light, elongated like the flame of a candle, without any stars,' he described it; only later did it get the designation M1 from his catalog, and much later the name 'Crab nebula.'

Within a year Messier had found his own comet; fame and promotions did follow, and eventually, in 1770, he gained the coveted election into the Académie Royale des Sciences in Paris. And what could be more fitting as his first major memoir for the Academy publications than a catalog of his nebulae and clusters, starting with the one he had found that summer night in 1758? He rounded up about 40 troublemakers, and then topped off his list with well-known objects such as the Orion nebula and Praesepe for a total of 45. In the years following he added one and then a second supplement to his original list, bringing the total to 103.

In our century, Messier's list is a guide to showpieces for professional astronomers and a delight to amateurs, and in fact, I began my own career as an amateur astronomer sweeping the sky for these sparkling star clusters and faint wispy nebulae. And so it was that one night in July of 1948 I stumbled on a nebula that moved! With quickened pulse I watched in my 8-inch reflector long enough to be sure it was a comet, and then dashed off a telegram to Harvard College Observatory, the western-hemisphere clearing house for comet discoveries. A week later I got back a postcard saying, 'Thank you for your observation of Comet Honda.'

In the long run, however, I would have to say that my investment in the telegram paid off, because one thing led to another and in 1949 I got a job as a summer assistant to Harlow Shapley, who was then director of Harvard College Observatory. In the dusty book stacks of HCO, I found all the original publications of Messier and his contemporaries, and there at HCO also (a few years later) I first met Helen Sawyer Hogg. Indeed, it did not take too long to discover among those musty almanacs and memoirs that I was close on her trail. Many of these long-forgotten antique reports of nebulae she had

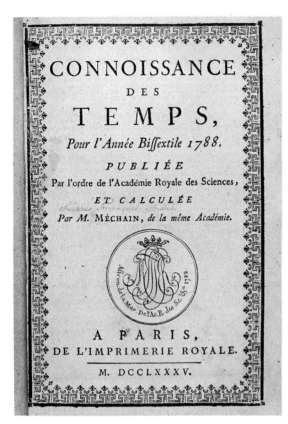

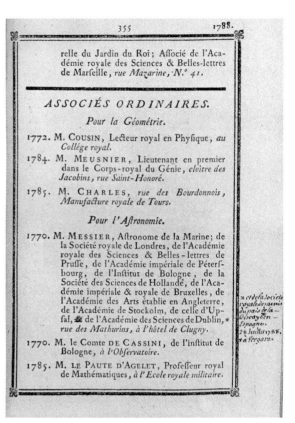

Figure 29.1 The title page of the HCO copy of Connoissance de Temps *for 1788 bearing the circular bookplate inscribed with Messier's name, and a page showing where he has updated his own list of honors.*

already brought to light in her wonderful series for the *Journal of the Royal Astronomical Society of Canada* called 'Out of Old Books,' which ran from 1946 to 1954. I took courage in hand and wrote to ask for her *Bibliography of Individual Globular Clusters*, which was rich in citations of early literature. Distribution of the bibliography was rather restricted, she responded, but in special cases like this it was available, and with that vote of confidence she sped me on my way to becoming a professional astronomer, and ultimately, a professional historian of astronomy.

I was fascinated with the old books still found on open shelves at HCO in those days, and I pursued Messier with a particular passion. My French was then singularly primitive, and Helen generously gave me a long translation that substantially aided my researches. One of the interesting games was to discover the patterns of nebular discovery in the eighteenth century. It was clear that Messier's colleague Pierre Méchain was the first to appreciate the richness of the Virgo region of galaxies, and that Méchain had found a number of faint nebulae that Messier never got around to adding to his own list. These latter discoveries were described in a letter by Méchain published in Bode's *Astronomisches Jahrbuch* for 1786. Dr Hogg had identified three of them, and now they are sometimes listed as M105, M106 and M107. I found that it was not too difficult to identify the other two, but I have always been a little uneasy by designations such as Messier 105 or Messier 109, since we have no specific evidence that Messier ever examined those particular nebulae. It would perhaps be more appropriate to call them Méchain 105 or Méchain 109, but even that would be rewriting history since Méchain never made any attempt to number them systematically.

Except for a large number of nebulae near the end of his catalog, Messier was apparently the independent discoverer of the objects in it. However, he did report a certain number of earlier sightings. He found his first object, M1, depicted on an atlas authored by John Bevis in England. Bevis had bad luck with his *Uranographia Britannica*, for after the plates were made but before even a title page had been set, around 1750, his printer went bankrupt. Apparently Bevis had a number of proof copies that had been struck from the plates, one of which Messier eventually saw. (Complete sets are now quite rare – fewer than 20 are known to exist.) Messier does not mention Bevis in the 1783 supplement to his catalog, but in the revised and extended edition of the following year he does include the information that Bevis had found the nebula in the horn of the Bull in 1731. Except for this printed note of Messier's, the 1731 date is completely undocumented. Since Bevis had died in 1771, Messier clearly did not get the information directly from the discoverer, and the source of that date remains a mystery, although, in the absence of a standard title page, Messier might have for some reason believed (erroneously) that 1731 was the date of the atlas.

In the popular literature, Messier is always referred to as a comet hunter who cataloged the nebulae primarily as nuisances to his primary goal. His original memoir of 1771 contains nary a hint of this, and certainly gives the impression that he was examining the nebulae for their own intrinsic interest. However, I think his true colors are revealed by a passage that I found in another of the old almanacs at Harvard, the *Connaissance des Tems* for 1801. (Note the rationalized spelling of the revolutionary period.) There he wrote:

What caused me to undertake the catalogue was the nebula I discovered above the southern horn of Taurus while examining the comet [of 1758]. . . . This nebula had such a resemblance to the comet, in its form and brightness, that it caused me to search for others, so that astronomers would not confuse these same nebulae with faint comets. I observed more of them with telescopes suitable for comet hunting, and this was the intention I had in forming my catalogue. After me the celebrated Herschel published in the *Philosophical Transactions* for 1786 and 1789 a catalog of 2000 that he has observed. This unveiling of the sky was made with instruments of great power, which are not useful for sweeping the sky in a search for comets just beginning to appear. Thus, my object was different from his.

In contrast, three decades earlier Messier remarked that each time he had observed a nebula he had made a careful drawing, 'so that in the future it would be useful for observing them to see if they had been subject to any changes.' Such a motivation placed him onto the cutting edge of speculative eighteenth-century sidereal astronomy, for the possible evolution of nebulae was an emerging question, and one that remains of interest to this day. That Messier did nothing

Figure 29.2 The belt and sword region of Orion from Galileo's Sidereus Nuncius *(1610), showing numerous additional stars but omitting the nebula.*

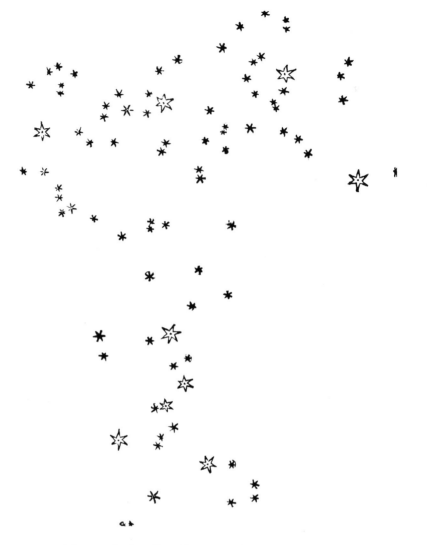

more with it undoubtedly reflects the observational difficulty of finding changes in the nebulae as well as his distaste for theoretical or mathematical astronomy.

Nevertheless, the idea of nebular evolution arches over the entire history of telescopic observations. Recently there has even been speculation concerning possible observed changes in the Orion nebula since the time of Galileo; I would like to argue that the case for a marked increase in brightness early in the seventeenth century is built on quicksand. The grounds for these (spurious) claims go back to 1610, when Galileo was first examining the starry realms with his newly developed refractor. Early that year he mapped the sword of Orion, without recording any nebulosity, at least as far as the published diagram in his *Sidereus Nuncius* is concerned. Does this mean that there was no milky nebulosity to be seen?

Here we must look carefully at the context of his published chart. The title page of Galileo's report proclaims (in part),

THE STARRY MESSENGER, revealing great, unusual, and marvellous sights,

Figure 29.3 Lord Rosse's greatly metamorphosed drawing of the Crab nebula, made with his 36-inch reflector around 1844, and the later drawing made by Secchi with a 9-inch refractor and published in 1856, showing obvious influence from Lord Rosse's erroneous depiction.

observed by Galileo Galilei, with the aid of a spyglass newly invented by him, on the face of the moon, the innumerable fixed stars, the Milky Way, the nebulous stars, and especially the four planets revolving at different distances around Jupiter. . . .

These discoveries had fallen very quickly into Galileo's hands after his first recorded observation of the moon on 30 November 1609 and the sequence of Jovian observations beginning 7 January 1610. He readied his small booklet for the printer in February and early March, and apparently decided as an afterthought to expand the section on the stars, for these observations come on two unnumbered leaves inserted between 16 and 17.

In this section Galileo writes,

I have observed the essence or substance of the Milky Way, with an ocular certainty to resolve all the disputes that have vexed philosophers throughout so many ages: the Galaxy is nothing but a congeries of innumerable stars packed together in clusters. And not only in the Milky Way are these whitish clouds seen, but several patches shine here and there in the aether with similar faint colour, and if a telescope is turned on any of them, we are confronted with a closely packed mass of stars. And what is even more surprising, the stars which astronomers have up till now called *nebulous* turn out to be groups of small stars wonderfully arranged.

This, then, is another discovery worthy of note on the book's title page.

To clinch his point more specifically, Galileo draws in detail two regions that have been called 'nebulous' both by Ptolemy and by Copernicus in their star catalogs. One is Praesepe in Cancer, the other is the head of Orion, and both are well resolved, into 38 and 21 stars respectively.

But Galileo also diagrams in rich detail the belt and sword regions of Orion, without the slightest indication of any milky nebulosity in the sword. Why not? Had Galileo noticed the nebulosity (and it would be remarkable if he had not), he surely would have supposed that this milkiness, too, would be resolved with some improved telescopic power. It would have diminished his important discovery to mention so quickly a counterexample, and in his mind it would surely have been unwarranted to conclude that a genuine nebulosity existed after the previously known examples had broken into stars under his telescopic scrutiny.

Recently an old and exceedingly rare printed account of nebulae by the Sicilian astronomer Hodierna has come to light. Hodierna had no ax to grind about the nature of nebulae, and in 1653 he depicts (for the first time) the Orion nebula. It seems to me to be an outright false deduction to suppose that the Orion nebula came into brilliance in the short interval between 1610 and 1653. In detective work, we must always remember that absence of evidence is *not* evidence of absence.

Nevertheless, the question of possible changes in nebulae tantalized astronomers increasingly in the nineteenth century. Because the Orion nebula had been so comparatively well observed, it became the inevitable focus for examination. Edward Holden, who had not yet become director of Lick Observatory, published in 1882 an entire monograph on this object. Lavishly illustrated, it included what was then the earliest known depiction, by Christiaan Huygens, as well as eighteenth-century drawings by Messier and Herschel. The role of nineteenth-century observers was extensive, and Holden included drawings by Struve, Bond, Lassell, Rosse, Secchi, d'Arrest and Trouvelot. These images proved little about any evolutionary changes in the Orion nebula, but they do illuminate interesting points about both the psychology and physiology of observation.

Beginning in the 1860s the drawings by G. P. Bond, William Lassell, and Lord Rosse depicted the nebula in increasingly cubistic terms. When observers with lesser telescopes followed this style, one is almost obliged to conclude that it was fashionable to sketch the Orion nebula this way, rather than to show what the nebula was really like.

The effect of a prior image influencing a depiction is even more striking with respect to M1, the Crab nebula. In 1844 Lord Rosse published in the *Philosophical Transactions* a figure of M1 made with his 36-inch reflector that showed the nebula with remarkable appendages. Not long afterward Angelo Secchi, using a much smaller refractor at the Vatican Observatory, produced a notoriously similar view. All later depictions, including Rosse's, were totally different, and in 1880 he remarked 'would have figured it different from drawing in *Philosophical Transactions* 1844.' A recent examination by David Dewhirst of the archival drawings from Birr Castle shows how the original telescopic drawing (perhaps made from memory the next morning) metamorphosed through a series of intermediate images to the 1844 published drawing, which no longer bore much resemblance to the celestial object.

But Father Secchi was not the only observer who brought to the eyepiece unwitting preconceptions. After Lord Rosse found the spiral structure in M51, calling it the 'Whirlpool nebula,' he began to see spirals everywhere in the sky – not just in genuine spiral galaxies, but in planetary nebulae, and even in the irregular companion to M51 itself. (See Lord Rosse's drawing, Figure 29.4.)

After the 1880s, *drawing* nebulae became obsolete. The introduction of dry plates made possible the photography of faint objects such as the Orion nebula – first in 1882 by Henry Draper in New York state, and then much more dramatically in 1883 by Andrew Ainslie Common in England. These early photographs in fact showed some of the

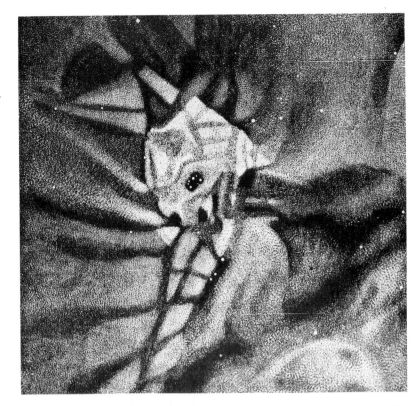

Figure 29.4 The cubistic representation of the central region of the Orion nebula made by Lord Rosse with the 72-inch reflector in 1865–7 should be compared with Malin's picture below.

Figure 29.5 David Malin's unsharp mask photograph of the central region of the Orion nebula is from plate V1650 of the 3.9-meter Anglo-Australian Telescope.

Figure 29.6 One of the drawings made of the Whirlpool nebula with Lord Rosse's 72-inch reflector, showing spiral structure not only in the main nebula but in the irregular satellite galaxy as well.

blocky structure of the visual drawings, but soon a very different, more homogeneous view of the Orion nebula imprinted itself on astronomers' minds. Not until the recent unsharp mask photographs by David Malin have we truly been able to appreciate the wonderful skills of the visual observers of a century ago. Perhaps in part they were psychologically influenced by the fashions of the day, but the fashions had a real physiological basis: they exaggerated contrast differences at the same time that they suppressed large intensity differences in order to produce their remarkable drawings, so astonishingly similar to the unsharp mask images.

The introduction of photography into astronomy gradually transformed the way nebulae were discovered and studied. When James Keeler became director of Lick Observatory in 1898, he took over the previously scorned 36-inch Crossley reflector and made it into a powerful tool for discovery of nebulae. By the turn of the century he could estimate that 120 000 nebulae were bright enough to be photographed by the Crossley, and a very large fraction of them appeared to be spirals.

Photography eventually showed that the sought-for large-scale evolutionary changes were virtually absent in the nebulae. However, it began to reveal more subtle alterations: the changing brightness of stars within the spirals, first novae, and then other variables. Keeler's success in photographing the spiral nebulae with the Crossley reflector at Lick Observatory paved the way for the building of the large reflectors at Mount Wilson Observatory under the astronomical entrepreneurship of George Ellery Hale. Soon these two mountaintop

Figure 29.7 Dr Harlow Shapley at his famous circular revolving desk, as photographed by Frank Hogg in the late 1920s.

observatories were locked into intense rivalry. The first novae were found in the spirals in March, 1917, by Heber D. Curtis with the Crossley reflector, but before he published his discovery, George Ritchey at Mount Wilson announced by telegram that he had found one in NGC 6946. Curtis, although scooped by the competing establishment to the south, at least had a clear idea of what to do with the discoveries, which he promptly used to infer that "the novae in spirals furnish weighty evidence in favor of the well-known 'island universe' theory of the spiral nebulae."

Meanwhile, Harlow Shapley, a young Missourian with a Princeton PhD who had been hired by Hale to study star clusters, used the 60-inch reflector at Mount Wilson to find his own nova, in M31. Shapley promptly published a discussion in which he stated that 'the minimum distance of the Andromeda nebula must be of the order of a million light years.' This was not the entire story, however, for photography was revealing (or so it seemed) yet another change in the spirals: a steady rotation. In his paper Shapley mentioned the work of his Pasadena colleague, Adriaan van Maanen, on the apparent rotations of the spirals, which if correct, and if the spirals were at extragalactic distances, would have led to rotational velocities approaching the speed of light. "Measurable internal proper motions, therefore, can not well be harmonized with 'island universes' of whatever size," Shapley concluded, but in October, 1917, he was prepared to dismiss van Maanen's findings on the evidence of the novae.

Within a few months, however, Shapley flip-flopped on his weighting of the conflicting evidence. At Princeton he had worked on the orbits of eclipsing binary stars, and among other things, he convinced himself that the Cepheids could not be interpreted (as was then commonly thought) as eclipsing binaries. Hence he had strong reasons to believe that the periodic variation arose not by geometric accident but by the intrinsic physical nature of each Cepheid. Thus Henrietta Leavitt's period–luminosity connection was more than just a 'remarkable relation' but a physical law, something that could be assumed to be the same everywhere and therefore suitable for distance measurements. Shapley calibrated the intrinsic brightnesses of the Cepheids statistically (on the rather shaky basis of proper motions of 11 galactic Cepheids), and, applying this result to his globular clusters, concluded that they lay at tens of thousands of light years from the sun.

By the beginning of 1918 Shapley took a mighty conceptual leap: the globular clusters must define a distant nucleus of our own extended Milky Way system. Writing to Eddington on 8 January, Shapley said:

I have in mind from the first that results more important to the problem of the galactic system than to any other question might be contributed by the cluster studies. Now, with startling suddenness and definiteness, they seem to have elucidated the whole sidereal structure. . . .

The equatorial diameter of the system is in the order of 300 000 light-years: the center is some 60 000 light-years distant. Our local cluster, very loose and perhaps ill-defined, is about half way out to the edge. . . . So far as I have gone everything seems to fit together beautifully. The above is sketchy and arrogant, I know; and I haven't much excuse for it. You will be able to see all the consequences I have, and probably many more, so there is no pressing need to summarize further. [Harvard University Archives.]

Among the other consequences of this 300 000-light-year diameter was that the Milky Way seemed to be so much larger than M31 (at a million-light-year distance) that Shapley abandoned the idea of island universes, and van Maanen's measurements of the internal motions in the spirals seemed to corroborate this judgment. The story of how Shapley debated with Curtis over the scale of the universe is now part of the standard lore of astronomy. The two men faced off at an evening *converzatione* at the National Academy of Sciences on 26 April, 1920. Shapley, not an experienced debater, was apprehensive. He devoted his talk primarily to an elementary exposition of the distance scale, coming eventually to the large size of the Milky Way, and he ended with a glowing account of an optical 'image intensifier' that he had invented. This account must have been intended for the benefit of a pair of Harvard representatives, who were there scouting for a new observatory director and who probably needed something like an image intensifier at the Cambridge observatory to overcome the handicaps of East Coast weather. Curtis, in a momentary quandary, decided to carry on with his more technical rebuttal of Shapley's use of Cepheids and his considerable doubts concerning van Maanen's spiral motions. Speaking from the floor, Henry Norris Russell defended Shapley's variable star techniques so brilliantly that the Harvard representatives decided straightaway that the vacant directorship should be offered to Russell!

The following day the *Kansas City Star* carried a 17-inch account of the debate. Since it is unlikely that they had a reporter present, and since Shapley himself had once been a Kansas reporter and was always a fanatical baseball enthusiast, this ephemeral account (here considerably abridged) from his vast archive of newspaper clippings probably gives Shapley's view of the debate (although he can hardly be held accountable for the headlines):

A LEAGUE OF UNIVERSES!

OUR UNIVERSE, INCLUDING THE MILKY WAY, IS A COMPARATIVELY SMALL AFFAIR, A TAILENDER IN A BUSH LEAGUE, ONE FACTION HOLDS

Washington, April 27. – Is there one great universe or a million? Is the great unmeasured nebulae [*sic*] of which the world is but one atom, one big unit or is there a league of universes, each independent of each other?

Undisturbed by the problems of reconstruction, heedless of the League of Nations war between the President of the United States and the senate, whether overalls will accomplish the undoing of high prices, and like pregnant issues of the earthly hour, the National Academy of Sciences in session here listened to a learned debate on the dimensions of the universe. . . .

On the one side ranged the supporters of Dr. Heber D. Curtis of the Lick Observatory, who defended the old theory that our universe is one of many universes, a theory until recently generally accepted by astronomers.

On the other side appeared the following of Dr. Harlow Shapley, who, in an exhaustive discourse, replete with terms of the higher and highest calculus, sought to demonstrate that there is but one universe, but a universe ten times larger than that hitherto conceived by the wildest astronomical calculation. . . .

In the confines of his comparatively cramped universe, Dr. Curtis touched off a series of pyrotechnic displays of cosines and tangents demonstrating to the satisfaction of his partisans his conception of a league of universes, leaving the question open whether our universe belongs to a major, minor or bush league in the cosmic game. . . .

Then Dr. Shapley came to bat with his theory of one big universe, which he did not attempt to bound. He began by postulating a galaxy of stars ten times greater than any conceived by Dr. Curtis. He descanted upon star spirals 'as presumably inter-galactic objects of nebular construction – that is, a part of the grand system and not individual galaxies or other great universes. . . .'

In the weeks immediately following the 'Great Debate' Henry Norris Russell turned down the offer of the HCO directorship, and eventually Harlow Shapley received the position. By April of 1921 he

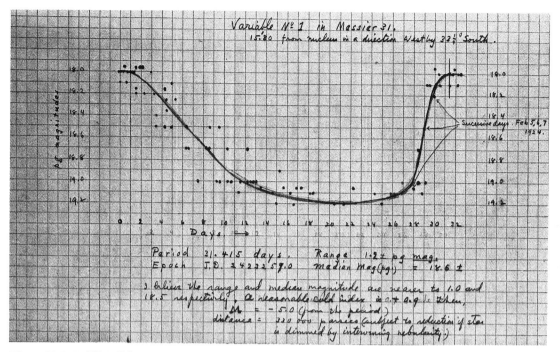

Figure 29.8 Hubble's light curve of the first Cepheid found in the Andromeda nebula, sent to Shapley in 1924.

had left Mount Wilson, and therefore he witnessed from afar the resolution of the island–universe controversy.

The resolution came through the researchers of Edwin Hubble, another Missourian brought to Mount Wilson by Hale. Hubble exploited the 100-inch reflector with great effect in his studies of spiral nebulae, and on a plate of the outer regions of M31 taken 5 October 1932 he recognized as a Cepheid a star that he had at first mistaken for yet another nova. After he had obtained a full light curve, he broke the news in a letter to Shapley. In response, Shapley admitted, 'Your letter telling of the crop of novae and of the two variable stars in the direction of the Andromeda nebula is the most entertaining piece of literature I have seen for a long time. . . . That second, fainter variable [the Cepheid] is, under the circumstances, a highly important object.' [27 February 1924.]

An opportunity to present these results came at the 1924 joint meeting of the American Astronomical Society and American Association for the Advancement of Science held in Washington, DC, where a substantial prize was being offered for the best scientific paper. Joel Stebbins, then the secretary of the AAS, later described to Hubble an amusing and memorable anecdote of the occasion:

On the first evening of the meeting, I happened to take dinner with Russell who arrived rather late, and one of the first things he inquired about was whether you had sent in any contribution. On my answering no, he then said, 'Well, he is an ass. With a perfectly good thousand dollars available he refuses to take it.' These remarks led to some discussion, and afterwards in a group in the hotel lobby we drafted a telegram urging you to send by night letter the principal results which Russell and Shapley could make up into a paper. After this message was drafted, Russell and I started to go over to the telegraph office to send it, but on the way we stopped at the desk and put it on a regular blank. Just as we were leaving, Russell's eye caught beyond him on the floor a large envelope addressed to himself, and at the same time I spied your name on the upper left corner. The clerk gave us the material, and we walked back to the group in the lobby saying that we had got quick service, and that the paper was in hand. At the time, the coincidence seemed a miracle. [Stebbins to Hubble, 16 February 1925, AAS Archives, American Institute of Physics.]

Russell read Hubble's paper for him, and it shared the prize for the best paper at the AAAS. Curtis had been at the meeting, but had already taken the train home before the paper was read. In any event, all the principals had got the news long before the session on New Year's Day of 1925 put it on the official record.

It was perhaps fortunate for Helen Sawyer Hogg that Shapley had come East, and was therefore only a peripheral player in the denouement of the island–universe debate. But in Cambridge Shapley set up the astronomy graduate school from which Helen and her husband Frank obtained two of the earliest doctorates. It was there, with Harlow Shapley, that she developed her love for variable stars in globular clusters, a field that Shapley had worked so energetically to promote. I am particularly pleased, therefore, that I could end this brief series of episodes from the early history of nebulae with Shapley and with variable stars.

Notes and references

Revised text of the first Helen Sawyer Hogg Lecture, delivered during a meeting of the Canadian Astronomical Society, in Toronto, on 28 May 1985.

Some of Helen Sawyer Hogg's 'Out of Old Books' in the *Journal of the Royal Astronomical Society of Canada* of particular interest to our topic are: 'Halley's List of Nebulous Objects,' **41** (1947), 60–71; 'Derham's Catalogue of Nebulous Objects from Hevelius' Prodromus,' **41**, 233–8; 'Catalogues of Nebulous Objects in the Eighteenth Century,' **41**, 265–73; and 'The Early Discoveries of Four Globular Clusters,' **43** (1949), 45–8. My early contributions on Messier are 'Messier and His Catalogue,' *Sky and Telescope*, **12** (1953), 255–8, 288–91 and 'The Missing Messier Objects,' **20** (1960), 196–9. The fullest information on Bevis' atlas is given by William B. Ashworth, Jr., 'John Bevis and His *Uranographia* (*ca.* 1750),' *Proceedings of the American Philosophical Society*, **125** (1981), 52–73. The dubious idea that the Orion Nebula has changed since Galileo's day is espoused by Thomas G. Harrison, 'The Orion Nebula: Where in History Is It?' *Quarterly Journal of the Royal Astronomical Society*, **25** (1984), 65–79. See also G. Fodera Serio, L. Indorato and P. Nastasi, 'G. B. Hodierna's Observations of Nebulae and His Cosmology,' *Journal for the History of Astronomy*, **16** (1985), 1–36. Much interesting material on the history of nebulae, including a transcription of Shapley's actual 'Great Debate' text, appears in Michael A. Hoskin's *Stellar Astronomy* (Chalfont St Giles, 1982). The Stebbins letter is quoted from Richard Berendzen and M. A. Hoskin, 'Hubble's Announcement of Cepheids in Spiral Nebulae,' *Leaflet of the Astronomical Society of the Pacific* No. 504 (1971). See also Chapter 30.

30 *Harlow Shapley and the Cepheids**

Now this work of Shapley's, which was done when he was a young astronomer at Mount Wilson, is really most remarkable. I have always admired how Shapley in a very short time finished this whole problem, ending up with a picture of our Galaxy that just smashed everything which the old school had said about it.

So recalled the distinguished observational cosmologist Walter Baade when he lectured on stellar evolution at Harvard in 1959. The 'whole problem,' of course, involved determining the size of our Milky Way galaxy and locating the sun's place within it. Early in this century one of the curious unsolved astronomical puzzles involved the very remarkable distribution of the globular clusters – they were all located in one hemisphere of the sky! While researching variable stars in globulars, Harlow Shapley stumbled onto the answer to this question, which at once provided the key to the arrangement and scale of our Milky Way.

Shapley was born near Carthage, Missouri, in November, 1885. As a youth he became a newspaper reporter and eventually decided to further his prospects by attending the University of Missouri. Shapley's first mentor was Missouri's professor of astronomy, Frederick Seares, who was to influence his career in two important ways. In 1911 Seares recommended Shapley for a fellowship at Princeton, and in 1913, when Shapley was completing his PhD thesis, helped him obtain a staff appointment at the Mount Wilson Observatory.

Shapley's second and more important scientific mentor was Henry Norris Russell, an enthusiastic new astronomy professor at Princeton. Under Russell's guidance he made nearly 10 000 observations of eclipsing binary stars and calculated 90 orbits, about 10 times as many as had been established previously.

At that time many astronomers supposed that Cepheid variables, which are now known to pulsate, were a form of eclipsing binary star. As Shapley wrote in a key paper in 1914, it was a 'misfortune for the progress of research' that the pulsation-induced Doppler shifts of a Cepheid's spectral lines mimicked those due to elliptical orbital motion in a binary system. Shapley demolished the binary idea by showing that any such pair of stars would have to lie *inside* each other to reproduce the observations. Having been led in this way to the Cepheids, Shapley went on to make them a centerpiece of his long and outstanding career.

At Mount Wilson he had access to the 60-inch reflector, then the world's largest telescope. There he began an observational attack on the globular clusters and solved the intriguing problem of their lopsided distribution. Shapley's solution led to a far larger Milky Way system than astronomers had ever before considered.

*Coauthored with Barbara Welther.

Figure 30.1 Harlow Shapley about 1920, roughly when he used Cepheid variable stars to determine the dimensions of the Milky Way galaxy.

His strategy to probe galactic structure rested on an important discovery made at Harvard College Observatory just a few years earlier. There Henrietta Leavitt had studied the variable stars in the Small Magellanic Cloud (SMC). In 1908 she published a list of 1777 variables, including 16 for which she had determined the periods of their light variations. Leavitt noted that their cycles ranged from 1.25 to 127 days and that the longer the period, the brighter the star on the photograph. Curiously, no one seemed to pay any attention to that paper.

She subsequently determined periods for nine more variables in

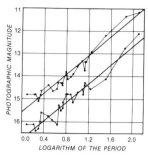

Figure 30.2 Henrietta Leavitt's 1912 plot showed that the brightest Cepheid variables in the Small Magellanic Cloud had the longest periods. Apparent blue-light magnitude is given vertically and the logarithm of the stars' periods horizontally (0.0 equals 1 day; 1.0, 10; 2.0, 100). For each star she plotted its maximum and minimum brightness. The straight lines through these extremes show Leavitt's now-famous period–luminosity relation and the fact that the Cepheids all have about a 1.2-magnitude change in amplitude.

the SMC and in 1912 published a second paper giving her results graphically (see Figure 30.2). In addition she mentioned that, if her photographic magnitude scale could be calibrated for absolute magnitude (representing the stars' intrinsic luminosities), the distance of the SMC could be determined. This article caught the attention of the eminent Danish astronomer Ejnar Hertzsprung, and he cited her result in a paper in which he derived the absolute magnitude of a typical Milky Way Cepheid.

Finding the intrinsic brightness of a Cepheid would have been easy if such a star existed close enough to the sun for its trigonometric parallax, and thus its distance, to be measured reliably. Yet even the nearest Cepheids are too far away for satisfactory triangulation from the earth's annual orbit, though distances could be fathomed using the much longer baseline swept out by the sun's motion through space over several decades. Unfortunately, the unknown space motions of the stars themselves spoil this procedure for individual objects. But by combining the data from a handful of Cepheids statistically, Hertzsprung managed to find an average distance.

However, according to Baade,

Hertzsprung made a decimal error in this case, deriving a distance that was too small by a factor of 10. And this passed completely unnoticed, and I remember when Hertzsprung, some 20 years later, violently protested that he had not made an error. It is psychologically so interesting because . . . it was the largest astronomical distance known at that time. It was so terribly large for these boys that a factor of 10 didn't make any difference. You see, today any student would have a feeling that it was wrong by a factor of 10 – it cannot be so close as our backyard. There was no feeling for these things.

Indeed, Hertzsprung's erroneous statement is right there in the *Astronomische Nachrichten* for November, 1913: 'One gets $p=0''.0001$, corresponding to a distance of about 3000 light years.' Whether he made a factor-of-10 error in calculating the distance, or whether it was an uncaught typographical error, is hard to establish.

Almost simultaneously, but entirely independently, Russell used the same data for the 13 Milky Way Cepheids and determined practically the same absolute magnitudes for them. However, as he wrote to Hertzsprung in the summer of 1913:

I had not thought of making the very pretty use you make of Miss Leavitt's discovery about the relation between period and absolute brightness. There is of course a certain element of uncertainty about this, but I think it is a legitimate hypothesis. If the analogy can be pushed a little farther, one may reason as follows. Miss Leavitt's stars, of period comparable with the [Milky Way] Cepheid variables, are of about the 15th photographic magnitude. The [Milky Way] Cepheid variables appear of about the 5th photographic magnitude. Therefore, if the two sets of stars are really similar, and there is no absorption of light in space, the *distance of the Small Magellanic Cloud* is 100 times of the Cepheid variables, or about 80 000 light-years! This is an enormous distance, but is not intrinsically incredible.

At this time Shapley was still Russell's doctoral student, and the two must have discussed this astonishing result. Furthermore, they must have realized that the Cepheids were not only intrinsically very bright but also very large. This information was essential for Shapley to show that, if treated as an eclipsing binary system, the two putative companions of a typical Cepheid had to be inside each other!

Thereafter, Shapley sought an alternative explanation, and he picked up the suggestion that the Cepheids must be pulsating stars. This new hypothesis made a great deal of difference in understanding the stars' characteristics. If the Cepheids were eclipsing binaries, then many of their features (such as the relation of light curves to periods) would just be an accidental result of the component stars' separations and orbital orientations. On the other hand, if Cepheids pulsated, then their light curves, luminosities, and periods would be genuine physical properties intrinsic to the stars.

Thus, when Shapley went to Mount Wilson and began studying variable stars in globular clusters, all this knowledge was part of his toolkit. Working with the 60-inch reflector, Shapley soon found a few Cepheid variables in the globular clusters. Already confident that the Cepheids represented a distinct physical type of star, Shapley noticed that those in a given cluster seemed to show the same period–luminosity relationship that Leavitt had found. Their apparent magnitudes differed from cluster to cluster, but he was sure this effect arose simply because of the globulars' different distances. Hence he could immediately establish the *relative* distances of the clusters.

What remained was to reestablish an absolute magnitude scale for the Milky Way Cepheids in order to determine the actual distances to the globulars. Here he followed in the footsteps of Hertzsprung and Russell and employed the same baker's dozen stars they had used. This sample was hardly enough to give good statistics, especially after he threw out two stars because their light curves looked suspect. In retrospect that was a good move, but it didn't add confidence to the statistics.

The results of Shapley's analysis in 1918 are now well known. Using his calibration of the Cepheids, he established that a typical globular cluster was perhaps 50 000 light years distant. And from the asymmetrical distribution of these clusters (concentrated in the direction of Sagittarius), he postulated that the invisible nucleus of the Milky Way lies in the center of the globular cluster system some tens of thousands of light years away.

Not only had our galaxy proved to be far larger than anyone had imagined, but the sun fell in its outskirts. "I can assure you these were exciting times when Shapley's results were published," Baade reported, "because the 'old boys' didn't take it sitting down. To the old group enamoured of the old picture, these distances seemed to be fantastically large. Every attempt was made to punch holes into Shapley's arguments."

One of the most famous attempts to discredit the large size of the Milky Way came a few years later. At the annual spring meeting of the National Academy of Sciences in April, 1921, the evening *conversazione* was devoted to a debate between Shapley and Lick Observatory's Heber D. Curtis. In this 'great debate' Shapley was certainly wrong on several counts. Here, however, we will look at just the treatment of the period–luminosity relation.

Figure 30.3 Over 40 percent of the 140 or so known Milky Way globular clusters are found in less than three percent of the sky's area. Explaining this asymmetry led to Harlow Shapley's discovery of the size of the Milky Way and our sun's place in it. The globulars are circled on this wide-field Harvard College Observatory photograph, though many are invisible here.

In the published (and much modified) version of the debate, Shapley downplayed the role of the Cepheids. He believed that good corroborative evidence for the distances to the globulars could be obtained, for example, from their bright blue stars. To see his period–luminosity curve for the Cepheids, we have to look at an earlier paper, one published in the November, 1917, *Astrophysical Journal*. Today, the most startling detail of his diagram, reproduced in Figure 30.4, is that his 11 Cepheids (the open circles) all sit exactly on the line. In fact not only these stars but also 25 in the SMC and 21 in five different globular clusters all line up like beads on a string. Shapley had deliberately patterned his curve after the one Leavitt had determined for the SMC Cepheids. And because he was convinced that the period–luminosity relation was an exact physical law, he had no qualms about smoothing the results for individual stars.

Shapley had calculated the absolute magnitudes for his 11 Milky Way Cepheids individually. Plotted against their periods, these magnitudes scattered widely. However, by averaging the values in groups of three, he defined a curve that could be superimposed on Leavitt's. Then he plotted the smoothed values of the absolute magnitudes for all his Cepheids so that they also fell neatly on the curve.

Curtis questioned Shapley's 'elaborate system of weighting' and replotted these 11 stars to show their actual scatter (circles with slashes in Figure 30.5). He also included 36 additional points, which

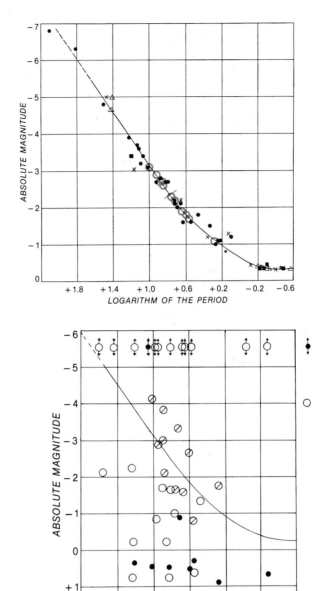

Figure 30.4 Perhaps to make his period–luminosity relation look different from Leavitt's, in 1917 Shapley published this mirror-image version. Also, though he used some of her data (dots), he corrected their blue-light magnitudes to yellow-light ones; consequently, the curve is steeper. Large open circles identify smoothed data for the Milky Way Cepheids Shapley used to calibrate the relation. The other symbols represent Cepheid variables in globular clusters.

Figure 30.5 In attempting to disprove the large size of our Milky Way galaxy, H. D. Curtis replotted Shapley's calibration stars (open circles with slashes) without manipulating the data. He also added Cepheids for which distances had been determined through trigonometric parallaxes (dots) or space motions (open circles). When all of this was combined, it resulted in a meaningless scatter diagram, which apparently invalidated the period–luminosity relation. The points at the top represent stars with negative parallaxes or otherwise indeterminate distances.

produced a hopeless mess. Curtis concluded that 'it would seem that available observational data lend little support to the fact of a period–luminosity relation among galactic Cepheids.' Clearly, there was no possibility of disputing Leavitt's results from the SMC. But Curtis had seemingly mounted a formidable challenge to the idea that Milky Way variables could be used to calibrate her relation, and he certainly tried to make it look as if Shapley had fudged his data.

Where did Curtis go wrong? The black dots in his diagram

represent direct trigonometric parallaxes, but we now know that the Cepheids are too distant for such results to be meaningful. In fact, due to random errors, observers should have measured about as many *negative* parallaxes as positive ones. But since a negative parallax implies that a star is farther away than infinity, observers were understandably reluctant to report such findings! Hence the data available to Curtis were naturally biased toward the closer distances, which made the stars intrinsically less luminous and caused them to cluster in the lower part of the diagram. As for parallaxes based on proper motions (open circles), the data were clearly much inferior to those used by Hertzsprung, Russell, Shapley, and others.

While Shapley's intuition was correct regarding the period–luminosity relation, his calibration actually erred by about a magnitude, and he was certainly pushing his luck by smoothing the sparse data in a rather arbitrary way. In this regard Robert Smith quotes effectively from J. R. Ravetz:

A problem under investigation grows in interaction with its materials, and . . . when it is completed it is necessarily rough-hewn. This will generally be even more so in the case of deeply novel results; for the conquering of new pitfalls, the forging of new tools, and the establishment of new objects of inquiry, require great talent, daring, and ruthlessness, and also a complete identification of the scientist with the result. . . . If every anomaly in experience, and every ambiguity in concept, were completely ironed out before the work was presented to the public, nothing new would ever appear.

Others before Shapley understood how to calibrate the period–luminosity relationship. But only he had the verve and energy to apply it imaginatively and effectively to an intriguing problem of the day. Thus he came up with a truly revolutionary conception of the structure of the heavens.

Indeed, within a few years Edwin Hubble would be consulting with Shapley as to the most up-to-date calibration of the period–luminosity relation. He did so to apply the relation to newly discovered Cepheids in the Andromeda nebula. Despite the still undetected calibration error, this work established that galaxy's enormous distance and paved the way to the discovery of the expanding universe. As Baade said in reminiscing about those days, 'Altogether, it was a very exciting time.'

Notes and references

Robert Smith's *The Expanding Universe: Astronomy's 'Great Debate' 1900–1931* (Cambridge, 1982) is essential reading for anyone deeply interested in the history of Milky Way and galaxy studies. We have taken the Russell–Hertzsprung letter from this source. Michael Hoskin's *Stellar Astronomy* (Chalfont St Giles, 1982) is also recommended. The Baade quotations are edited from a transcript prepared by Richard B. Rodman. Hertzsprung's paper appears in *Astronomische Nachrichten*, **196** (1913), 201–10 and the Shapley–Curtis debate is in *Bulletin of the National Research Council*, **2** (1921), 171–217. When

Shapley published his *Source Book in Astronomy 1900–1950* (Cambridge, Mass., 1960) he included his paper on the pulsation hypothesis of Cepheid variability; in private conversations with us he explained how significant this physical idea was when he began using these variations to probe the Milky Way structure.

31 *A search for Russell's original diagram*

One of the most familiar pictorial tools of modern astronomers is the Hertzsprung–Russell diagram, in which the luminosities or magnitudes of stars are plotted against their spectral types or colors. This representation did not spring full-grown like Minerva from the brow of Zeus; there are various antecedent tabulations and plots.

The earliest ones that astronomers easily recognize were first displayed by Henry Norris Russell in 1913. The Princeton astronomer spoke about 'Relations between the Spectra and Other Characteristics of Stars' before a joint meeting of the American Astronomical Society and the American Association for the Advancement of Science in Atlanta, Georgia, on 30 December 1913. On that occasion he illustrated his paper with lantern slides of the now-famous diagram.

The accumulating knowledge of stellar parallaxes had made it possible to know for the first time the distances, and hence the absolute magnitudes, of a good many stars. In his first diagram Russell plotted more than 200 stars, all the ones for which directly measured parallaxes were available in the spring of 1913. In the second diagram he showed data for about 150 stars in four 'moving clusters,' whose distances were deduced from their convergent points on the sky. This independent group completely confirmed the distinctive 'backwards 7' pattern of the first diagram. In a third diagram Russell plotted about 550 stars for which what we now call spectroscopic parallaxes had been determined; that is, components of binary systems whose masses were approximated as being equal to the sun's. In the final slide he exhibited a diagram for 80 eclipsing stars.

As it happened, rather few astronomers heard Russell's trailblazing presentation at the Atlantic session. In fact, the AAS meeting was so poorly attended that a quorum could not be mustered for an election. Thus, president E. C. Pickering, treasurer Annie J. Cannon, and others continued in office for another eight months. Apparently, the dozen members present took turns reading the 29 other papers of their mostly absent colleagues!

Five months after the meeting, in the May issue of *Popular Astronomy*, Russell published the first installment of his paper, followed a month later by the final section. The four diagrams reproduced in that journal are rather miserable, muddy half-tones which, unlike many of the more picturesque and glossy views in the same volume, are printed on dull book stock. As a result, they are impossible to reproduce in any aesthetic fashion; hence, many historical accounts of the diagrams contain redrawn figures.

Several years ago I decided to try to find out if the original figures survive, with the hope of making slides of them for my class lectures. At that time the Russell papers in the Princeton University library were still largely unsorted, so it was almost as bad as looking for the proverbial needle in a haystack, and I found nary a trace of his plots. However, I knew that Russell had given a preliminary version of his

Figure 31.1 A group of astronomers at the December, 1913, Atlanta meeting of the Astronomical and Astrophysical Society of America (as the AAS was then named). In the front row are, left to right, G. Comstock, E. C. Pickering, and P. Fox, and behind them are F. Slocum, F. Moulton, and H. N. Russell. The photograph may have been taken by the Popular Astronomy editor, C. H. Gingrich, who was present at the meeting but not in the picture; it appears as Plate XII in the 1914 volume of that journal.

paper before the Royal Astronomical Society in London on 13 June 1913. Along with other Americans, he was en route to a meeting of the International Union for Solar Research in Bonn, and the RAS took advantage of the opportunity to hear from Annie Cannon, E. C. Pickering, and at the end of a very crowded program, Russell himself. The Princeton astronomer spoke on " 'Giant' and 'Dwarf' Stars," giving what he called 'a hurried account,' but showing the four slides that eventually became figures in his *Popular Astronomy* paper.

This earlier presentation gave another possibility for finding the elusive manuscripts. As I searched through the miscellany of memorabilia from his European trip, I suddenly encountered the dinner menu for that Friday the 13th, 1913. But, alas, that was as close as I came to the missing materials.

Today, thanks to the efforts of Peggy Kidwell and David DeVorkin, the archive has been carefully arranged and catalogued. Unfortunately, the end result is still the same: no original copies of the diagrams. However, Russell's correspondence is now more accessible, and there one can read a request from Arthur Eddington for copies of the diagrams that Russell had shown in his RAS talk.

'I am glad you reminded me of my promise to send you copies of my diagrams of the giant and dwarf stars,' Russell replied in February, 1914. 'I enclose prints of all four of my diagrams, the ones I showed at the RAS meeting last June. They are on a rather small scale, but are sharp and clear, and I have printed the negatives on the best paper I could get here. I hope you will find them useful. They could probably be enlarged a little if necessary for reproduction in your book.' Russell added that he was sending the paper to *Nature* because he had promised it to the editor, Sir Norman Lockyer, but since all the AAS papers were to be published in *Popular Astronomy*, it was also being submitted there. Hence, identical versions of the paper appeared simultaneously on each side of the Atlantic.

Apparently, both publications used very similar photographic prints for the figures, with identical lettering, but there are subtle differences in the grid lines and borders. Although the linecuts in

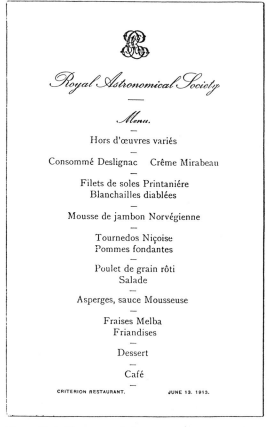

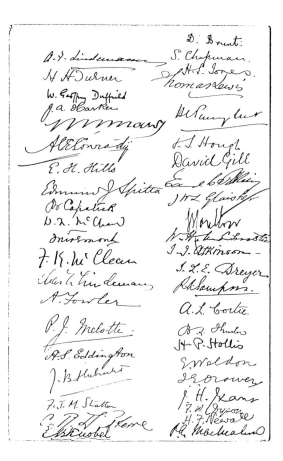

Figure 31.2 The menu of the Royal Astronomical Society Club on the night Russell gave the first version of his paper. On the back of Russell's copy are the autographs of the other guests and members present.

Nature are considerably clearer than the gray half-tones, we can sympathize with the decision made by *Popular Astronomy*'s editor, C. H. Gingrich, because his reproductions include several small stars lost in the British publication. Close on the heels of these diagrams came the one in Eddington's *Stellar Movements and the Structure of the Universe*, clearly the best reproduction of the lot. Eddington, however, used only the first of Russell's four plots.

But now I return to the dinner (whose menu I found in the Russell archive) following the London presentation of Russell's dwarf-and-giant-stars paper. It was the 649th dinner of the Royal Astronomical Society Club, a venerable institution that antedates the RAS itself by some months, and whose members still continue to dine together on the evenings of the Society meetings. Because the club has kept impeccable records, it is possible to know that this was a particularly large affair with 45 in attendance. We even know that its president, J. W. L. Glaisher from Trinity College, Cambridge, was flanked by E. C. Pickering and Lord Moulton, and that Russell sat across the table with H. H. Turner of Oxford, the club treasurer and notetaker, at his side. Among the other visitors was J. L. E. Dreyer, who had some years previously completed the *New General Catalogue* as well as its two supplementary *Index Catalogues*, and who was currently at work on the multivolume *Opera* of Tycho Brahe.

Not many months after my initial visit to the Russell archives, I

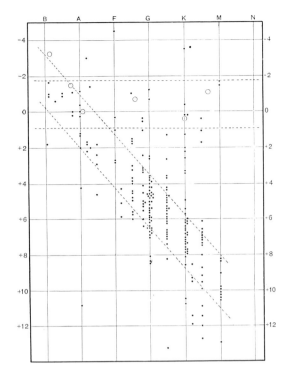

Figure 31.3 In Eddington's carefully redrawn version of Russell's first diagram, the original's various levels of reliability indicated for the plotted points are no longer distinguished, and horizontal dashed lines have been added to emphasize the giant branch. The large circles represent mean values for bright stars of small proper motions and parallax.

found myself a guest at the RAS Club, and in accord with their time-honored custom, they asked me to make a few remarks. I took the opportunity to report that I had found the printed menu of their June, 1913, dinner while searching in vain for Russell's original copy of what we now call the H–R diagram.

A few weeks later, Cdr Derek Howse of the National Maritime Museum in Greenwich sent me a photocopy of the minute book of the RAS Club for that earlier occasion. It was interesting to see that Russell spoke not about his topic of the afternoon but about the 'triangular cooperation' in photography of the moon between Harvard, Yale, and Princeton. Far more fascinating to me, however, was the five-line abstract of Dreyer's remarks that evening: 'Dreyer is doing works of Tycho Brahe and finds Kepler mistaken in his diagnosis of Tycho's method of inferring that Mars was nearer Earth than Sun at opposition – He worked it out from retrograde motion.' By chance I had been working through the same historical problem, which had puzzled historians of science for many years, and it was quite astonishing to see that Dreyer had come to a conclusion 60 years earlier even though he had apparently never published it.

In the next chapter I shall explain the circumstances of this problem, which relates to the way in which Tycho arrived at his own cosmological system.

Notes and references

Ejnar Hertzsprung plotted colors against apparent magnitudes for the Hyades cluster in the *Publikationem des Astrophysikalischen*

Observatoriums zu Potsdam, **22** (1911), but his diagram was certainly not as striking as those of Russell. Both Hertzsprung and Russell attended the International Union for Solar Research meeting at Bonn in August, 1913, and stood near each other in the group photograph – undoubtedly they must have discussed giants, dwarfs, and the diagram!

For details of the complex history of the H–R diagram, see A. V. Nielsen, 'Contributions to the History of the Hertzsprung–Russell Diagram.' *Centaurus*, **9** (1963), 219–53, (*Meddelelser fra Ole Roemer-Observatoriet i Aarhus*, No. 30) and David DeVorkin, 'Steps Toward the Hertzsprung–Russell Diagram,' *Physics Today*, March, 1978, 32–9, with its numerous references. DeVorkin has also described the evolutionary studies leading up to the diagram in Vol. 4A, Chapter 6, of *The General History of Astronomy* (ed. O. Gingerich, Cambridge, 1984).

I wish to thank the Princeton University library for permission to reproduce the RAS Club menu (Russell Collection, Box 89, Folder 12) and to quote from the letter to Eddington (Box 7, Folder 87).

There is a Persian fairy tale in which the three princes of Serendip have such a marvelous propensity for making accidental discoveries that, ever since the mid-eighteenth century, the word *serendipity* has referred to the aptitude for, or process of, making happy, unexpected discoveries. Not only in science, but also in historical research, an unexpected answer can turn up to a question not even being asked!

In the last chapter I described my frustrated endeavor to find Henry Norris Russell's original version of his famous diagram, a search that led ultimately to the 694th dinner of the Royal Astronomical Society Club in London on 13 June 1913, and to the after-dinner remarks by Russell and J. L. E. Dreyer. What caught my attention was Dreyer's rather casual discussion of a scholarly problem concerning Tycho Brahe's own special cosmology, a problem that I had found perplexing and which, though I didn't know it then, Dreyer had solved to his satisfaction.

John Louis Emil Dreyer was born in Denmark, and from his youth on he found a particular fascination in his countryman, Tycho Brahe. Although Dreyer made his astronomical reputation by editing the *New General Catalogue of Nebulae and Clusters*, he repeatedly returned to the original texts and manuscripts of his illustrious Danish predecessor. In 1890 he published *Tycho Brahe, a Picture of Scientific Life and Work in the Sixteenth Century*, which remains to this day the standard biography of Tycho.

One of the puzzles facing any Tycho scholar is to decide why Tycho rejected both the ancient Ptolemaic system and the radical new Copernican system, only to settle on a composite scheme. In the Tychonic system the earth remained safely at rest while the sun, orbited by all the other planets, cycled around the earth. Part of the reason for adopting this new arrangement was that Tycho, a self-assured egotist, wanted to promulgate a system of his very own. But Tycho was also an excellent scientist, and he sought some satisfactory observational reasons as well.

In the sixteenth century, there was no unambiguous observational way to decide between the Ptolemaic and Copernican systems. In these alternative models, predictions concerning the *directions* of planetary motion were essentially the same, but they differed a great deal with respect to the *distances* of the planets. For example, in the Ptolemaic arrangement Mars always lay beyond the sun (see Figure 32.2). Quite the contrary was true for the Copernican system.

The problem was that no one had been able to measure the distances to the planets with any accuracy. Tycho realized that a good distance measurement could distinguish between the two systems, and he set out to do this in an ingenious way, by measuring the positions of Mars both in the morning and in the evening. According to the Copernican system, the earth would rotate in those 12 hours,

Figure 32.1 J. L. E. Dreyer (1852–1926). In addition to being a respected historian of science and compiler of the famous New General Catalogue, *he was director of Armagh Observatory from 1882 to 1916. After graduating from Copenhagen University, he worked at Lord Rosse's observatory at Birr Castle where he observed with the great 6-foot speculum mirror.*

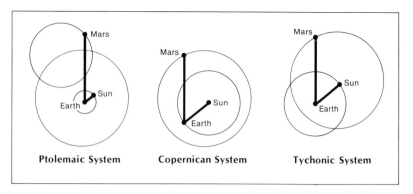

Figure 32.2 In each of the solar system models depicted here, the directions *(angles) to the sun and Mars are identical, but the* distances *are quite different. In Ptolemy's epicyclic system, earth is always closer to the sun than to Mars; this is not the case according to Copernicus' concept.*

carrying the observer an earth-diameter around and thereby giving a new position from which to sight Mars.

Tycho set to work and soon claimed to find a measurable parallax for Mars, making it closer than the sun, and hence ruling out the possibility of the Ptolemaic system. To a modern reader who knows that the so-called horizontal parallax of Mars can at most reach 23 arc seconds, Tycho's result was, in Dreyer's words, 'a surprising statement,' because the quantity was much too small for Tycho to measure with his naked-eye instruments. The statement was equally puzzling to Johannes Kepler, who vainly sought to find the Martian parallax from Tycho's observational data.

Kepler, in fact, proposed another and stronger geometrical configuration for finding the distance to Mars, but unfortunately Tycho had not provided any measurements under these circumstances. Consequently, Kepler decided to attempt the observations himself: 'This will treat you to a ridiculous spectacle,' he wrote in describing his attempts. 'In 1604, when I was considering the parallax, an occasion presented itself, which would have been more convenient in another climate.'

Kepler said this because the February night was so cold that his bare hands stuck to the iron of the sextant, and with gloves it was impossible to get the clamps set securely. Furthermore, strong winds ruled out the use of lamps – the scales could be read only by the light of a glowing coal. 'I would be insane to rely on these observations for anything so subtle,' Kepler admitted. 'I hope that my readers, on account of their distaste for these uncertain things, will all the more seek out the surety of Tycho's observations.'

However, neither Kepler's nor Tycho's observations could settle such a delicate matter, and Kepler sought some explanation as to why Tycho had thought that his data had established Mars' distance. In searching through Tycho's record books, Kepler found that Tycho had assigned the calculations to his students. 'But behold an unexpected thing!' wrote Kepler in Chapter 11 of his *Astronomia nova*. The student assistants, instead of computing the effect of the rotation of the earth in establishing a baseline for a triangulation to Mars, had actually made an elaborate Copernican calculation and necessarily discovered that Mars at planetary opposition is closer than the sun! 'Thus Brahe had intended one thing, but his assistants carried out another. What was desired was for the morning and evening

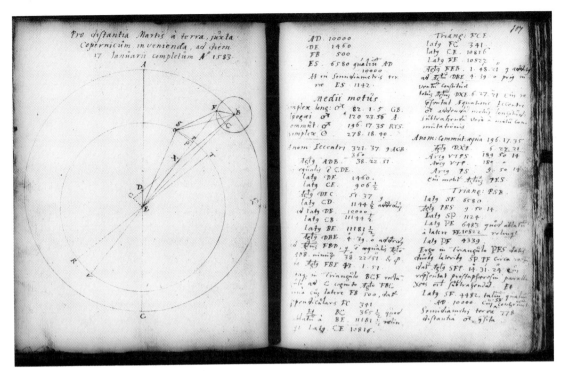

Figure 32.3 The Copernican diagram for finding the distance to Mars. Kepler saw this was not in Tycho's hand, and he therefore assumed that Tycho did not know the manner in which the calculations were being carried out.

observations to be compared among themselves in order to investigate the parallax of Mars, but what they really investigated was how much parallax the Copernican diagram had. Whether it was just from trust in his assistants that Brahe made his pronouncement about the parallax, I do not know.'

In his biography of Tycho, Dreyer paraphrased Kepler's remarks and let the matter rest. The obvious conclusion would be that Tycho rejected the Ptolemaic system for totally spurious reasons. However, by the time of the 1913 RAS Club dinner, Dreyer had examined Tycho's manuscripts more thoroughly and had found something overlooked by Kepler. Dreyer discovered among the Tycho papers in the Austrian National Library in Vienna another copy of the Copernican calculations, but they were in Tycho's own hand, not that of an assistant. In other words, Tycho himself had also done the Copernican calculation, and contrary to Kepler's deduction, it was not an unwarranted trust in his assistants that had led Tycho to a rejection of the Ptolemaic system. In his own personal copy of the Tycho biography, Dreyer wrote 'Nonsense' alongside his citation of Kepler's opinion.

Why, then, did Tycho reject the Ptolemaic system? Popular accounts often credit the comet of 1577, on the grounds that its movement through the heavens would have smashed the crystal spheres that held the Ptolemaic arrangement in place. The chronology of events suggests, however, that this was an afterthought, because Tycho himself indicates that he got the idea for his own planetary arrangement in 1583, at the time he was intensively studying the opposition of Mars.

Figure 32.4 Dreyer found this manuscript in Tycho's own hand; beginning at the top of the page it says, 'Here follows the investigation of the distance to Mars from the Earth on 17 January 1583.'

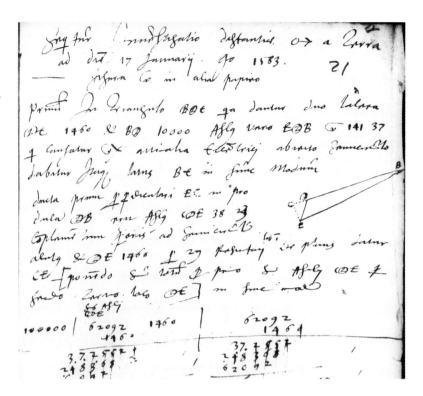

In investigating the question, Dreyer concluded that the apparition of Mars was indeed the catalyst for Tycho's thinking, and he reasoned that the speed of Mars' retrogression provided the answer. The sun moves across the sky at about 1° per day, Mars at about $\frac{1}{2}$° on the average. In the first few days of 1583, however, Mars was moving *retrograde* by nearly $\frac{1}{2}$° per day. Tycho's Copernican notes, which are labeled 'Concerning the distance to Mars from the Earth around an opposition, according to Copernicus,' and which derives the distance for 17 January, 1583, showed Tycho that this motion would be possible only if Mars came closer than the sun.

Dreyer also knew that the great Danish astronomer repeatedly mentioned in his correspondence that he had tried hard to find the distance to Mars, and in a 1588 letter to the Wittenberg professor Caspar Peucer, he explicitly mentioned the daily motion of Mars as leading to the solution. Dreyer described his findings in the Latin prologue to his magnificent multivolume set of Tycho Brahe's '*Complete Works.*' The prologue was dated 'Armagh, May, 1913,' which means that he had just completed the first volume of this *Opera omnia* when he spoke at the RAS Club.

This explanation of the road to the Tychonic system still leaves one crucial point hanging. In any model of the planetary system in which the orbit of Mars goes around the sun – either the Copernican or Tychonic arrangement – the ratio of the average sun–Mars and sun–earth distances must be about 1.52:1. Therefore, at opposition Mars will, on the average, approach the earth to a proportional distance of 0.52, and thus be closer than the sun. But the Ptolemaic scheme, in which Mars moves on a large epicycle forever beyond the

Figure 32.5 The RAS Club secretary's minute book for the 694th dinner. Note at the bottom his annotation that Dreyer had found Kepler 'mistaken in diagnosis of Tycho's method of inferring that Mars was nearer Earth than Sun at opposition,' as described in the text. The top part of the page lists the members and guests present.

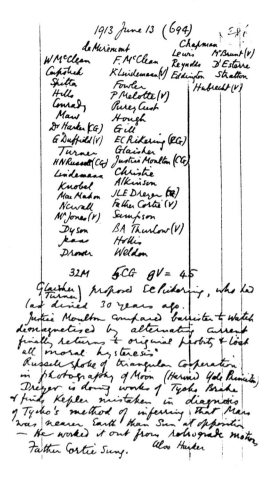

sun, can also reproduce the observed positions, and none of Tycho's observations could disprove this model. Not until Galileo's telescopic examination of Venus, made about 30 years later, was there any direct observational proof against the Ptolemaic arrangement. Hence, despite Tycho's apparent belief that he had established his cosmological system on observational grounds, he had actually taken a major unproved intuitive leap in accepting a circumsolar orbit for Mars.

Although Kepler erred in interpreting the Copernican diagram he found among Tycho's papers, he was correct in noticing that indeed Tycho had not been able to establish the distance to Mars empirically; hence, Tycho did not have a logical observational basis for rejecting the Ptolemaic system. But I would maintain that such intuitive leaps, often not strictly logical, are probably the most natural way in which creative science progresses. As a bold and imaginative scientist, Tycho had dropped the Ptolemaic system in favor of a scheme more unified by its circumsolar orbits. And while his geo-heliocentric arrangement did not stand the test of time, it actually helped to bridge the way to the New Astronomy of Kepler.

Notes and references

In 1916 Dreyer moved from Armagh to Oxford, and there he continued working on the Tycho Brahe *Opera Omnia* until his death in 1926. Twelve volumes were published during his lifetime, and he left essentially complete manuscripts for the final two volumes. His remarkable astronomical library was sold by the London bookseller Henry Sotheran (*Catalogue* 804), and Dreyer's annotated copy of his biography *Tycho Brahe* was item 231. The book passed into the hands of the English physicist E. N. da C. Andrade and subsequently into my own collection. *Tycho Brahe* was reprinted by Dover and, more recently, by Peter Smith; it has now been superseded by the biography by Victor Thoren, *Lord of Uraniborg* (Cambridge, 1991).

33 Robert Trumpler and the dustiness of space

Figure 33.1 Robert J. Trumpler and his wife Augusta at the 1916 meeting of the American Astronomical Society, held at Sproul Observatory.

In our modern cities with their flood of artificial light there is little opportunity to view the Milky Way. The observer has to be on a moonless night in the open country, or better still on an isolated mountaintop in order to appreciate fully the mysterious splendour of that pale, diffuse band of light.

Thus began a slide lecture presented in 1929 for the Graduate Council of the University of California. The speaker, chosen for the importance of his research, was an experienced mountaintop observer, an astronomer greatly fascinated by the Milky Way. For a decade Robert Julius Trumpler had been working at the Lick Observatory, primarily with the 36-inch Crossley reflector, on the star clusters of the Milky Way. When he lectured in Berkeley that March, he was on the threshold of making one of the great discoveries of twentieth-century astronomy, but apparently he then hadn't the slightest hint of what was to come, according to his personal papers.

Trumpler was born in Zurich, Switzerland, in 1886. As a teenager, his interest in astronomy was aroused by a talk given by one of his classmates, but he did not enter a scientific career easily. The elder Trumpler, a successful businessman, looked forward to the day when young Robert would enter the family firm. Apparently a stint in a Swiss bank proved so boring that Robert was finally allowed to study astronomy at the University of Zurich and later at Göttingen, where he received his doctorate in 1910.

'At the outbreak of the European War in August 1914,' Trumpler said in applying for a fellowship at the Lick Observatory, 'I was called to the colors of my country in order to protect its neutrality. I had previously done my regular military training and been promoted to First L[i]eutenant. In this commission I served until April 1915, when I obtained a leave of absence with permission to leave the country in order to follow an engagement as assistant of the Allegheny Observatory.'

At Allegheny in Pittsburgh, the young Swiss astronomer began to investigate star clusters as a class of objects. Three years later W. W. Campbell, director of Lick Observatory, gave him the opportunity to pursue these and other studies with the larger instruments on Mount Hamilton.

While Campbell quickly recognized Trumpler's ability as an observer, chosing him as a collaborator in the exacting attempts to determine the relativistic deflection of starlight by the sun during a total eclipse, the stern, no-nonsense director was less enamoured with the investigations of star clusters. Trumpler still had not finished the report of his Allegheny observations, and Campbell wrote to him in 1921:

I do not want to establish the precedent of publishing papers embodying work done at other universities. I therefore strongly recommend that you proceed soon to comply as well as possible with Dr Curtis's suggestions and get the manuscript to him promptly, in acceptance of his definite offer to publish under these conditions. [Heber D. Curtis, then director at Allegheny, had urged Trumpler to shorten his paper for publication there.] I think that five or six full days of work on your part would see this through, and the Lick Observatory will contribute gladly this much of your time to the problem.

In fact, Trumpler's *Allegheny Publications* paper on star clusters is pretty non-memorable. His chief competitor was Harlow Shapley, an astronomer just one year older than himself, who by 1921 had published over a score of papers on star clusters and who had used the globular clusters in a dramatic fashion to delineate the vast structure of the Milky Way system. From one of Shapley's contributions, Trumpler had gleaned data on the globular cluster M13. He went on to suggest that 'It may be questioned, if two objects so entirely different as the Pleiades and M13 should be compared at all.' Trumpler attempted rather lamely to justify the comparison, but in future papers he restricted himself more and more to open clusters while Shapley emphasized the globulars.

Throughout the 1920s Trumpler continued to work on open star clusters along with other observing activity, and by 1925 he could outline a system of classification based on the Hertzsprung–Russell diagrams of the clusters. He noticed, for example, that some clusters had intrinsically brighter stars than others. From the H–R diagrams he could deduce the absolute magnitudes of the brightest stars in the clusters, and from the observed apparent magnitudes he could then obtain a photometric distance based on the principle 'faintness means farness.' (We shall examine this procedure in somewhat more detail presently.)

A second way to obtain a cluster's distance might be expressed as 'smallness means farness' – in other words, the more distinct a cluster, the smaller it appears in the sky. This, of course, assumes that all clusters have the same physical size. Trumpler attempted to sort out the bewildering variety of clusters according to their degrees of central concentration and richness as well as by their H–R diagrams, thereby hoping to find intrinsically similar clusters.

In this he was luckier than he could have expected. With 20–20 hindsight we now realize that the shearing forces of our galaxy (due to its differential rotation) trim the majority of clusters down to a fairly narrow range of diameters, from about 3 to 7 parsecs (10–22 light years).

By 1929 Trumpler had enough open-cluster distances to map out their space distribution. They fell in a lens-shaped, sun-centered system about 35 000 light years in diameter. He further believed that he could detect spiral structure emanating from the solar neighborhood. The open-cluster system was much like the distribution of stars that had been measured statistically by the Dutch astronomer J. C. Kapteyn and stood in marked contrast to the grand scheme proposed by his rival, Shapley.

In his 30 June 1929 annual report, Trumpler mentioned that he had reliable estimates of distance for 80 clusters, which outlined such a lens-shaped system. If he saw implications of anything further, he dropped nary a clue. Then, within the next six months, something new and unexpected fell into place with startling clarity. For several decades astronomers had been asking if interstellar space is indeed perfectly transparent, and always the answer had come back yes, there seems to be no absorption except in the obvious black obscuring clouds such as the southern Coalsack or the Great Rift in the summer Milky Way.

In his 1930 *Star Clusters*, Shapley had written, 'As far as it goes, the result for globular clusters indicates the essential transparency of space up to a distance of 100 000 light years.' But what Trumpler discovered was convincing evidence that the space within the plane of the Milky Way was not transparent. He found a fogginess or dustiness of space that obscured 0.7 magnitude for every 1000 parsecs (or 0.2 magnitude per 1000 light years). Furthermore, he noticed that the more distant clusters were redder, presumably because the interstellar dust obscured the blue light more efficiently than the red.

How Trumpler's discovery was made can be illustrated by picking two clusters of the same intrinsic size; the nearby Hyades and M103. (Of course, Trumpler did not know what sizes these clusters really were. He had to get his result by considering a large sample of 100 clusters in order for the varying sizes to average out.)

First and simplest, we can determine the relative distances of the Hyades and M103 by the 'smallness means farness' criterion. If both clusters have the same linear diameter, then their angular diameters are inversely proportional to their distances. The angular sizes can be estimated from the photographs reproduced here, which are both to the same scale.

The Hyades cluster is over 400 arc minutes wide while M103 is about 7. Hence, M103 is about 60 times farther away than the Hyades. The Hyades stars are sufficiently nearby for their distance to be found by the so-called 'moving cluster' method, based on the apparent convergence of the individual stars' paths in space, which places them at about 140 light years. We then infer a distance for M103 of about 8000 light years.

Now let us determine the relative distance photometrically, based on the 'faintness means farness' criterion. For this we must match up the main sequences on the H–R diagrams for the two clusters. Note that M103 is so far away and small that only its brightest members have been measured for the diagram, but as Trumpler knew, these brightest stars had spectral type *B*3, one spectral class earlier than the brightest stars of the Hyades cluster (*A*2).

The result of sliding the two H–R diagrams over each other vertically to match the main sequences is shown in Figure 33.3. Note that an M103 star of apparent magnitude, say, 14.0, matches a star of 3.5 in the Hyades. In other words, the greater distance of M103 causes a star of the same intrinsic brightness as one in the Hyades to appear 10.5 magnitudes fainter.

The formula $\log (D_1/D_2) = \Delta M/5$ gives the distance ratio D_1/D_2 corresponding to the magnitude difference ΔM. A slide rule or pocket calculator gives a distance ratio of 125 for a magnitude difference of

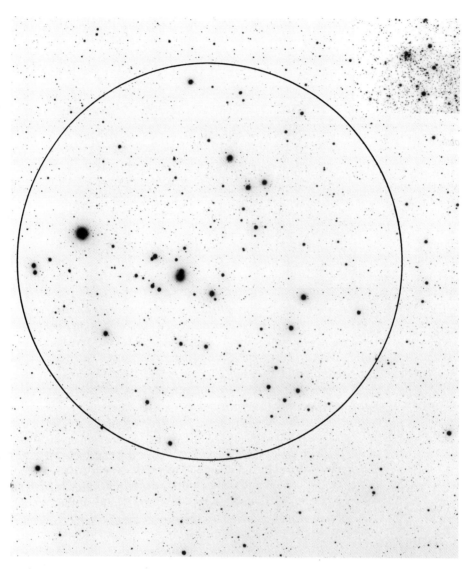

Figure 33.2 The Hyades star cluster in Taurus extends to the edges of the large circle, and actually a bit beyond, while M103 in Cassiopeia is the clumping of stars in the center of the small circle. Both are reproduced at the same scale of 0.6 inches per degree from the Lick Sky Survey. North is up. Aldebaran, the brightest star in the Hyades picture, is not a member of the cluster.

10.5. In other words, given that the Hyades are 140 light years away, the uncorrected photometric distance for M103 (that is, assuming transparent space) comes out as 17 500 light years!

But suppose that there are 1.5 magnitudes of obscuration between us and M103. If 1.5 magnitudes arise from the fogginess, then M103 is only 9 magnitudes fainter on account of its distance. With 9 plugged into the formula, we get a distance ratio of 60, the same that we found earlier from the linear diameter. *Provided the clusters have the same physical size*, then the discrepancy in the two procedures is resolved by postulating 1.5 magnitudes of obscuration over the distance to M103.

By analyzing many clusters so that the variations in physical size averaged out, Trumpler made his important discovery concerning the

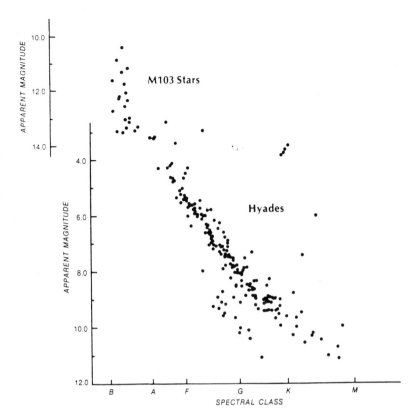

Figure 33.3 Superimposed H–R diagrams for the Hyades (below apparent magnitude 4.7) and M103 (above apparent magnitude 14.2). These are plots of the apparent magnitudes of the stars against their spectral types. Due to distance and interstellar obscuration, only the brightest stars in M103 can be measured, but these hot B and A stars define the uppermost part of the main sequence. Reddening displaces the M103 stars to the right compared to the Hyades.

existence of general interstellar absorption. There remained, however, the serious discrepancy between this result from open clusters and Harlow Shapley's previous conclusion from the globular clusters. Trumpler suggested that the disparity arose because the absorption lay in a thin layer along the plane of the Milky Way where the open clusters were found. The globulars, which were largely observed away from the Milky Way itself, escaped this galactic obscuration for the most part.

Trumpler's discovery was embedded in a major paper which appeared in 1930 as *Lick Observatory Publication* 420. It was entitled 'Preliminary Results on the Distances, Dimensions and Space Distribution of Open Star Clusters.' Trumpler, like Shapley, used his cluster studies in an attempt to delineate the structure of the Milky Way itself. He had already formulated his conclusion (that the Milky Way was a comparatively small, lens-shaped, sun-centered system) before he stumbled onto the evidence for galactic absorption. Unfortunately, he did not immediately recognize the implications of his discovery both for Kapteyn's model of the galaxy as determined by star counts and for his own belief that he had rather fully sampled the open clusters of the Milky Way. In both instances the fogginess in the plane of the Milky Way obscured the more distant vistas and created an artifically low limit to the observed size of our galaxy.

Trumpler must have rather quickly realized his error, for in a *Leaflet* of the Astronomical Society of the Pacific published the following year he astutely avoided mention of the sun's place in the system or his speculations on the spiral structure of the Milky Way. He concluded,

Astronomers are only beginning to realize that highly-rarefied dark matter, as free atoms or fine cosmic dust, plays an important role in the structure of the Milky Way system and seems to spread through vast portions of interstellar space. The study of the Milky Way is yet in its infancy. A tremendous amount of observational work on the distance and composition of individual star clouds, and on the nature and distribution of dark matter spread through them, has to be mastered before we can reach as adequate knowledge of our galaxy.

Modern astronomers could not agree more!

Notes and references

I wish to thank Donald Osterbrock of the Lick Observatory and Dorothy Schaumberg of Lick's Mary Lea Shane Archives for facilitating my access to the Trumpler papers and for permission to quote from the manuscript sources. Besides the archival materials a useful biographical source is Harold Weaver and Paul Weaver, 'Robert Julius Trumpler 1886–1956,' *Publications of the Astronomical Society of the Pacific*, **69** (1957), 304–307. The ASP *Leaflet* cited is No. 36. The H–R diagrams have been adapted from Gretchen Hagen, 'An Atlas of Open Cluster Colour-Magnitude Diagrams,' *Publications of the David Dunlap Observatory*, **4** (1970).

34 *The discovery of the Milky Way's spiral arms*

Although the general form of our Milky Way galaxy – a great watch-shaped configuration of stars – was already understood in the eighteenth century, not until the 1920s did astronomers begin to appreciate either its vast scale or the fact that the entire assemblage rotates around a very distant center. At about the same time astronomers also began to realize that the well-known spiral nebulae were distant systems comparable to our own.

Thus, by the 1930s a key research program for any Milky Way specialist was the challenge of finding whether our galaxy, like those distant pinwheels, had a spiral structure. At that time Dutch astronomers, many trained in Groningen by the pioneering stellar statistician J. C. Kapteyn or by his student P. J. van Rhijn, were at the forefront of Milky Way studies.

One of the leading experts was Bart J. Bok. His interest had been stimulated by Harlow Shapley's model of a Milky Way system far larger than anyone had previously contemplated and with the sun far from the galactic nucleus. For his doctoral research, Bok emigrated to Harvard College Observatory, where Shapley had taken up the directorship, but he actually obtained his degree from the Kapteyn Astronomical Laboratory in Groningen. More than anyone else he developed and applied the numerical methods originated by Kapteyn to the problem of Milky Way structure and, in particular, to the delineation of its spiral features.

Bok hoped to prove the existence of spiral arms through the analysis of extensive star counts in selected areas of the Milky Way. From the numbers of stars of each apparent magnitude in a given direction in space, it was possible to reconstruct the variations in stellar concentration along the line of sight. The method worked very well for locating clouds of dark absorbing matter. However, more than a decade of devoted counting and analysis, particularly at Harvard, failed to disclose the expected stellar density concentrations that could be identified with spiral arms.

The history of astronomy shows repeatedly that well-defined problems are often solved by totally unexpected lines of research, and so it was in deducing the spiral structure of our galactic system. The solution to this puzzle lay in the observational analysis of the Andromeda nebula (M31) and other nearby galaxies. This work was carried out by Walter Baade at Mount Wilson, and it took particular advantage of the dark skies produced by the wartime blackout of Los Angeles and Hollywood.

Like Bok, Baade had been inspired by the work of Shapley, in his case by the physical nature of pulsating stars. In 1931 Baade received an offer to join the staff of the Mount Wilson Observatory, and he immediately accepted. He eventually applied for American citizenship but lost the papers, and with a characteristic disdain for bureaucracy Baade never reopened the matter. Thus, when the United States entered World War II, he was classified as an enemy alien unfit for war work, and consequently he had free rein with the 100-inch reflector during those dark years. Baade pushed the telescope to its limits to resolve the inner portions of M31 and its satellite galaxies M32 and NGC 205.

The results were published in 1944 in a famous article in the *Astrophysical Journal* in which he distinguished between two stellar populations and introduced the now-familiar terms Type I and Type II. Characteristic of the first type were highly luminous *O* and *B* stars and members of open clusters. The second type was represented in globular clusters and by short-period Cepheid variable stars.

Baade added that the same two types of stars had already been recognized by Jan Oort in 1926 from their differing motions in space. This Dutch astronomer, like Bok, had studied with van Rhijn and had played a leading role in the discovery of the rotation of our Milky Way. Because of their common interests in galactic structure, a lively correspondence ensued between Oort and Baade after the publication of the latter's classic paper. Among the topics they discussed was the possibility of using the Type I *O* and *B* stars as spiral-arm indicators and of surveying the Milky Way with Schmidt telescopes for such highly luminous stars.

Although Baade himself did not get involved with such a program at Palomar, he appreciated that it was the proper approach for solving the problem of spiral structure, rather than by the less discriminating method of star counting. His opportunity to discuss this vision in the context of the new, wide-field telescopes came at the dedication of the Curtis Schmidt at Michigan in June, 1950. Baade gave both the opening lecture and the first symposium paper, on 'Galaxies -- Present Day Problems.' He addressed a wide variety of issues before he came to his final topic, 'Our Galaxy as a Spiral Nebula.' Baade wrote, 'We have, I think, convincing evidence now that our galaxy is an Sb spiral, because it has a nucleus similar to that of the Andromeda nebula.' Then he turned to the spiral structure itself as another problem ready for attack: 'The procedure in this case is obvious. Since the supergiants of the population I are restricted to the spiral arms, we have to study their spatial arrangement in the solar neighborhood. The most promising stars for a first test are undoubtedly the *O* and early *B* stars, on account of their high frequency in spiral arms.'

Baade showed a very suggestive illustration, of the ionized-hydrogen (H II) regions in M31. These are huge spheres of gas internally excited by the highly luminous *O* and *B* stars. Around this time Baade must have formulated the analogy that the spiral arms in M31 are much like the candles on a birthday cake – all show and little substance. In other words, there was not much of a stellar density difference in the spiral arms; hence little was to be found by star-counting methods. Instead, we had to look for bright and conspicuous spiral indicators, either the *OB* stars or their associated H II regions.

Figure 34.1 Chief participants at the Michigan symposium in June, 1950. Front row (left to right): F. D. Miller, K. G. Henize, H. Shapley, W. Baade, J. J. Nassau, and L. Goldberg. Back row: G. Abetti, B. Lindblad, W. W. Morgan, A. N. Vyssotsky, N. U. Mayall, R. Minkowski, J. Stebbins, and S. W. McCuskey.

But to delineate the spiral structure of the Milky Way itself, it was also necessary to get the distances of the spiral indicators. This crucial determination could only be made if the absolute magnitudes were known for these highly luminous objects as well as the amount of interstellar absorption along the line of sight. 'W. W. Morgan's spectroscopic luminosity criteria for O and B stars should fill this gap,' reported Baade, 'and it is, I think, no secret that Morgan and Nassau are now engaged in a large program of determining the absolute magnitudes of O and B stars by this method.'

No secret indeed! William W. Morgan of Yerkes Observatory and Jason J. Nassau of Warner and Swasey Observatory reported their results at the same symposium. With respect to the spiral arms, however, their paper took a very conservative stance. It is fascinating to examine the report of the Nassau-Morgan paper given in the *Sky and Telescope* report of the symposium (August, 1950, page 243).

The search yielded over 900 OB stars, but for the majority of them the distances are undetermined. However, for 49 relatively nearby OB stars and for three [other] groups ... Dr. Morgan has collected the required data. Combining the results with already existing knowledge of many facts about the galaxy and other galaxies, these astronomers suggested that the sun is located near the outer border of a spiral arm. The arm extends roughly from the constellation Carina to Cygnus. The fact that many faint and hence distant OB stars are found toward Cygnus indicates that we are observing the stars in the extension of this arm beyond the clustering in that constellation, that is, beyond 3000 light years. ...

Dr. Nassau cautioned, however that the evidence is insufficient at present to preclude the hypothesis that a great disorganization exists in the galaxy and the star groupings do not trace definite spiral arms.

A far more positive conclusion to these studies came approximately 18 months later at the 1951 Christmas meeting of the American Astronomical Society in Cleveland. There Morgan presented new

Figure 34.2 The original plot by J. J. Nassau and W. W. Morgan of 49 OB stars and three OB groups for which distances were determined. The position of the sun is shown by S, and cross-hatching designates the limit of the survey.

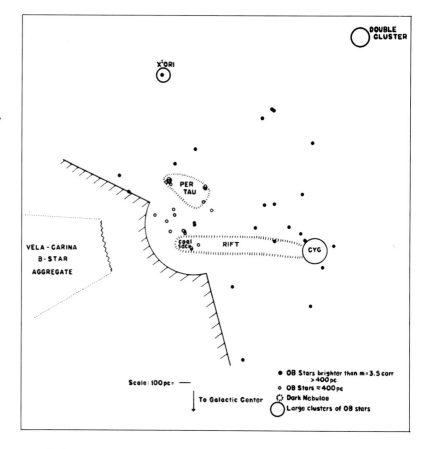

results based partly on the *OB* stars but largely on an investigation of the H II regions. By examining the distribution of such emission nebulae, he, together with Stewart Sharpless and Donald Osterbrock, was able to delineate segments of two spiral arms. One passes through the sun, and the other is over 6000 light years away in the direction opposite to the galactic center that is, about twice as far as the limits of the earlier Nassau–Morgan work.

What had changed between the Michigan symposium and the Cleveland AAS meeting? Although, as Baade had stated, the procedure was obvious, carrying it out was not. Finding the 900 *OB* stars was only the first stage. Determining luminosity criteria (to know the absolute magnitudes and hence the distances) and finding the really faint specimens was more difficult. Morgan had long appreciated that for studies of galactic structure accurate spectral types with luminosity classifications would be required. To that end he had produced in 1943 with P. C. Keenan and E. Kellman the MKK *Atlas of Stellar Spectra*. Thus, Morgan had correctly analyzed the approach for finding the large-scale structure of our galaxy and had established the basis for a successful program. What remained was to find the truly distant high-luminosity objects.

In an oral history interview taped a few years ago at Yerkes University by David DeVorkin of the American Institute of Physics, Morgan remarked that he had *two* papers at the Michigan symposium, a joint one given by Nassau on the arrangement of the *B* stars in space, 'which at the time had not gone far enough to show anything but a beautiful Gould belt . . .' and another of his own containing a description of what he called 'natural groups' in stellar spectra. In that second paper Morgan had coined the expression '*OB* stars' and had described these as a natural group with little spread in luminosity. This property made it possible to determine easily the absolute magnitude by a glance at the spectrum of such stars.

> I used to go to Cleveland for a week or so every few months, for a number of years. Nassau and I did all of the classifying. . . . We had a belt I believe 10 degrees wide, as far south as we could get around the sky, and this [furnished] the basic catalogue that was used here [at Yerkes] for taking slit spectrograms of as many of those stars as possible.
>
> Anyway, in the fall of 1951 I was walking between the observatory and home, which is only 100 yards away. I was looking up in the northern sky, just looking up in the region of the Double Cluster, and it suddenly occurred to me that the Double Cluster in Perseus and then a number of stars in Cassiopeia and even Cepheus, that along there I was getting distance moduli of between 11 and 12. Well, 11.5 is two kiloparsecs, and so I couldn't wait to get over here and really plot them up. It looked like a concentration. . . . But the hardest thing is to know what's going on if you're in the middle of something. So when I plotted out the Perseus arm, I then plotted out the other stars, and it turned out through the sun there was this narrow lane parallel to the other one. So that's the way it happened. It was a burst of realization. It was not a question of a reasoned process of steps.

In addition to the classification work with Nassau, Morgan had been searching for the more distant H II regions, which, he was convinced, would be the pointers for the faint distant *OB* stars and hence the key for tracing the spiral structure.

Concerning the discovery of the spiral arms Osterbrock told me,

> Morgan wanted to do it; he wanted to do it himself, and I guess part of his fear was that somebody else would tumble to the idea before he got it done in what he considered the right way. And I must say that Morgan involved Sharpless and me in every stage of it. Yet our major contribution was really in taking the plates. Most of the H II regions that were found he found, although we had looked very hard for them too. I've always felt that he gave us an awful lot of credit for two young graduate students whose contribution was quite minor. Many other investigators, I think, would have written the paper themselves. It was an idea that he had had for many, many years; he was asked to go to a meeting to deliver a paper on it, and yet he felt it was right to include us as authors.

Thus it was that Morgan took to Cleveland a joint paper that proved to be the sensation of the December, 1951, AAS meeting. Concerning the Morgan–Sharpless–Osterbrock paper, Otto Struve wrote,

> Astronomers are usually of a quiet and introspective disposition. They are not given to displays of emotion. Moreover, they tend to be cautious – more often than not they take plenty of time to weigh the evidence of any new and startling development before they accept it. But in Cleveland, Morgan's paper on galactic structure was greeted by an ovation such as I have never before witnessed. Clearly, he had in the course of a 15-minute paper presented so convincing an array of arguments that the audience for once threw

Figure 34.3 (a) The model of the three spiral arms, as shown by Morgan on a lantern slide at the Cleveland AAS meeting in December, 1951. Features are identified in (b).

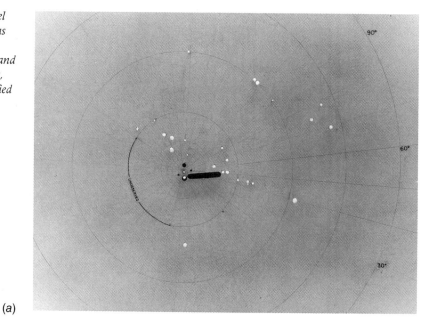

(a)

(b)

caution to the wind and gave Morgan the recognition which he so fully deserved.

Morgan described the situation:

Oort had introduced me, and when he sat down to listen, he sat down in my seat. It was one of those steeply sloping classrooms at Case with the seats all the way up high. Well, when I got through, the first thing was that I had no place to sit down. The second thing was people started to applaud by

Figure 34.4 By 1958 the spiral arms had been delineated by radio astronomers using the 21-cm line of neutral hydrogen gas, as shown in this diagram by J. H. Oort, F. Kerr, and G. Westerhout published in the Monthly Notices of the Royal Astronomical Society. *The more detailed pattern between 340° and 210° had been established by the group at Leiden, while the part on the left came from the Sydney astronomers. Problems in making a full reconciliation between the two hemispheres and with the optical work were already becoming apparent.*

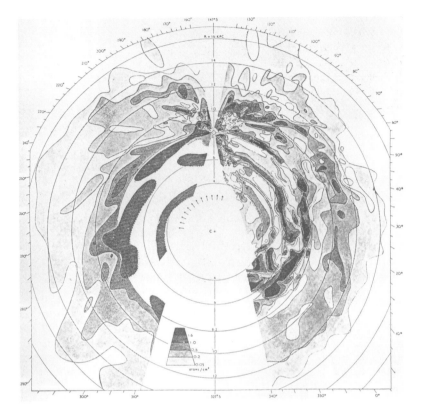

clapping their hands, but then they started stamping their feet. It was quite an experience.

Remarkably, the full paper describing the first discovery of the Milky Way's spiral arms was never published, and it is necessary to go to the April, 1952, issue of *Sky and Telescope*, page 138, to find the best account of it. In a poignant letter to me Morgan wrote, 'The reason for this was that I had a collapse in the spring of 1952 and spent the summer in Billings Hospital in Chicago in a helpless condition. When I returned to Yerkes in October, I had my partially written paper waiting for me, begun in the early part of the year; I was unable to work on it and complete it; instead, I wrote the UBV paper with Harold Johnson. The rapid growth of radio astronomy resulted in my never finishing and publishing the original paper.'

In this way a problem that had stood at least three decades, and which had been one of the primary targets of galactic-structure studies, finally found its solution by a quite different avenue from the numerical star-counting procedures. Yet the analysis that grew out of the earlier Dutch studies soon found another and even more powerful application in the interpretation of the 21-cm radiation from neutral hydrogen. This radio emission had been discovered in the same year, 1951, as the optical discovery of the spiral structure from the ionized hydrogen. Almost immediately Oort and Bok and their students began a vigorous investigation of the Milky Way's radio structure, and by 1952 the spiral pattern had been confirmed and extended.

Nevertheless, I think there is some larger justice in that the optical studies, on which so many decades of effort had been spent, narrowly won the race against the new and powerful radio astronomy to establish that our galaxy really did have spiral arms, as had long been conjectured.

Notes and references

This chapter is based on a longer account presented at IAU Symposium No. 106, Groningen, the Netherlands, in May, 1983 (and published in *The Milky Way, Proceedings of IAU Symposium 106*, Dordrecht, 1985, pp. 59–70). I wish to thank W. W. Morgan, Jan Oort, Leo Goldberg, and Donald Osterbrock for providing a variety of relevant historical materials and permission to quote from them, as well as the Niels Bohr Library of the American Institute of Physics for permission to use the oral history interview with Morgan. The Struve quotation is from *Leaflet* No. 285 of the Astronomical Society of the Pacific, January, 1953. The abstract of the Morgan–Sharpless–Osterbrock paper appeared in *Astronomical Journal*, **57** (1952), 3.

35 *The Great Comet of 1965*

Of all the memorable comets that have excited astronomers and stirred men's imaginations, not one had more impact on our concepts of the universe than the Great Comet of 1577. Discovered in November of that year, the comet stood like a bent red flame in the western sky just after sunset. The celebrated Danish astronomer Tycho Brahe was among the early observers: he caught sight of the brilliant nucleus while he was fishing, even before the sun had set. As darkness fell, a splendid 22° tail revealed itself. Tycho's precise observations over the 10-week span before the comet faded away were to deal the deathblow to ancient cosmogonies and pave the way for modern astronomy.

In the sixteenth century nearly everyone accepted Aristotle's idea that comets were meteorological phenomena, fiery condensations in the upper atmosphere. Or, if not that, they were burning impurities on the lower fringe of the celestial ether, far below the orbit of the moon. In 1577 most astronomers still subscribed to the ancient belief that the moon and planets were carried around the earth on concentric shells of purest ether. Tycho, by comparing his careful measurements of the comet's position with data from distant observers, proved that it sped through space far beyond the moon. The comet of 1577 completely shattered the immutable crystalline spheres, thereby contributing to the breakdown of Aristotelian physics and the acceptance of the Copernican system.

But the most renowned and most thoroughly studied of all comets is the one associated with Edmond Halley. It was the first to have a periodic orbit assigned, thus securing for comets their place as members of the solar system. Halley had matched the comet of 1682, which he had observed, with those of 1531 and 1607. Assuming these to be different appearances of the same celestial object, he predicted another return in 1758. Although he was ridiculed for setting the date beyond his expected lifetime, the comet indeed returned, and Halley's name has been linked with it ever since.

On its return, in 1910, Halley's comet put on a magnificent display, reaching its climax several weeks after perihelion passage in mid-April. During the early part of May it increased until the brilliance of its head equaled the brightest stars and its tail extended 60° across the sky. Later in May, the earth grazed the edge of the tail. The thin vacuous tail caused no observable effect on earth, except for such human aberrations as the spirited sale of asbestos suits. That no terrestrial consequences were detected is not surprising when we learn that 2000 cubic miles of the tail contained less material than a single cubic inch of ordinary air.

If prizes were offered for cometary distinctions, then 1965's Comet Ikeya–Seki would have won a medal as the most photographed of all time, and it might have won again for the range of astrophysical observations carried out. As it swung around the sun, its brilliancy

Figure 35.1 Comet Ikeya–Seki, photographed with one of the Baker–Nunn satellite tracking cameras of the Smithsonian Astrophysical Observatory.

outshone that of the full moon, and within ten days its tail extended almost as far as the distance from the earth to the sun. The behavior of the comet was neatly explained by the 'dirty snowball' theory. According to this widely accepted picture, a comet's nucleus is a huge block of frozen gases generously sprinkled with dark earthy materials. Occasionally the gravitational attraction of nearby passing stars can perturb a comet from its cosmic deep freeze in the distant fringes of the planetary system beyond Neptune; the comet then can penetrate the inner circles of the solar system, where it develops a shining gaseous shroud as its surface vaporizes under the sun's warming rays. Hence, the closer a comet approaches the sun, the more it vaporizes and the larger and brighter it becomes. Comet Ikeya–Seki passed unusually close to the sun, becoming possibly the brightest comet of the century; the resulting tail was the fourth longest ever recorded.

Today I look back with a wry smile to the Sunday morning in September when I decoded the telegram bringing the first word of the new comet. Early that morning in Benten Jima, Japan, a youthful comet hunter, Kaoru Ikeya, had discovered a fuzzy glow not charted on his sky maps. At the same time, another young amateur 250 miles away, Tsutomu Seki, had independently detected the new celestial visitor. Both men had used simple, homemade telescopes for their discovery, and both had sent urgent messages of their find to the Tokyo Astronomical Observatory.

News of the comet's appearance was quickly relayed from Tokyo to my office at the Smithsonian Astrophysical Observatory. Here the name 'Comet Ikeya–Seki' was officially assigned, as well as the astronomical designation 1965f. Throughout that day, 19 September, the communications center at Smithsonian alerted observatories and astronomical groups all over the world – Flagstaff, Rio de Janeiro, Johannesburg, Prague, Peking, Canberra – in all, more than 120. Included were the 12 astrophysical observing stations of the Smithsonian Observatory, whose specially designed satellite-tracking cameras are ideal for comet photography. Within hours a confirmation of Ikeya–Seki arrived from the Woomera, Australia, station.

By Tuesday afternoon, half a dozen approximate positions were in hand, more than enough for us to try for a crude preliminary solution of the comet's orbit. Unfortunately, the positions from the observing stations were only approximate 'eyeball' measurements obtained by laying the film onto a standard star chart with marked coordinates. Furthermore, the observatory's computer program had not been fully checked out. When the rough observations were used in different combinations, the computer produced two orbits in wild disagreement. Nevertheless, Prof. Fred L. Whipple, director of the Smithsonian Astrophysical Observatory and author of the 'dirty snowball' comet theory, noted that the second of the preliminary orbits closely resembled the path of a famous family of sun-grazing comets. The agreement was too close to be coincidence, he reasoned, and therefore the second solution must be correct.

Prof. Whipple's astute suggestion provided the first hint of the excitement that was to come. Several of the previous sun-grazers had

Figure 35.2 The staff of the Central Bureau for Astronomical Telegrams in 1965 included Richard Southworth, Barbara L. Welther, Brian Marsden (seated), and Owen Gingerich, who was then Director of the Bureau.

Figure 35.3 Professor Fred Whipple holds a teletype sheet of artificial satellite predictions in the early days of optical tracking. He stands in the Smithsonian's communications center, part of the Smithsonian Observatory he then directed.

been spectacular objects. Notable among them was the Great Comet of 1843, whose 70° tail stretched 200 million miles into space, setting an all-time record, and whose brilliance induced the citizenry of Cambridge to build a 15-inch telescope for Harvard equal to the largest in the world. And the second comet of 1882 achieved such brilliancy as it rounded the sun that it could be seen in broad daylight with the naked eye.

In the few days following the first computer solutions three 'precise' positions were reported to the Central Telegram Bureau, one from Steward Observatory in Tucson, Arizona, and two from the Skalnate Pleso Observatory in Czechoslovakia. When these new positions were fed by themselves into the computer, the result indicated an ordinary comet, and not a sun-grazer at all. But our programmers noticed that something was seriously wrong. When positions from the satellite-tracking cameras were included in the calculations, the computer gave different answers. Among them was the interesting possibility that Comet Ikeya–Seki might die by fire, plunging directly into the sun.

Then, suddenly, the mystery vanished. Six accurate positions from veteran comet observer Elizabeth Roemer at the Flagstaff, Arizona, station of the US Naval Observatory established the path with great precision. One of the earlier 'precise' observations had been faulty, and with its elimination, the others fell into place. Comet Ikeya–Seki was accelerating along a course that would carry it within a solar radius of the sun's surface. And since a comet's brightness depends on its closeness to the sun, there was every indication that Comet Ikeya–Seki would become a brilliant object.

Armed with predictions of Comet Ikeya–Seki's sun-grazing path, the Smithsonian staff set out to forewarn space scientists and radio astronomers whose attention does not normally encompass comets.

Figure 35.4 Tsutomu Seki with his comet hunter, taken in September, 1965.

We called a press conference to describe the magnificent view hoped for as the comet swung around perihelion, its nearest approach to the sun. First discovered in the morning sky, the comet would cross into the evening sky for only a few hours on 21 October. If a tail of this comet were to appear in the evening, it would sweep across the western sky after sunset on that evening. Afterward it would reappear in the morning twilight. Such a prediction was hazardous, because although the comet's trajectory was well established, its brightness and tail length resisted astronomical forecasting since no one knew just how much material would be activated as it sped past the sun.

Had we examined more carefully the historical records of Comet 1882 II, we might have been more cautious in telling the public to look for the tail of Comet Ikeya–Seki sweeping across the western sky after sunset on 21 October. Each new observation of the 1965 comet confirmed that it was a virtual twin of the Great Comet of 1882; thus, by looking at the observations from the last century, we should have guessed that the comet's enormous velocity as it rounded the sun – one million miles per hour – would dissipate the tail so widely that it could not be seen in the dark sky. On the other hand, we hardly dared publicize what the computer's brightness predictions showed: that Comet Ikeya–Seki would be visible in full daylight within a few degrees of the sun!

And thus it happened that thousands of would-be observers in the eastern United States maintained a cold and fruitless search in the early morning hours of 21 October. Thousands of others, especially in the American southwest, had the view of a lifetime – a bright comet with its short silvery tail visible next to the sun in broad daylight. Simply by holding up their hands to block out the sunlight, they could glimpse the comet shining with the brilliance of the full moon. Hazy, milky skies blocked the naked-eye view for observers in the eastern United States and much of the rest of the world; even in New England, however, telescopes revealed the comet with a sharp edge facing the sun and the beginnings of a fuzzy tail on the other side. Professional astronomers were excited by the opportunity to photograph the object at high noon. For the first time, the daylight brilliance of a comet permitted analysis from solar coronagraphs. Airborne and rocket-borne ultraviolet detectors examined features never before studied in comets.

The spectrum observations ended eight decades of controversy. In most comets, the reflected spectrum of sunlight is seen, combined with the more interesting bright molecular spectrum from carbon and carbon compounds. The molecules are excited by the ultraviolet light from the sun, and glow in much the same way that certain minerals fluoresce under an ultraviolet lamp. But back in 1882, when spectroscopy was in its infancy, the great sun-grazing comet yielded an entirely different spectrum. Scientists at the Dunecht Observatory in Scotland thought they saw emission lines from metal atoms such as iron, titanium, or calcium, but a similar spectrum was never found in subsequent comets. Some observers expressed their disbelief in this unique record.

Astronomers did not get another chance to examine a comet so close to the sun until 20 October 1965. On that morning at the Radcliffe Observatory in South Africa, Dr A. D. Thackeray obtained spectrograms of the nucleus of Comet Ikeya–Seki, then only 8 million miles from the sun. These showed bright lines of both iron and calcium. The telegraphic announcement, again relayed by the Central Bureau, set other spectroscopists into action. Within days, there were reports of nickel, chromium, sodium, and copper.

Though fully expected from a theoretical point of view, these observations confirmed that the impurities in comets had a chemical composition similar to that of meteors. The connection is not fortuitous; for many years astronomers recognized that those ephemeral streaks of light in the night sky, the meteors, were fragile cometary debris plunging through the earth's atmosphere. As the gases boil out of a cometary nucleus, myriads of dirty, dusty fragments are lost in space. In time, they can be distributed throughout a comet's entire orbit, and if that path comes close to the earth's own trajectory, a meteor shower results.

The Leonid meteors are a splendid example of 'falling stars' closely related to a comet. A meteor swarm follows close to Comet Tempel–Tuttle. Every 33 years, as the comet nears the earth's orbit, a particularly good display of Leonids appears around 16 November. The recovery of this same comet in 1965 was followed by a November shower in which hundreds of brilliant meteors flashed through the sky within a period of a few hours. Nonetheless, the 1965 Leonids provided a sparse show compared with the hundreds of thousands seen in 1833 and 1866. In 1899, astronomers predicted yet another fireworks spectacular. The prognostication proved to be a great fiasco, for gravitational attraction from the planet Jupiter had slightly shifted the orbit of the comet and its associated meteor swarm. Ever since, astronomers have been wary of alerting the public to meteors or comets. Our enthusiasm in predicting the greatness of Comet Ikeya–Seki on 21 October was indeed risky.

Nevertheless, the daylight apparition of Comet Ikeya–Seki was but a prelude to a more spectacular show. Its surface thoroughly heated by its passage through the solar corona, the comet developed a surrounding coma of gas and dust some thousands of miles in diameter as it left the sun. As it slowed its course and receded from the hearth of our planetary system, the solar wind drove particles from that coma into a long stream preceding the comet.

As soon as Comet Ikeya–Seki could once again be seen in the early morning sky, its long twisted tail caused a sensation. Standing like a wispy searchlight beam above the eastern horizon, the tail could be traced for at least 25°. Its maximum length corresponded to 70 million miles, ranking it as the fourth longest ever recorded. Only the Great Comets of 1843, 1680, and 1811 had tails stretching farther through space. (Quite a few comets have spanned greater arcs of the sky because they were much closer to the earth. Their actual lengths in space could not compare with that of the Great Comet of 1965.) At its peak brightness, Comet Ikeya–Seki was about equal to the sungrazers of 1843 and 1882. Even after it receded from the sun, its nucleus shone brilliantly through the morning twilight. By all accounts, Comet Ikeya–Seki compared favorably with the great comets of the past. Those portentous sights, compared to giant swords by many a bygone observer, had little competition from city lights, smog, and horizon-blocking apartment buildings.

Comet Ikeya–Seki surprised most astronomers by developing a strikingly brilliant tail on its outward path from the sun, especially when compared with the poor show on its incoming trajectory. Had they looked in Book III of Newton's *Principia*, however, they would have seen another sun-grazing comet neatly diagrammed with a short, stubby tail before perihelion passage and the great flowing streamlike tail afterward. Newton spent many pages describing that Great Comet of 1680. Especially interesting to American readers is the generous sprinkling of observations reported from New England and 'at the river Patuxent, near Hunting Creek, in Maryland, in the confines of Virginia.'

In the New World not only astronomers were interested in the comet. From the Massachusetts pulpit of Increase Mather came the warning,

Figure 35.5 Hideo Ikeya's reflector telescope with which he discovered what was to become the 'Great Comet of 1965.'

As for the SIGN in Heaven now appearing, what Calamityes may be portended thereby? . . . As *Vespasian* the Emperour, when There was a long *hairy Comet* seen, he did not deride at it, and make a Joke of it, saying, That it concerned the Parthians that wore long hair, and not him, who was bald: but within a Year, Vespasian himself (and not the Parthian) dyed. There is no doubt to be made of it, but that God by this *Blazing-star* is speaking to other Places, and not to *New England* onely. And it may be, He is declaring to the generation of hairy Scalps, who go on still in their Trespasses, that the day of Calamity is at hand.

Superstitions concerning comets reached their highest development and received their sharpest attacks at this time. For centuries comets had been considered fearsome omens of bloody catastrophe, and Increase Mather must have been among the great majority who considered the comet of 1680 as a symbol fraught with dark meanings. The terrors of the superstitious were compounded when a report came that a hen had laid an egg marked with a comet. Pamphlets were circulated in France and Germany with woodblocks of the comet, the hen, and the egg. Even the French Academy of Sciences felt obliged to comment:

Last Monday night, about eight o'clock, a hen which had never before laid an egg, after having cackled in an extraordinarily loud manner, laid an egg of an uncommon size. It was not marked with a comet as many have believed, but with several stars as our engraving indicates.

In a further analysis of this comet, Newton's *Principia* reported that a remarkable comet had appeared four times at equal intervals of 575 years beginning with the month of September in the year Julius

Caesar was killed. Newton and his colleague Halley believed that the Great Comet of 1680 had been the same one as seen in 1106, 531, and in 44 BC. This conclusion was in fact false, and the Great Comet of 1680 had a much longer period. Within a few years, however, Halley correctly analyzed the periodicity of the famous comet that now bears his name.

Is Comet Ikeya–Seki periodic like Halley's? If so, can it be identified with any of the previous sun-grazers? The resemblance of Comet Ikeya–Seki to Comet 1882 II has led many people to suppose that these objects were identical. The orbits of both these comets take the form of greatly elongated ellipses, extending away from the sun in virtually identical directions. Nevertheless, even the earliest orbit calculations scuttled the possibility that the comets were one and the same, since at least several hundred years must have passed since Ikeya–Seki made a previous appearance in the inner realms of the solar system. On the other hand, it is unlikely that Comet Ikeya–Seki, Comet 1882 II, and a half dozen others would share the same celestial traffic pattern and remain unrelated. The only reasonable explanation is to suppose that some single giant comet must have fissioned into many parts hundreds of years ago.

Indeed, the Great Comet of 1882 did just that. Before periphelion passage, it showed a single nucleus; a few weeks afterward, astronomers detected four parts, which gradually separated along the line of the orbit. The periods for the individual pieces are calculated as 671, 772, 875, and 955 years. Consequently, this comet will return as *four* great comets, about a century apart.

It was, therefore, not at all unexpected when the Central Bureau was able to relay the message on 5 November that Comet Ikeya–Seki had likewise broken into pieces. The first report suggested the possibility of three fragments, but later observers were able to pinpoint only two. One of these was almost starlike, the other fuzzy and diffuse. Though first observed two weeks after perihelion passage, the breakup was probably caused by unequal heating of the icy comet as it neared the sun.

If the Great Comet of 1965 was itself merely a fragment, what a superb sight the original sun-grazer must have been. Appearances of comets with known orbits total 870, beginning with Halley's in 240 BC, but the earliest known sun-grazer of this family is the comet of 1668. In medieval chronicles and Chinese annals, and on cuneiform tablets, hundreds of other comets have been recorded, but the observations are inadequate for orbit determinations. Undoubtedly, that original superspectacular sun-grazer was observed, but whether it was recorded and whether such records can be found and interpreted are at present unanswerable questions.

A similar search of historical records, which holds more promise of success, is now under way at the Smithsonian Astrophysical Observatory. The comet with the shortest known period, Encke, cycles around the sun every $3\frac{1}{3}$ years. Inexorably, each close approach to the sun further erodes Comet Encke. The size of its snowball has never been directly observed, but a shrewd guess based on the known excrescence of gaseous material places it in the order of a few miles.

By calculating ahead, Prof. Whipple has predicted the final demise of Comet Encke in the last decade of this century. By calculating backward in time, he has concluded that it might once have been a brilliant object. Its $3\frac{1}{3}$-year period would bring a close approach to the earth every third revolution, so that a spectacular comet might appear in the records at 10-year intervals. In the centuries before Christ, the Chinese and Babylonian records show remarkable agreement, but the register is too sketchy, and so far, Comet Encke's appearances in antiquity have not been identified.

In addition to Encke there are nearly 100 comets whose periods are less than 200 years. Like Comet Encke, they face a slow death, giving up more of their substance on each perihelion passage. On an astronomical time scale, the solar system's corps of short-period comets would be rapidly depleted if a fresh supply were unavailable. On the other hand, there is apparently an unlimited abundance of long-period comets that spend most of their lifetime far beyond the planetary system. Astronomers now envision an extensive cloud of hundreds of thousands of comets encircling the sun at distances well beyond Pluto. Originally there may only have been a ring of cometary material lying in the same plane as the earth's orbit – the leftover flotsam from the solar system's primordial times. Perhaps the density of material was insufficient to coalesce into planetary objects, or perhaps at those great distances from the sun the snowballs were too cold to stick together easily.

Gravitational attractions from passing stars presumably threw many of the comets out of their original orbits into the present cometary cloud. These gravitational perturbations still continue, and a few comets from the cloud reach the earth's orbit every year. Their appearances are entirely unexpected, and their discoveries are fair game for professional and amateur alike. But since most professional astronomers are busily engaged in more reliable pursuits, persistent amateurs manage to catch the majority of bright long-period comets. Devotees such as Ikeya and Seki have spent literally hundreds of hours sweeping the sky with their telescopes in the hope of catching a small nebulous wisp that might be a new comet. The great sun-grazer was the third cometary find for each man. Within a week of its discovery, a British schoolteacher, G. E. D. Alcock, also found a new comet – his fourth. Alcock started his comet-finding career in 1959 by uncovering *two* new comets within a few days.

How does an amateur, or a professional, recognize a new comet when he finds one? Most new-found comets are as diffuse and formless as a squashed star, completely devoid of any tail. In this respect they resemble hundreds of faint nebulae that speckle the sky, with this difference: nebulae are fixed, but a comet will inevitably move. Consequently, a second observation made a few hours later will generally reveal a motion if the nebulous wisp is indeed a comet. However, most comet hunters compare the position of their suspected comet with a sky map that charts faint nebulae and clusters. Then the discovery is quickly reported to a nearby observatory or directly to the Central Bureau.

Today the chief reward for a comet find lies in the tradition of attaching the discoverer's name to the object, but in times past there have been other compensations. Jean Louis Pons, who discovered 37 comets during the first quarter of the nineteenth century, rose from observatory doorkeeper to observatory director largely as a result of his international reputation for comet finding. And the Tennessee astronomer E. E. Barnard paid for his Nashville house with cash awards offered by a wealthy patron of astronomy for comet discoveries in the 1880s. Barnard has recorded a remarkable incident relating to the great sun-grazing comet of 1882:

My thoughts must have run strongly on comets during that time, for one night when thoroughly worn out I set my alarm clock and lay down for a short sleep. Possibly it was the noise of the clock that set my wits to work, or perhaps it was the presence of that wonderful comet which was then gracing the morning skies, or perhaps, it was the worry over the mortgage and the hopes of finding another comet or two to wipe it out. Whatever the cause, I had a most wonderful dream. I thought I was looking at the sky which was filled with comets, long-tailed and short-tailed and with no tails at all. It was a marvelous sight, and I had just begun to gather in the crop when the alarm clock went off and the blessed vision of comets vanished. I took my telescope out in the yard and began sweeping the heavens to the southwest of the Great Comet in the search for comets. Presently I ran upon a very cometary-looking object where there was no known nebula. Looking more carefully I saw several others in the field of view. Moving the telescope about I found that there must have been ten or fifteen comets at this point within the space of a few degrees. Before dawn killed them out I located six or eight of them.

Undoubtedly Barnard's observations referred to ephemeral fragments disrupted from the Comet 1882 II then in view.

A great majority of the comets reaching the earth's orbit go back to the vast comet cloud, never to be identified again. Occasionally, however, a comet swings so close to the great planet Jupiter that its orbit is bent, and it is 'captured' into a much shorter period. A 'Jupiter capture' has never been directly observed, because most comets are still too faint when they reach Jupiter's orbit. Nevertheless, in 1965 astronomers came almost as close as they ever will to witnessing the aftermath of this remarkable phenomenon.

In January that year the press reported the discovery of two new comets by the Chinese, a rather unexpected claim inasmuch as it has been centuries since the Chinese discovered even one comet, not to mention two. To everyone's astonishment a pair of telegrams eventually reached our Central Bureau via England confirming the existence of the objects. At the same time, the Chinese managed to flout the centuries-old tradition of naming comets after their discoverer. In the absence of the discoverer's name, our bureau assigned to both comets the label Tsuchinshan, which translated means 'Purple Mountain Observatory.'

Tsuchinshan 1 and Tsuchinshan 2 have remarkably similar orbits, whose greatest distances from the sun fall near the orbit of Jupiter. As these faint comets swung around that distant point in 1961, Jupiter was passing in close proximity. Quite possibly the gravitational attraction from Jupiter secured the capture of a long-period comet in that year, simultaneously disrupting it into the two Tsuchinshan fragments. However, it is more likely that the capture occurred at a somewhat earlier pass, a point that will eventually be established by a computer investigation. In any event, the observation of a comet pair with such a close approach to Jupiter is without precedence in the annals of comet history.

The complete roster of comets for 1965 included not only the Tsuchinshan pair, Comet Alcock, and the once-in-33-years visit of Tempel–Tuttle, but the recoveries of four other faint periodic comets and another new one, Comet Klemola, which was accidently picked up during a search for faint satellites of Saturn. Of this rich harvest, Comet Ikeya–Seki received more attention than all the others combined. Day after day, the Smithsonian observing stations around the world kept a continual photographic watch as the long twisted tail developed and faded.

By now the Great Comet of 1965 has faded beyond the range of either Ikeya's or Seki's small telescope, and has apparently vanished from the larger instruments of professional astronomers as well. Perhaps in a millennium hence an unsuspecting amateur, never imagining that he has caught a sun-grazer, will find it on its next return.

'When discovered, the comet was only a white spot in the moonlit sky,' Seki recently wrote to us. 'I did not even dream that it would later come so close to the sun and become so famous.'

Notes and references

The quotation from Increase Mather is found in his KOMETO-ΓΡΑΦΙΑ *or A Discourse concerning Comets* (Boston, 1683) and the one from E. E. Barnard is in Robert Hardie's 'The Early Life of E. E. Barnard,' *Leaflets of the Astronomical Society of the Pacific*, Nos. 415 and 416 (1964). For more on the IAU telegram bureau, see my 'The Central Bureau for Astronomical Telegrams,' *Physics Today*, **22**, no. 12 (1966), 37–40 – at that time I was Director of the Bureau.

Albert Einstein: a laboratory in the mind

The long conical shadow of the moon was scheduled to sweep from Brazil to Africa on 29 May 1919. Near each end of the path of the eclipse, astronomers waited in tense expectancy to test a bold new theory of gravitation that challenged the time-honored laws of the English mathematician Sir Isaac Newton.

This new theory had been evolving for nearly a decade in the mind of a young German-born physicist, Albert Einstein. When it was finally published in 1916 as the 'general theory of relativity' – to distinguish it from his earlier 'special theory of relativity' – Einstein was only 37, eight years younger than Newton was when he wrote his *Principia* with its famous law of universal gravitation. Einstein described gravitation as a natural property of curved space. Just as a marble placed on a stretched rubber membrane would deform the rubber surface, so would a massive object bend or warp the space around it. Thus, Einstein predicted the sun bends the space around it, and the distortion could be detected by measuring the deflection of starlight passing near it. Of course, the blinding glare of the sun normally masks the distant stars behind it. But during an eclipse, stars whose line of sight lies near the sun can be photographed.

Waiting to test Einstein's theory on the tiny Portuguese island of Príncipe were Arthur S. Eddington, a great enthusiast of the theory of relativity, and his young assistant, E. T. Cottingham. Before embarking on the expedition, Cottingham had asked Sir Frank W. Dyson, director of Great Britain's Royal Greenwich Observatory and official organizer of the expedition, 'What will it mean if we get double the [predicted] Einstein deflection?' Dyson replied, 'Then Eddington will go mad and you will have to come home alone.' But when Eddington had completed the measurements of his first photograph plate on Príncipe, he turned to his companion and said, 'Cottingham, you won't have to go home alone.'

Eddington's formal announcement of the expedition's results the following September provided dramatic verification of Einstein's esoteric theory. Overnight, 'Einstein' became a household word, and the man was plunged into a fame from which he never escaped.

Einstein knew that his theory of gravitation would have several consequences that could be verified by experiments. One was that the planets' elliptical orbits should slowly change their orientations beyond what Newtonian theories accounted for. Most conspicuously, the long axis of Mercury's orbit should rotate around the sun by 43 arc seconds (43/1296000 of a circle) more than the 500 arc seconds per century predicted by Newton's laws. Indeed, astronomers had measured such a rotation and had been puzzled by it.

Another consequence was that in the warped space-time near a massive object, clocks would run more slowly. One form of clock is a vibrating atom, and the color of any spectral line radiated by the atom indicates the clock's rate. Since the atoms on the surface of the sun or a star act like tiny clocks in a strong gravitational field, they should run slightly slower than their counterparts on earth. As a result, their spectral lines should be slightly redder. This effect is slight on the sun and on most stars, and measurements were not sensitive enough to confirm it when the general relativity theory was published. The slowing of time has since been confirmed. Extremely accurate maser clocks carried in high-flying aircraft (where the gravity is less than at the Earth's surface) run faster than identical clocks in the laboratory. Another experiment testing this theory is so sensitive that it has detected the difference in the gravity between the first and the fourth floors of a Harvard University physics laboratory.

Who was this remarkable physicist so abruptly cast into the limelight by the scientific events of 1919? Although Einstein had received a prestigious research professorship at the University of Berlin at the comparatively early age of 34, as a youth he did not appear to be precocious. In fact, his parents had worried that he might be retarded because he did not begin talking until he was three years old. Nevertheless, young Albert was fascinated by physical and mechanical phenomena. Later, when he was in his mid-20s and working in the Swiss Patent Office in Bern, he could understand the intricacies of the submitted inventions so quickly that he finished a normal day's workload in a few hours. This left him plenty of time to ask himself 'simple' questions about the nature of space, time, and matter.

Many years later, when Einstein set down some autobiographical notes, he did not bother with such details as, 'I was born on 14 March 1879 in the German town of Ulm.' Nor did he document his moves to Munich; Milan, Italy; or Switzerland. Instead he recorded his sense of wonder about things when, for example, as a child of 4 or 5, his father showed him a magnetic compass.

Einstein also described how he developed a deep religiosity as a boy, although his Jewish parents did not practice any religion. He abruptly lost this in 'an orgy of freethinking' at age 12. Much later, as a result of his experiences in Germany between World Wars I and II, he identified with his Jewish heritage and became an active Zionist. But it was the contemplation of the universe that induced a deep reverence in Einstein. 'The road to this paradise was not as comfortable and alluring as the road to the religious paradise,' he wrote, 'but it has shown itself reliable, and I have never regretted having chosen it.'

In the fall of 1895, the teen-aged Albert took the exams at the renowned Swiss Polytechnic Institute in Zurich – and flunked them. A year later, however, he was admitted and in 1900 was awarded a degree. But his independent spirit had so alienated his teachers that the door to a university career seemed closed. Thus, Einstein took the job in the Swiss Patent Office in 1902. During the seven years he worked there, Einstein developed the concept most irrevocably linked with his name – the special theory of relativity. This theory rests on two postulates, or assumptions: first, the laws of light and optics are the same for any reference frame in which Newton's laws also hold; second, the velocity of light in empty space is always the

Figure 36.1 Swiss citizen Albert Einstein (far right) was among the cream of European physicists who met in Brussels, Belgium, in 1911. Other notables included the French Marie Curie (seated at the table) and – standing – the German Max Planck (second from left), the French Louis de Broglie (sixth from left), and the British Ernest Rutherford (third from right).

same, regardless of the motion of its source. Einstein published these postulates in 1905 in a paper entitled 'On the Electrodynamics of Moving Bodies.'

At first glance, the two postulates seem innocent enough. Yet they contradicted the two simplest models for the behavior of light. First, if light waves were like sound waves, there must be a medium through which they travel. The theory of electromagnetic waves, worked out in the 1860s by British mathematician James Clerk Maxwell, assumed the existence of an ethereal medium in which light waves traveled and which provided a preferred, fixed reference frame for them. But this concept of light traveling in an invisible ether is denied by the first postulate of relativity, which declares that all observers must use exactly the same law of light and optics. Even if two observers are moving with respect to each other, the laws apply for each one as if each were at rest. According to relativity and in contradiction to Maxwell's theory, there can be no preferred reference frame – no ether – for light that is 'really' at rest.

Many experiments in the late 1800s had been designed to detect an ether in which light might move. The most notable one was carried out by American physicist Albert A. Michelson, who won the 1907 Nobel prize for physics, and his colleague Edward W. Morley. Without exception, these experiments failed to detect any ether.

A second simple model pictures light as particles shot out from its source. In this case, we might expect the particle of light to travel

Figure 36.2 Einstein at his stand-up desk when he was an examiner at the Swiss patent office in Bern around 1905.

with the combined speed of the source of light itself. An analogy would be a baseball thrown forward on a moving train; another, a spacecraft that races toward Mars with the combined velocity of the rocket's thrust and the earth's orbital motion. But for light, such a combined velocity is denied by the second postulate of relativity. The truth of this postulate is verified by observations of binary stars – pairs of stars that orbit around each other. As a companion star orbits its primary, it alternately approaches and recedes from the earth. According to the model, light should be alternately speeded up and slowed down so that its rays in traveling the vast distance to the earth would arrive out of their original sequence. But they arrive in sequence. Analysis of X-ray binary stars in 1978 has confirmed this second postulate of relativity to an accuracy of 1 part in 10 billion.

In exploring the consequences of the two postulates of special relativity, Einstein realized that our notion of time was intimately bound up with our concept of space, and that he had to be exceedingly careful in judging two events to be simultaneous. One can imagine the impatience of the German physics professors in their stiff starched collars when Einstein elaborated these details in his paper: ''If, for instance, I say, 'That train arrives here at 7 o'clock,' I mean something like this: 'The pointing of the small hand of my watch to 7 and the arrival of the train are simultaneous events.' ''

By pressing his careful reasoning to its logical conclusion, Einstein found a startling result: in trying to synchronize two clocks that are far apart from each other, we get different readings when we do it by carrying a third clock between the two than we do if we try to synchronize them with light signals. Moreover, time appears to run more slowly for a moving clock than for a clock at rest. Einstein's theory yielded additional astonishing results. Moving rods appear to contract, compared with identical rods at rest, and moving objects become more massive than similar objects at rest. But all of these concepts are relative, because it is impossible to find absolute rest or absolute motion. Whether a clock or rod is at rest or in motion depends on the point of view of the observer.

Einstein's conclusions appear to fly in the face of common sense – at the very least, they are contrary to ordinary experience. Conspicuous changes in clocks, rods, or masses occur only at velocities near the speed of light, something rarely encountered in everyday life. After scientists began working with objects at very high speeds, it became apparent that nature really does work according to Einstein's theory of relativity.

For example, as atomic particles are accelerated in synchrotrons and linear accelerators, they become more massive and, beyond a certain point, additional energy can produce almost no further increases in speed. The atomic particles resist being accelerated past the velocity of light.

Another example is provided by the quasars, whose spectral lines are spectacularly shifted toward the red. According to the ordinary Doppler effect, which explains why a radiating object moving away from an observer appears to be radiating at a lower frequency, these far-distant objects must be rushing away from us at two or three times the speed of light. A relativistic Doppler effect, however, shows that while they are moving rapidly, they do not exceed the speed of light.

In 1922, only a few years after relativity had entered the public vocabulary, Einstein won the 1921 Nobel prize for physics. Ironically, the citation did not specifically mention relativity. Instead, it honored another discovery he made in the same wonder year of 1905. The prize was awarded 'for his services to theoretical physics and, in particular, for his discovery of the law of the photoelectric effect.'

In the photoelectric effect, light shining on a metal releases electrons from the surface. Einstein proposed that the phenomenon could be explained quantitatively if light was treated as a series of discrete, or separate, packets of energy, or quanta. Because well-known optical phenomena, such as diffraction and polarization, could only be described by picturing light as a wave, Einstein's quanta at first seemed contrary to established physical principles.

Indeed, American physicist Robert A. Millikan found Einstein's photoelectric hypothesis so unbelievable that he conducted detailed experiments beginning in 1913 to settle the matter. But instead of disproving the theory, Millikan's work confirmed it in every detail. Millikan himself won the Nobel prize for physics in 1923. So when Einstein won his prize in 1922, the photoelectric effect had been verified experimentally, while much of his relativity theory had not.

Although Einstein placed relativity foremost among his achievements and chose to speak on this topic when he accepted his Nobel prize, he made numerous other advances in the quantum theory of

Figure 36.3 Danish physicist Niels Bohr and Einstein developed a major disagreement on the nature of the quantum theory.

atomic structure. Quantum theory had its foundation in the work of German physicist Max Planck, who, in 1900, discovered the law that described how radiant energy is distributed at different temperatures – the so-called black body radiation. This work earned him the Noble prize for physics in 1918. Planck had to assume that radiating atoms could not vibrate with any arbitrary energy, but only with specific discrete amounts, or quanta. In explaining the photoelectric effect, Einstein noticed that he could apply these same quantized energy levels to light. Two years later, by a further brilliant recourse to quantum theory, he explained unresolved discrepancies in experiments on the heat capacity of solids.

The quantum view of nature received another boost in 1913 with Danish physicist Niels Bohr's new interpretation of the atom. Bohr, who won the 1922 Nobel prize for physics for his work, theorized that electrons traveling about an atomic nucleus would move only in certain definite, or quantized, orbits. In Bohr's picture, a specific quantum, or photon, of light could be emitted or absorbed when an electron changed from one allowed orbit to another. Einstein realized that Planck's radiation law would not work on the basis of only absorption or spontaneous emission. A third process called stimulated emission was required.

Stimulated emission provides the basis for two important modern inventions, the microwave maser and its optical counterpart, the laser. These devices use photons, created when excited atoms in the material drop back to their unexcited state, to stimulate other excited atoms to release identical photons. This rapidly multiplies the number of photons, increasing the intensity of microwave or light signals.

Laser beams have many uses, from drilling tiny holes in sapphires for watch bearings to transmitting telephone and television signals on glass fibers. Masers, operating by the same principle in the microwave region of the electromagnetic spectrum, are used as exceedingly accurate atomic clocks.

In the early 1920s, Einstein supported the 'crazy' ideas of the French physicist Louis de Broglie, winner of the Nobel prize for physics in 1929. De Broglie had proposed that the specific atomic orbits suggested by Bohr could be pictured as standing waves. A permitted orbit could contain only complete multiples of its wave-form. Thus, electron particles were intimately bound up with waves, just as light itself exhibited these dual particle and wave aspects that seemed to be contradictory.

Einstein's recommendations had a catalytic effect in the physics community. Almost overnight, the field of wave mechanics was born. Shortly thereafter, the waves were reinterpreted as probability waves, and German physicist Werner Heisenberg (who won the 1932 Nobel prize in physics) introduced his famous 'uncertainty principle.' According to this view, complete information concerning the position and motion of atoms was forever unobtainable. Einstein was repelled and distressed at this loss of determinism. At two major meetings of physicists in 1927 and 1930, he argued vehemently with Bohr and Heisenberg against the emerging interpretation of atomic physics. Failing to convince them, he retired from the field he had done so much to create, remonstrating, 'God does not play dice.'

In the early 1930s, political events in Germany took an ominous turn. Einstein was in Pasadena, California, when Adolf Hitler came to power, and he realized at once that he could not go back to Germany. He resigned from the Prussian Academy of Sciences in March, 1933, while his writings were being burned by the Nazis in their anti-Semitic frenzy. Einstein went to Belgium for a short period, and then briefly visited Oxford University in England. In October, he accepted an appointment at the newly founded Institute for Advanced Study in Princeton, New Jersey, where he stayed until his death in 1955.

While he was at Princeton, the wide-ranging implications of yet another of his 1905 papers became clear. In a sequel to his special relativity article, Einstein had shown that a further consequence was the equivalence of mass and energy. Ironically, his first derivation of this relationship contained a flaw in logic and, in effect, he assumed the answer he intuitively sought. In 1906, he gave a correct derivation. The following year, he wrote what has become the most famous equation of physics: $E=mc^2$ – the energy of a body is equal to its mass times the square of the velocity of light.

How the mass of atomic nuclei supplied the prodigious energy that powers the stars became known through the work of Eddington in the 1920s. In 1919, Eddington wrote: 'If, indeed, the subatomic energy in the stars is being freely used to maintain their great furnaces, it seems to bring a little nearer to fulfillment our dream of controlling this latent power for the well-being of the human race – or for its suicide.'

By 1939, physicists in the United States and Germany had discovered the specific nuclear reactions that power the stars. Almost simultaneously, other physicists realized that a quite different set of reactions, involving the *fission* (splitting) of uranium atoms, could also release vast amounts of energy. Uranium held not only the promise of a vast new energy source, but also the grim specter of an atomic bomb. Einstein's colleagues convinced him to write to President Franklin D. Roosevelt about building a nuclear weapon.

Many of the physicists who worked on the bomb really hoped that the task would prove impossible, but such was not the case. On 6 August 1945 a nuclear bomb was dropped on Hiroshima, Japan. One of Einstein's biographers, Banesh Hoffmann, writes in *Albert Einstein, Creator and Rebel*: "Einstein's secretary heard the news on the radio. When Einstein came down from his bedroom for afternoon tea, she told him. And he said, 'Oh weh,' which is a cry of despair whose depth is not conveyed by the translation 'Alas.' "

Einstein's $E=mc^2$, like most scientific knowledge, is morally neutral. It has indeed opened a Pandora's box before which we stand with much trepidation. The dreadful, haunting fear of nuclear warfare casts a sobering pall over all international relations. Similarly, the fear of a catastrophic nuclear accident poisons the potential for abundant power for a civilization rapidly depleting its conventional energy resources. But $E=mc^2$ also forms the basis of a beautiful understanding of the evolution of stars, laying open the past and future of our sun, and revealing the physical reasons for the wide variety of stars spanning the night skies. It is we, and not Einstein, who must be held responsible to use $E=mc^2$ wisely.

During the 1970s, scientists began to search for yet another phenomenon predicted by Einstein's general theory of relativity – gravitational waves. If a massive object that bends the space in its vicinity also changes its arrangement – for example, in the collapse of a supernova, or exploding star – surrounding space will be deformed by the propagation of gravitational ripples that travel with the speed of light. Sensitive machines are being built in several laboratories throughout the world to detect gravitational waves. Meanwhile, analysis of the acceleration of pulses from a binary pulsar by radio astronomers at the University of Massachusetts in Amherst in 1978 provided indirect evidence of gravity waves.

Astronomers are currently in hot pursuit of yet another phenomenon described by general relativity – the black hole. In the 1700s, the British astronomer John Michell and French astronomer Pierre Simon Laplace independently considered the possibility of an object so dense and massive that the escape velocity from its surface would equal or exceed the velocity of light. Light itself could leave from such an object, but gravity would eventually pull it back again.

Relativistic calculations show, however, that light could not even leave from such an object – hence the name 'black hole.' Although black holes can never be observed directly, their gravitation can betray their presence by the intense warp they create in space. Likely candidates for black holes include the collapsed companion of an X-ray binary, Cygnus X-1; the cores of several compact globular star clusters; and a massive condensed object within our own galaxy.

Just as objects bend the space in their vicinity, the cumulative effects of the many masses embedded in space can curve it on a large scale. In his early work on general relativity, Einstein realized that over cosmic distances space might not conform to the principles of Euclidean geometry – that is, the sum of the angles of a very large triangle might not equal exactly 180°.

When Einstein announced his general theory of relativity in 1916, astronomers had little conception of distances beyond a few thousand light years. By the 1920s, they began to realize that what appeared as the spiral nebulae consisted of remote stellar systems millions of light years away. In the 1930s, they realized that these far-flung galaxies were rushing away from one another at immense velocities and that the universe was rapidly expanding.

Will the universe expand forever, or will it eventually coast to a stop and then begin to contract? If the universe is dense enough, gravitation can overcome the expansion and pull it back until it ultimately collapses. Such a universe is said to be 'closed,' and the curvature of its space produces triangles containing more than 180°. If the density is too low, however, gravity will be insufficient to halt its outward rush. Such a universe is said to be 'open,' and its space is curved in such a way that triangles contain less than 180°.

So far, the observations point toward an open universe, providing a real challenge for a small but dedicated group of scientists who believe on aesthetic grounds that the universe should be closed. In 1978, many of them hoped that the discovery of a faint background of X-ray radiation would furnish evidence for previously overlooked matter – mass that would raise the measured density past the minimum needed for closure. In 1979, these hopes were dashed by results from a new orbiting X-ray telescope appropriately named the *Einstein* satellite. The new data show that the background X-rays

Figure 36.5 A portion of a famous 1913 letter from Einstein to American astronomer George Ellery Hale, showing how the sun would bend starlight passing nearby.

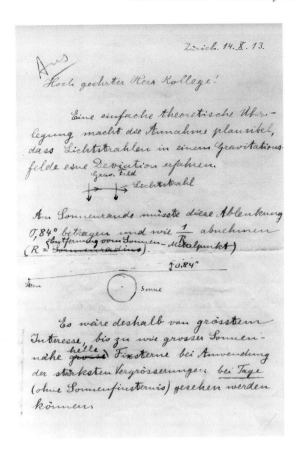

come from distant quasars that were already counted in the mass totals and not from previously overlooked hot gas.

Part of Einstein's genius was his incredibly well-tuned intuition and his remarkable sense of the aesthetic. He believed his general theory of relativity had to be right because it was so beautiful mathematically. It is fascinating to note that many competing gravitational theories have been postulated in the past 20 years, and that virtually all of them have been eliminated by more refined experiments. At the same time, Einstein's relativistic theory of gravitation has come through every test unscathed.

One of Einstein's University of Berlin students, Ilse Rosenthal-Schneider, reported that one day in 1919 when she was studying with him, "He suddenly interrupted the discussion of the book, reached for a telegram that was lying on the window sill, and handed it to me with the words, 'Here, this will perhaps interest you.' It was Eddington's cable with the results of the eclipse measurements. When I was giving expression to my joy that the results coincided with his calculations, he said, quite unmoved, 'But I knew that the theory is correct'; and when I asked, what if there had been no confirmation of his prediction, he countered: 'Then I would have been sorry for the dear Lord – the theory is correct.'"

Einstein's aesthetic sense drove him to search even further for a harmonic unity between the gravitation of space and the electrical structure of matter. Although he failed in this quest for a unified field theory, several leading theoretical physicists in several parts of the world are continuing to search for it.

Einstein is a towering figure in science because he found and attacked a broad range of crucial problems and discovered deep underlying harmonies that other physicists missed. But even more important, while many of the other giants of his age won their laurels with a dazzling display of new discoveries in the laboratory, Einstein's workshop was in the mind. His experiments were almost always thought experiments. Like Copernicus before him, who had begun a revolution with theories 'pleasing to the mind,' Einstein astonished his contemporaries with entirely new ways of looking at phenomena that had been known for decades, or even centuries.

Einstein's work, like a shaft of sunlight in the great cathedral of science, suddenly illuminates the structure and lets us realize that this beautiful edifice is man-made. Scientific theory is not an absolute, but a human construction, molded according to our taste and perspectives. Einstein himself clearly summarized this understanding of science when he wrote, 'The sense experiences are the given subject matter. But the theory that shall interpret them is man-made. It is the result of an extremely laborious process of adaptation: hypothetical, never completely final, always subject to question and doubt.'

Relativity and quantum mechanics have radically altered our picture of nature. But Einstein's insight into these concepts reminds us of the remarkable impact that a single gifted individual can make in our scientific view of the world.

Notes and references

Gerald Holton's several essays and lectures on Einstein have guided my interest; two from which I have profited especially are "Einstein, Michelson, and the 'Crucial' Experiment" and 'On Trying to Understand Scientific Genius,' both reprinted in *Thematic Origins of Scientific Thought: Kepler to Einstein* (Cambridge, Massachusetts, revised edition, 1988). Holton gives the Ilse Rosenthal-Schneider quotation about the eclipse telegram on pp. 254–5; I once had occasion to question Einstein's long-time secretary, Helen Dukas, about it, and she replied that it seemed quite authentic, as Einstein had more than once expressed himself similarly. Two excellent biographies of Einstein are Abraham Pais, *Subtle is the Lord: the Science and Life of Albert Einstein* (Oxford, 1982) and Banesh Hoffmann, *Creator–Rebel: Albert Einstein* (New York, 1972). The Eddington incident at Princípe in 1919 is described in A. Vibert Douglas' biography, *Arthur Stanley Eddington* (London, 1957), p. 40.

Figure acknowledgements

Chapter 1. Ancient Egyptian sky magic
Sky and Telescope 65, 418–20, 1983
1.1. Author's collection.
1.2. Photo by Owen Gingerich.
1.3. Photo by Owen Gingerich.
1.4. © 1974 John Ross.

Chapter 2. The origin of the zodiac
Sky and Telescope 67, 218–20, 1984
2.1. Tafel XVII in *Archiv für Orientforschung,* **16** (1953).
2.2. Supplied by the author.
2.3. *Sky and Telescope.*

Chapter 3. The basic astronomy of Stonehenge
Technology Review 80 #2, 64–73, December, 1977
3.1. Supplied by the author.
3.2. Supplied by the author.
3.3. Invented by the author.
3.4. Photo © 1971 Owen Gingerich.
3.5. Photo © 1958 Owen Gingerich.
3.6. Photo © 1971 Owen Gingerich.

Chapter 4. Some puzzles of Ptolemy's Star Catalog
Sky and Telescope 67, 421–3, 1984
4.1. Supplied by the authors.
4.2. Supplied by the authors.
4.3. Andre Thevet, *Vrais pourtraits et vies des hommes illustres* (Paris, 1584), Burndy Library, photo by Owen Gingerich.

Chapter 5. Ptolemy and the maverick motion of Mercury
Sky and Telescope 66, 11–13, 1983
5.1. Supplied by the author.
5.2. *Sky and Telescope.*
5.3. Supplied by the author.
5.4. Supplied by the author.

Chapter 6. How astronomers finally captured Mercury
Sky and Telescope 66, 203–5, 1983
6.1. *Sky and Telescope.*
6.2. Author's collection.
6.3. Permission of Houghton Library, Harvard University.
6.4. Harvard College Library.
6.5. Author's collection.

Chapter 7. Islamic astronomy
Scientific American 254 #4, 74–83, April, 1986
7.1. Al-Sabah Collection, Kuwait National Museum.
7.2. Photo by Owen Gingerich, Cairo National Library.
7.3. © 1981 Scientific American, Inc. All rights reserved.

7.4.   © 1981 Scientific American, Inc. All rights reserved.
7.5.   (a) Thomas Photos, Oxford
      (b) © 1981 Scientific American, Inc. All rights reserved.
7.6.   Bodleian Library, Oxford.

Chapter 8.   The astronomy of Alfonso the Wise
Sky and Telescope **69**, 206–8, 1985
8.1.   Supplied by the author.
8.2.   Harvard College Observatory Library.
8.3.   Permission of Houghton Library, Harvard University.

Chapter 9.   From Aristarchus to Copernicus
Sky and Telescope **66**, 410–12, 1983
9.1.   Courtesy of *Aramco World* magazine.
9.2.   Jagiellonian Library, Cracow, reproduced from
Nicholas Copernicus Complete Works, vol. 1 (Warsaw, 1972).

Chapter 10.   The Great Copernicus Chase
American Scholar **49**, 81–8, 1979
10.1.   City Museum, Torun, Poland.
10.2.   Photo by Owen Gingerich.
10.3.   Photo by Owen Gingerich.
10.4.   Photo by Owen Gingerich, Leipzig Universitätsbibliothek.

Chapter 11.   The Tower of the Winds and the Gregorian Calendar
Sky and Telescope **64**, 530–3, 1982
11.1.   Photo by Gordon Moyer.
11.2.   F. Manicinelli and J. Casanovas, *La Torre dei Venti in Vaticano*, courtesy of Martin McCarthy.
11.3.   F. Manicinelli and J. Casanovas, *La Torre dei Venti in Vaticano*, courtesy of Martin McCarthy.
11.4.   Author's collection.

Chapter 12.   Tycho Brahe and the Great Comet of 1577
Sky and Telescope **54**, 452–8, 1977
12.1.   Courtesy Baron Otto Reedtz-Thott, Gavno Castle, Sweden.
12.2.   Crawford Collection, Royal Observatory Edinburgh, Permission of Astronomer Royal of Scotland.
12.3.   Crawford Collection, Royal Observatory Edinburgh, Permission of Astronomer Royal of Scotland.
12.4.   Wikiana Collection, Zurich Zentralbibliothek.
12.5.   Tycho Brahe *Opera Omnia*, Vol. 4 (Copenhagen, 1922).
12.6.   Tycho Brahe *Opera Omnia*, Vol. 4 (Copenhagen, 1922).
12.7.   Crawford Collection, Royal Observatory Edinburgh, Permission of Astronomer Royal of Scotland.
12.8.   Wikiana Collection, Zurich Zentralbibliothek.
12.9.   Wikiana Collection, Zurich Zentralbibliothek.

Chapter 13.   Galileo and the phases of Venus
Sky and Telescope **68**, 520–2, 1984
13.1.   National Maritime Museum.
13.2.   *Sky and Telescope*.
13.3.   Supplied by the author.
13.4.   Supplied by the author.

Chapter 14.   The Galileo affair
              Scientific American 247 #2, 132–43, August, 1982
   14.1.   Author's collection.
   14.2.   Biblioteca Nazionale, Florence.
   14.3.   Supplied by the author.
   14.4.   Courtesy Vatican Archives.
   14.5.   Biblioteca Nazionale, Florence.
   14.7.   Courtesy Vatican Archives.
   14.8.   Courtesy Vatican Archives.

Chapter 15.   Johannes Kepler and the *Rudolphine Tables*
              Sky and Telescope 42, 328–33, 1971
   15.1.   Author's collection.
   15.2.   Photo by Owen Gingerich.
   15.3.   Supplied by the author.
   15.4.   Kepler Commission Archives, Weil der Stadt.
   15.5.   Supplied by the author.
   15.6.   Author's collection. Photo by Owen Gingerich, Russian Academy of
            Sciences Archives, St. Petersburg.
   15.7.   Kepler Commission Archives, Weil der Stadt.
   15.8.   Photo by Owen Gingerich.

Chapter 16.   An astrolabe from Lahore
              Sky and Telescope 63, 358–60, 1982
   16.1.   Author's collection.
   16.2.   Author's collection.
   16.3.   *Sky and Telescope*, adapted from *Scientific American*.
   16.4.   Author's collection.
   16.5.   Author's collection.

Chapter 17.   Fake astrolabes
              Sky and Telescope 63, 465–8, 1982
   17.1.   Courtesy Adler Planetarium, Chicago.
   17.2.   Courtesy Freer Gallery of Art, Washington, DC.
   17.3.   *Sky and Telescope*, adapted from *Scientific American*.
   17.4.   Author's collection.
   17.5.   Courtesy of Stanley Epstein.
   17.6.   Permission of Houghton Library, Harvard University.

Chapter 18.   Newton, Halley and the comet
              Sky and Telescope 71, 230–2, 1986
   18.1.   Author's collection.
   18.2.   Cambridge University Observatories.
   18.3.   Cambridge University Press.
   18.4.   Author's collection.
   18.5.   Uppsala University Observatory.

Chapter 19.   Eighteenth-century eclipse paths
              Sky and Telescope 62, 324–7, 1981
   19.1.   Permission of Houghton Library, Harvard University.
   19.2.   Permission of Houghton Library, Harvard University.
   19.3.   *Memoirs of the American Academy of Arts and Sciences*, Vol. 1, 1781.
   19.4.   Courtesy American Academy of Arts and Sciences.
   19.5.   Photo by Paul Raila.
   19.6.   Harvard Map Room and *Sky and Telescope*.

Chapter 20.   The 1784 autobiography of William Herschel
               Sky and Telescope **68**, 317–19, 1984
    20.1.   Harvard College Library.
    20.2.   Permission of Houghton Library, Harvard University.

Chapter 21.   The Great Comet that never came
               Sky and Telescope **65**, 124–6, 1983
    21.1.   Author's collection.
    21.2.   Harvard's College Library.
    21.3.   Author's collection.
    21.4.   Author's collection.
    21.5.   Author's collection.
    21.6.   Author's collection.

Chapter 22.   Unlocking the chemical secrets of the Cosmos
               Sky and Telescope **62**, 13–15, 1981
    22.1.   Author's collection.
    22.2.   Harvard College Observatory.
    22.3.   Harvard College Observatory.

Chapter 23.   The discovery of the satellites of Mars
               Vistas in Astronomy **22**, 127–32, 1978
    23.1.   *Harper's Weekly*, **38** (1894), p. 1144.
    23.2.   Courtesy US Naval Observatory.

Chapter 24.   The first photograph of a nebula
               Sky and Telescope **60**, 364–6, 1980
    24.1.   *Sky and Telescope.*
    24.2.   Courtesy of Gerard de Vaucouleurs.
    24.3.   E. S. Holden, *Monograph of The Central Parts of the Nebula of Orian*
               (Washington, 1882).
    24.4.   Harvard College Observatory, *Annals*, **5** (1867).
    24.5.   Agnes Clerke, *A Popular History of Astronomy during the Nineteenth Century*
               (London, 1887).

Chapter 25.   The Great Comet and the *'Carte'*
               Sky and Telescope **64**, 237–9, 1983
    25.1.   (a) Agnes Clerke, *A Popular History of Astronomy during the Nineteenth Century*
               (London, 1893).
               (b) *Sky and Telescope.*
    25.2.   Admiral E. Mouchez, *Photographie astronomique* (Paris, 1887).
    25.3.   Harvard College Observatory.
    25.4.   Harvard College Observatory.

Chapter 26.   James Lick and the founding of Lick Observatory
               Pacific Discovery **31** #1, 1–10, 1978
26.1.–26.11.   Mary Lea Shane Archives, Lick Observatory.

Chapter 27.   Atget's eclipse watchers
               Sky and Telescope **61**, 215–16, 1981
    27.1.   Print by Bernice Abbot, author's collection.
    27.2.   *L'Astronomie*, 1912.
    27.3.   *Sky and Telescope* from *L'Astronomie*, 1912.

Chapter 28. Faintness means farness
Proceedings of the Royal Institution **58**, 201–13, 1987
28.1. Supplied by the author.
28.2. Supplied by the author.
28.3. *Philosophical Transactions*, **75** (1785), 213–66.
28.4. Supplied by the author.
28.5. DeSitter's *Kosmos*, (Cambridge, Mass, 1932).
28.6. Mt Wilson and Palomar Observatories, supplied by Kenneth Glyn Jones.
28.7. Isaac Roberts, supplied by Kenneth Glyn Jones.

Chapter 29. The mysterious nebulae, 1610–1924
Journal of the Royal Astronomical Society of Canada **81**, 113–28, 1987
29.1. Harvard College Observatory.
29.2. Supplied by the author.
29.3. Harvard College Observatory.
29.4. Birr Castle Archives, permission of Lord Rosse.
29.5. Courtesy of David Malin.
29.6. Birr Castle Archives, permission of Lord Rosse.
29.7. Photograph by Frank Hogg, courtesy of Helen Sawyer Hogg.
29.8. Permission of Harvard University Archives.

Chapter 30. Harlow Shapley and the Cepheids
Sky and Telescope **70**, 540–2, 1985
30.1. Author's collection.
30.2. Adapted from *Harvard Circular No.* 173, (1912), by *Sky and Telescope*.
30.3. *Sky and Telescope*.
30.4. Adapted from *Astrophysical Journal*, **48** (1918), 104, by *Sky and Telescope*.
30.5. Adapted from *Bulletin of the National Research Council*, **2** (1921), 205, by *Sky and Telescope*.

Chapter 31. A search for Russell's original diagram
Sky and Telescope **64**, 36–7, 1982
31.1. *Popular Astronomy*, **22** (1914), Plate XII.
31.2. Princeton University Library.
31.3. Eddington, *Stellar Movements and the Structure of the Universe* (London, 1914), p. 171.

Chapter 32. Dreyer and Tycho's world system
Sky and Telescope **64**, 138–40, 1982
32.1. Mary Lea Shane Archives, Lick Observatory.
32.2. *Sky and Telescope*.
32.3. Austrian National Library, Vienna, Codex 10658.1, folio 107.
32.4. Austrian National Library, Vienna, Codex 10686.84, folio 21.
32.5. Royal Astronomical Society Club, courtesy of Derek Howse.

Chapter 33. Robert Trumpler and the dustiness of space
Sky and Telescope **70**, 213–15, 1985
33.1. Mary Lea Shane Archives, Lick Observatory.
33.2. Lick Sky Survey.
33.3. Adapted from Gretchen Hagen, 'An Atlas of Open Cluster Colour-Magnitude Diagrams,' *Publications of the David Dunlap Observatory*, **4** (1970).

Index

Abbasid dynasty, 43
Abbott, Bernice, 208–9
Abd al-A'imma, 140–3
Abetti, G., picture, 265
Achernar, 27
Adler Planetarium, 139–40
Afghanistan, 132
al-Sufi, etc. – see Sufi, etc.
Albategnius, 98
Alcock, G. E. D., 279
Alfonsine Tables, 57, 59, 61–2, 84;
    picture, 61
Alfonso X, 57–62; picture, 58
algorithm, 45
Alhazen, 53
alidade, 136–7
Allegheny Observatory, 205,
    257–8
Allis, E. H., 189
Almagest (Ptolemy), 7, 24–5, 31, 34,
    43–4, 49, 69
Almagestum novum (Riccioli), 120–1
almucantars, 136
Alnath, 24
Alpetragius (al-Bitruji), 54
Alphecca, 134–5
Altair, 43, 134
Amadeus, 110
American Academy of Arts and
    Sciences, 155–6, 158
American Association for the
    Advancement of Science, 236–7,
    246
American Astronomical Society,
    151, 236–7, 246–7, 257, 265,
    267–8
American Philosophical Society, 156,
    159
American Scholar, xi, 79
anagram, 98, 102–3, 110
Andrade, E. N. da C., 256
Andromeda nebula, 221–3, 233,
    236, 263–4; picture, 222
Anglo–Australian telescope, 231
Ann Arbor, 75
Antares, 134–5
Apollonius, 52
Arabic star names, 43, 49, 134–5
Aratus, 11
Arcetri, 117, 120
Archimedes, 63–6
Arcturus, 218
Aristarchus, 63–8
Aristotelian logic, 111
Aristotle, 89, 105, 271; picture, 106
Armagh Observatory, 251, 256
armillary sphere, picture, 60
Arrest, Heinrich Louis d', 180, 230
Ashbrook, Joseph, xi, 187–8, 194
Ashworth, William B., 237
Assayer (Galileo), 115
Astrographic Catalogue, 191, 193
astrolabe, 49–51, 57–9, 132–45;
    pictures, 44, 50–1, 133, 135–6,
    140–4
astrology, 51

Astronomia Carolina (Streete), 38,
    40–1
Astronomia Nova (Kepler), 125, 252
Astronomiae pars Optica (Kepler), 130
Astronomical Society of the Pacific,
    261, 270
Astronomische Nachrichten, 240
Astronomisches Jahrbuch, 226
Astrophysical Journal, 206, 242, 264
Atget, Eugène, 208–10, 212
Atkinson, R. J. C., 19
Aubrey, John, 14, 21–2, 40
Audubon's *Birds*, 77
Aurifaber, Andreas, 74, 78–9
Austrian National Library, 253
Averroës (Ibn Rushd), 54

Baade, Walter, 238, 240, 244,
    263–6; picture, 265
Babinet, Jacques, 166, 170
Baghdad, 43
Baily's Beads, 159
Baker–Nunn camera, 272
Ballochroy, 21
Barberini, Maffeo (= Urban VIII),
    108, 113, 115
Barnard, E. E., 201, 205, 280
Baronius, Cardinal, 111, 118
Barrow, Isaac, 147–8
Basel, 78
Bath, 160, 162, 164, 215
Battani, al-, 52
Bayer, Johannes, 8
Bellarmino, Cardinal Roberto,
    113–15, 117, 119–20
Berendzen, Richard, 237
Berkeley, 257
Berlin Academy, 173
Bern, 283
Bernoulli, Johann, 146
Betelgeuse, 49, 134
Bevis, John, 227, 237
Bible, 107, 110
Bibliothèque Nationale, 71, 76
Big Dipper, 3–4, 6, 10
Birr Castle, 251
Bitruji, Abu Ishaq al-, 54
black hole, 290
Bode, J. E., 168, 226
Bohr Library, Niels, 270
Bohr, Niels, 287–8; picture, 287
Bok, Bart J., 263, 269
Bomme, B., 165
Bond, George Phillips, 179, 184,
    186, 188, 230
Bonn, 247
Bonner Durchmusterung, 190
Book of Nature, 108–9, 111, 113,
    116
Book of Scripture, 107–8, 111, 113
Boston Athenaeum, 77
Boston Museum of Fine Arts, 139
Boulliau, Ismael, 39
Boyle, Robert, 39
brachistochrone, 146, 148
Brahe – see Tycho Brahe
Brecher, Kenneth, 30